PROBABILISTIC SYSTEMS ANALYSIS

PROBABILISTIC SYSTEMS ANALYSIS

AN INTRODUCTION TO PROBABILISTIC MODELS, DECISIONS, AND APPLICATIONS OF RANDOM PROCESSES

ARTHUR M. BREIPOHL

Oklahoma State University

JOHN WILEY & SONS, INC.

NEW YORK · LONDON · SYDNEY · TORONTO

Library of Congress Catalogue Card Number: 77-94920

SBN 471 10181 8

Printed in the United States of America

10 9 8 7 6 5 4 3 2

This book is dedicated to my father,
WALTER L. BREIPOHL

PREFACE

Probabilistic models, I am convinced, are basic in the analysis of almost all kinds of engineering problems; hence a course in applied probability should be included in the undergraduate curriculum. The courses in probability or statistics that undergraduate engineers often take are very useful but do not emphasize applied engineering problems. The application of probability and statistics to engineering is usually limited to an introductory course in communication theory which stresses noise and communication circuits. In my opinion such a course does not properly represent the breadth of engineering applications of probability.

I believe that there can be an efficient compromise between teaching the breadth of engineering applications as a terminal course and teaching a solid introduction to the theory that is needed in graduate courses in communication theory and system studies. I hope this book is such a compromise. It is designed to present the fundamentals of probability theory and to show the student how to develop probabilistic models that describe the variation inherent in the physical elements with which he works.

In teaching a course with these objectives, I have found detailed mathematical treatment of such subjects as Borel fields and Lebesgue integration, as well as detailed limiting arguments, to be unnecessary and have not included it in this text. I intend the developments to be logical, but they are not rigorous.

PREVIEW AND POSSIBLE USES OF THIS BOOK

This is an undergraduate textbook and is intended to form the basis of a course for engineering students. I hope that engineers in industry will find it useful for self-study, but it is not meant to serve as a reference book.

The book consists of an introductory chapter and eleven other chapters. Chapter II introduces probability. Chapters IV and V present random variables and expected values. Except for an emphasis on conditional distributions and expectation, these chapters cover the usual material. The stress on modeling engineering problems is their distinguishing characteristic.

Chapter VI deals with transformation of random variables, a problem of primary importance in engineering. Applications are discussed in more detail in Chapters III and VII. Chapters VIII, IX, and X introduce statistics. Although classical statistics are discussed, the emphasis in these chapters is on Bayesian statistics, because I think that this view will prove to be the prevailing one for engineers. The last two chapters discuss random processes, with the material chosen to emphasize the concepts that are needed in future courses in communication theory and stochastic control.

Just as I am convinced that there is variation in physical elements, so I believe, a fortiori, that there is variation in people; therefore, the book is designed to be used in more than one way. It contains more material than can be taught in one semester to students who have not had a prior introduction to probability. Thus depending on the background of students, intent of the course, and interests of the instructor, various parts of the text may be used. The relation between the parts is shown in the diagram.

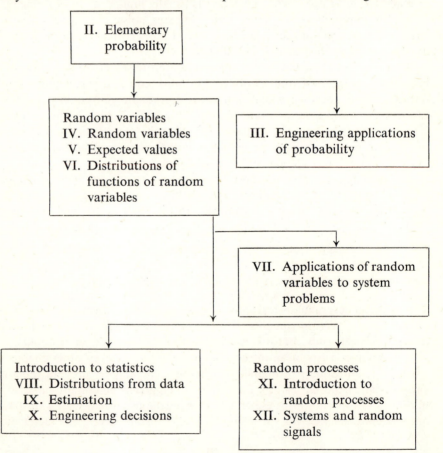

Chapters III and VII can be skipped without loss of logical continuity. However, I have found that these two chapters taught in sequence are very valuable in motivating engineering students. Also, either the statistics or the random process section can be postponed or omitted, as the instructor chooses.

I hope that this book will be useful in the following types of courses.

1. An introduction to applied probability for engineers who have not had a course in probability. [Chapters II, IV, V, and VI and selected material to emphasize (a) immediate applications (Chapters III and VII), (b) estimation and decisions (Chapters VIII, IX, and X), and (c) random processes (Chapters XI and XII).]

2. Applications of probability for electrical engineers who have had a course in probability. (Chapter VI and Chapters VIII through XII.)

Although the book is so designed that the instructor can select the material that he considers to be the most important, I hope that the examples and problems will not be slighted. The book is certainly not designed to be used without the examples and problems.

ACKNOWLEDGMENTS

I am greatly indebted to previous authors, many of whom are listed in the References at the end of the book. I am also in debt to several persons who taught probability to me: Dr. Bernie Ostle, Dr. George Steck, Dr. Arnold Koschmann, Dr. Bill Zimmer, and Dr. Jim Abbott. Of course, these teachers and authors are responsible only for what I learned correctly; the errors I managed in spite of them.

Encouragement provided by three colleagues—Bill Hughes, Bill Blackwell, and Bennett Basore—made this book a reality. Professor Basore taught from an earlier version of the manuscript, and his comments significantly clarified the text. In addition, many students at Oklahoma State University who studied from earlier manuscripts have suggested improvements. David Cunningham provided a very helpful final reading of the manuscript.

The publisher's reviews by several reviewers are responsible for a better book than would have resulted without their suggestions. I sincerely thank Professors John Thomas, Rudolph Drenick, and Abraham Haddad.

I wish to thank also Mrs. Barbara Adams, Mrs. Dixie Jennings, and Mrs. Toni Hadley who carefully typed various versions of the manuscript.

My special thanks to Shirley, Gary, and Diane for time I might otherwise have spent with them.

Finally, I will be thankful to any reader who sends me corrections or constructive criticisms.

Arthur M. Breipohl

Stillwater, Oklahoma
July, 1969

CONTENTS

IV. RANDOM VARIABLES

V. EXPECTED VALUES

VI. DISTRIBUTION OF FUNCTIONS OF RANDOM VARIABLES

X. ENGINEERING DECISIONS

XI. INTRODUCTION TO RANDOM PROCESSES

XII. SYSTEMS AND RANDOM SIGNALS

PROBABILISTIC SYSTEMS ANALYSIS

CHAPTER I

....

INTRODUCTION

1-1 VARIATION

There is a discontinuity in an engineer's academic training and the jobs he is required to perform as a practicing engineer. This fact will probably always be true and may even be desirable. However there is one rather general area that needs a better match between college and company. This area is the recognition of the variation of physical parameters and the study of methods of analyzing this variation. We mention some of the ways variation enters engineering problems.

Variation of physical quantities is apparent to anyone who has made measurements in a laboratory. No matter how well conditions are controlled and no matter how well instruments are designed, laboratory measurements are not exactly reproducible. In this case the engineer must analyze variation. Even if he models the parameter being measured as if it were fixed, the variations observed in the laboratory must be "averaged out" by taking more than one reading and combining the results (e.g., averaging or fitting a smooth curve to the data).

Any engineer who is responsible for the design of a product that is to be mass produced must take into account the variation of parts and specify their allowable variation. For example, he must decide whether he wants $\pm 20\%$ or $\pm 5\%$ resistors, what parameters of the transistor must be specified, and what their allowable variations are, etc.

Undesirable variation in signals is usually called noise. Design of communication and control systems is based on analyzing the noise, then reducing its effect on the system.

1

The purpose of this book is to present a theory for handling the types of problems described above. The book is based on the premise that all of these problems result from the fact that variation enters nearly all engineering problems and that this variation is best modeled by probabilistic means.

1-2 PROBABILISTIC AND DETERMINISTIC MODELS

In this book we are not concerned with how things "really" are.* Rather, we take the engineering view that we are interested in adopting a model that will best enable us to solve the problem at hand. For example, we are not concerned whether a resistor "actually" has a "true" resistance or is "actually" described by statistical properties of the atom; rather we attempt to show that by adopting a probabilistic model, an effective methodology can be developed for describing, say, the resistance of a circuit composed of $\pm 10\%$ resistors.

The student is familiar with models such as $E = IR$. We intend to use such models and identify some of the symbols in such a model with quantities that vary. One kind of variation results when the resistor used is selected from a class of, say, $\pm 10\%$ resistors and all members of the class are not identical. Another kind of variation is caused by the different performance of the same piece of equipment when used at different times. This type of variation is illustrated by the performance of a radio receiver.

Given that such variation is observed, why adopt a probabilistic model? There are two reasons for adopting such models. Probabilistic models enable us to state new fruitful questions, and they enable us to obtain more meaningful answers to questions which were previously phrased in deterministic terms.

The first reason is illustrated by the question, "How reliable is that computer?" Unless we want to use qualitative words like good or bad, we need probability in order to define reliability.

The second reason is illustrated by the question, "If twenty lengths of metal each 1.0 ± 0.1 in. are placed end to end, what is the total length?" In this case a deterministic answer, 20 ± 2.0 in. is possible; however, as we will see later, with practical assumptions the probability of being below 18.1 in. or above 21.9 in. is so near zero that the answer 20 ± 2.0 in. is not satisfactory from an engineering viewpoint. That is, if one did not take the probabilistic approach to this problem, his ultraconservative answer might be prohibitively expensive.

---

* The controversy between determinism and probability is a philosophical argument that has been the subject of many extensive discussions.

When designing communications systems (e.g., radio, TV, radar), the major problem is to receive the signal and reject the noise. Probabilistic models are necessary to describe the noise meaningfully; thus the design of modern communication equipment is based on probabilistic models. This application illustrates both reasons for adopting probabilistic models and has been a major motivation for engineers to study probabilistic models.

CHAPTER II

....

ELEMENTARY PROBABILITY

2-1 WHAT IS PROBABILITY?

Most people first encounter probability in games of chance or luck; in fact the theory of probability was originated to analyze a gambling problem. In 1654 Chevalier de Méré proposed a gambling problem to Blaise Pascal, and the principles of probability were founded by Pascal and Fermat in an interesting correspondence.* Since then the theory has grown to be a respected mathematical discipline and, from our point of view, more importantly, has many applications to practical engineering problems.

The intuitive notions of probability are rather simple, and these notions are useful in analyzing probabilistic models. However it is not easy to formulate a precise definition of probability based on this intuition.

The first approach to defining probability is an extension of gambling ideas. Consider the throw of a die. There are six possible outcomes that can occur: (a) one spot is on the up face, (b) two spots are on the up face, ..., (f) six spots are on the up face. It seems natural to define the probability of each outcome to be 1/6. This leads to what is usually referred to as the classical or equally likely definition of probability.

Equally Likely

If there are N possible equally likely and mutually exclusive outcomes (happenings) and if N_A of these outcomes correspond to event A, then the probability of event A, $P(A)$ is

$$P(A) = \frac{N_A}{N}. \tag{2-1}$$

* This correspondence except for the first letter is available in *Oeuvres de Fermat*, P. Tannery and C. Henry, Vol. 2 (1904). See also Ref. D2.

This definition seems to be adequate for the example of the die given above. Also consider the following example:

EXAMPLE 2-1

Suppose a name is picked at random (by chance) from a list of 50 names containing 30 electrical engineers, 15 mechanical engineers, 4 physicists, and 1 statistician. Then

$$P(\text{electrical engineer}) = \frac{\text{number of electrical engineers}}{\text{number of names}} = \tfrac{30}{50} = \tfrac{3}{5}.$$

The practical difficulties in applying this definition are illustrated by the following problem: list the number of equally likely possible ways in which a resistor can fail. There are also practical problems in deciding what equally likely means. Suppose a spaceship from Mars lands outside the building. The only thing we know about the inhabitants of Mars is that they are type A or type B. The door of the ship begins to open, and your friend wants to bet as to the type of the first occupant to emerge. What *is* the probability that a type A will emerge first? Using the principle of *equal likelihood*, one might reason that there are two possibilities; therefore the probability of type A is 1/2. Of course, this would be wrong if 90% of the inhabitants were of type A.

There is also a philosophical objection to such a definition. What does equally likely mean? After some thought one is led to say something about equally probable, which makes the definition circular.

As a summary of the use of this first definition, the following are the salient points:

1. The definition is often useful when it is reasonable to list N equally likely and mutually exclusive outcomes. Examples are found in games of chance such as card games, "craps," drawing balls from urns, etc.
2. In many engineering problems this definition is not useful because it is impossible to list N equally likely and mutually exclusive outcomes.
3. The definition is circular.

Relative Frequency

The second definition, often called the relative frequency definition, is quite popular among engineers and physicists. An experiment is performed under identical conditions n times. If the event A occurs n_A times, then its probability $P(A)$ is

$$P(A) = \lim_{n \to \infty} \frac{n_A}{n}. \qquad (2\text{-}2)$$

5

Note that n and n_A in (2-2) are quite different from N and N_A in (2-1). In this second definition n and n_A are experimental results, while in the first definition the quantities were determined before the experiment and are called a priori numbers.

This second definition is very useful in applications, but once again there are difficulties. In actuality, experiments are only repeated a finite number of times; therefore $P(A)$ cannot be found, and, in fact, the existence of a limit cannot be shown. Therefore this definition has practical limitations.

The relative frequency definition of probability is summarized by the following points:

1. The definition is a useful guide in considering many engineering problems.
2. Since one cannot perform an infinite number of experiments, probability cannot be measured by this definition. However, with a large number of experiments, probability can be estimated.

Personal Probability

A third definition of probability is based upon degree of belief. It is a numerical measure of your (the person assigning the probability) belief about the occurrence of a certain event. Assuming you are a reasonable person, your measure should correspond to the classical or relative frequency definition when you feel they are applicable. However they are not always applicable. Sometimes we are faced with situations where equally likely outcomes cannot be identified, and the situation will not be repeated. However probabilities based on experience and reason are often expressed in terms of a wager.

For example, consider the big football game. If you are willing to bet 3 to 2 that State will beat Tech and you are just as willing to bet 2 to 3 that State will not beat Tech, then your personal probability that State will beat Tech is 3/5. Note that there is no way to use an equally likely or relative frequency definition, but this and other problems more closely related to engineering seem worth considering even if the previous two definitions are not applicable.

Personal probabilities are subjective in the sense that different people will assign different probabilities to the same event because of different backgrounds, experience, and knowledge of the situation. Personal probability was cast into disrepute during the nineteenth century when science was believed to be absolute truth, because, with this definition, the results will depend upon the person solving the problem. However this objection to subjectivity has recently been countered quite effectively by Savage,* Jaynes, and other

* See for instance, L. J. Savage, "Bayesian Statistics" in *Recent Developments in Information and Decision Processes*, edited by Machol and Gray, Macmillan, 1962.

physical scientists. This definition is used and discussed more fully in Chapter VIII. References J1 and S2 contain more thorough discussions.

A More General Framework

In order to have the advantages of the three previous and other definitions of probability, we establish an axiomatic system of mathematics which is general enough to encompass all these definitions. From this point the development follows deductively. There will still be problems when the mathematics are applied to a physical problem; but within the mathematics there are no conceptual ambiguities. This is a standard procedure in engineering; for example, there is always a question of the validity of a differential equation as a model of a system, but there is no question of the solution of the differential equation.

The next section begins a more general theory of probability by introducing set theory. This more general theory is applicable no matter which of the above definitions of probability is chosen for an application.

2-2 ELEMENTARY SET THEORY

A *set* is defined to be a *collection of elements*. A *collection* is well defined if, given any element (object), this element is either a member of the collection or is not a member of the collection. An element cannot be both a member and not a member of the collection.

The concept of a set is very general. Examples of sets are the following:

1. All the students in this class.
2. All the students in this university.
3. All the students in this world.
4. The real numbers.
5. The complex numbers.
6. All even integers less than 99.
7. The resistors used in a radar.

Usually a set will have a rather natural composition, but the definition allows very general descriptions.

8. All pigs' tails and all movie actors.
9. The real numbers and girls.

The general practical procedure for describing a set is to describe the elements of the set. This may be done either by listing the elements or by describing the typical elements. Notationally capital letters A, B, ... will designate sets; and the small letters a, b, ... will designate elements or members of a set. The symbol $\in$ is read "is an element of," and the symbol $\notin$ is read "is not an element of." Thus $x \in A$ is read x is an element of A. When a set is described by listing the elements, the elements will be enclosed by brackets, for example,

$$A = \{2, 4, 6, 8\},$$

$$B = \{\text{John, Henry, Pete}\}.$$

When a set is described by describing the typical element, the notation will consist of the typical element listed and defined within brackets; for example, $C = \{z : z$ is a complex number$\}$, and this is read "C is the set of z such that z is a complex number."

Two special sets are of some interest. A set that has no elements is called the *empty set* or *null set* and will be denoted by $\varnothing$. The empty set is analogous to a committee that has no members. A set having at least one element is called *nonempty*. The *whole* or *entire space* S is a set that contains all other sets under consideration in the problem.

Set Inclusion

Given two sets A and B, the notation

$$A \subset B$$

or equivalently

$$B \supset A$$

is read A is contained in B, or A is a subset of B, or B contains A. (The notation might be remembered by thinking of a sloppy inequality sign.)

$A \subset B$ if and only if every element of A is an element of B.

EXAMPLE 2-2

Let

$$A = \{2, 4, 6, 8, 10\},$$

$$B = \{1, 2, 3, 4, 5, 6, 7, 8, 9, 10\}.$$

Then $A \subset B$, because every element of A is also an element of B.

EXAMPLE 2-3

Let

$$A = \{x : x \text{ is a male student of this class}\},$$

$$B = \{x : x \text{ is a student of this class}\}.$$

Then $B \supset A$ or $A \subset B$ because if $x \in A$ then surely $x \in B$.

There are three results that follow from the definitions given above. For any set A

$$A \subset S, \tag{2-3}$$

$$\varnothing \subset A, \tag{2-4}$$

$$A \subset A. \tag{2-5}$$

These results are obvious by noting that in each case every element of the left set is an element of the right set. In (2-4) this is trivially true since $\varnothing$ has no elements.

Many results in set theory can be seen by use of a Venn diagram. This diagram is shown as Figure 2-1 for a case where $A \subset B \subset C$. S will be represented by a rectangle and A, B, and C by circular curves within the rectangle. A is crosshatched and B is shaded.

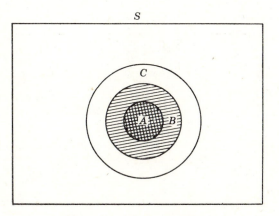

Figure 2-1. Venn diagram.

Set Equality

Two arbitrary sets A and B are called equal if and only if they contain exactly the same elements, or equivalently,

$$A = B \quad \text{if and only if} \quad A \subset B \text{ and } B \subset A.$$

Union

The union of two arbitrary sets A and B is written

$$A \cup B,$$

and is the set of all elements that belong to A *or* belong to B (or to both). $A \cup B$ is sometimes written $A + B$.

EXAMPLE 2-4

$$A \cup A = A.$$ (2-6)

This is shown by noting that $A \cup A \subset A$ and that $A \subset A \cup A$.

The union of the sets A and B is the shaded area in Figure 2-2.

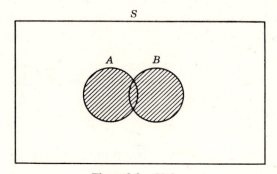

S

A B

Figure 2-2. Union.

EXAMPLE 2-5

If

$$D = \{2, 4, 6, 8\},$$
$$E = \{1, 2, 3, 4\},$$

then

$$D \cup E = \{1, 2, 3, 4, 6, 8\}.$$

Intersection

The intersection of two arbitrary sets A and B is written

$$A \cap B$$

and is the set of all elements that belong to both A *and* B. $A \cap B$ is also written AB.

EXAMPLE 2-6

If

$$D = \{2, 4, 6, 8\},$$
$$E = \{1, 2, 3, 4\},$$
$$D \cap E = \{2, 4\}.$$

EXAMPLE 2-7

$$A \cap A = A.$$ (2-7)

The intersection of two sets A and B is the shaded area in Figure 2-3.

10

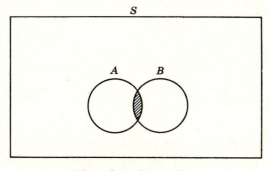

Figure 2-3. Intersection.

Mutually Exclusive

Two sets are called mutually exclusive (or disjoint) if they have no common elements; that is, for two arbitrary sets A, B; A and B are mutually exclusive if

$$A \cap B = AB = \varnothing,$$

where $\varnothing$ is the null set.

Two disjoint (mutually exclusive) sets are shown in the Venn diagram, Figure 2-4.

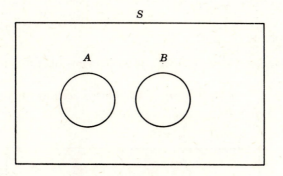

Figure 2-4. Disjoint sets.

Note that A and A are disjoint if and only if $A = \varnothing$.

The n sets $A_1, A_2, \ldots, A_n$ are called mutually exclusive if

$$A_i \cap A_j = \varnothing \text{ for all } i, j, i \neq j.$$

EXAMPLE 2-8

The set $\{2, 4, 6\}$ and the set $\{1, 3, 5\}$ are disjoint.

11

Complement

The complement $\bar{A}$ of a set A relative to S is defined as the set of all elements of S that are not in A.

In Figure 2-5 the shaded area is the complement $\bar{A}$ of a set A. Note that if $A \subset B$, then $\bar{A} \supset \bar{B}$. This may be seen by drawing a Venn diagram. Also if $A = B$, then $\bar{A} = \bar{B}$.

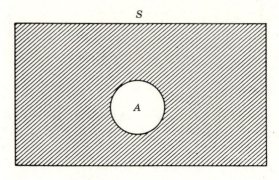

Figure 2-5. A complement.

EXAMPLE 2-9

If $S = \{2, 4, 6, 8, 10\}$ and $A = \{2, 4, 6\}$, then $\bar{A} = \{8, 10\}$.

Difference

The difference $A - B$ is the set of all elements of A that are not elements of B.

Thus $A - B = A \cap \bar{B} = A - AB$. This is seen in Figure 2-6 where the shaded area is $A - B$.

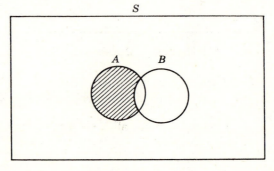

Figure 2-6. $A - B$.

Note that $S - A = S \cap \bar{A} = \bar{A}$. The difference $A - B$ is often called the relative complement of B in A.

EXAMPLE 2-10

Let

$$A = \{1, 2, 3, 4\},$$
$$B = \{2, 4, 6, 8\}.$$

Find $A - B$, $B - A$, $(A - B) \cup B$.

$$A - B = \{1, 3\},$$
$$B - A = \{6, 8\},$$
$$(A - B) \cup B = \{1, 2, 3, 4, 6, 8\}.$$

Results and Examples

This section shows some of the common results which can be used in solving problems and also allows the reader to use the definitions presented on the preceding pages.

Let S be the whole space and let A, B, C be arbitrary subsets of S. As before, $\varnothing$ is the null set and $\varnothing$ is a subset of all sets.

Commutative laws:

$$A \cup B = B \cup A, \tag{2-8}$$

$$A \cap B = B \cap A. \tag{2-9}$$

These are verified either by applying the definition and verifying that each is a subset of the other or more simply by observing the Venn diagram shown in Figure 2-7.

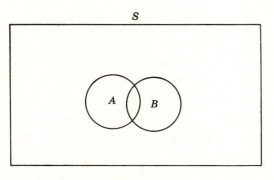

Figure 2-7

13

Associative laws:

$$(A \cup B) \cup C = A \cup (B \cup C) = A \cup B \cup C, \qquad (2\text{-}10)$$

$$(A \cap B) \cap C = A \cap (B \cap C) = A \cap B \cap C. \qquad (2\text{-}11)$$

These results are easily verified in the same manner as (2-8) and (2-9).
Distributive laws:

$$A \cap (B \cup C) = (A \cap B) \cup (A \cap C), \qquad (2\text{-}12)$$

$$A \cup (B \cap C) = (A \cup B) \cap (A \cup C). \qquad (2\text{-}13)$$

In order to check (2-12) consider the Venn diagram of Figure 2-8. $B \cup C$ is the /// shaded set. $A \cap (B \cup C)$ is represented by the crosshatched area.

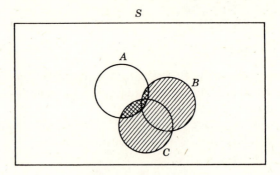

Figure 2-8. $A \cap (B \cup C)$.

In Figure 2-9 $A \cap B$ is the /// shaded set and $A \cap C$ is the \\\ shaded set. The union is seen to be the same area as that crosshatched in Figure 2-8.

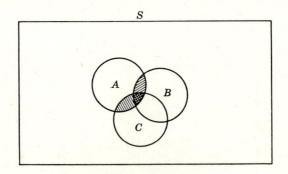

Figure 2-9. $(A \cap B) \cup (A \cap C)$.

As an exercise the reader may verify (2-13) by Venn diagram. Both sides of the equality are represented by the shaded area of Figure 2-10.

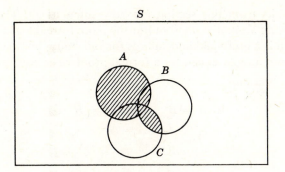

Figure 2-10. $A \cup (B \cap C)$.

De Morgan's law:

$$\overline{(A \cup B)} = \bar{A} \cap \bar{B}, \tag{2-14}$$

$$\overline{(A \cap B)} = \bar{A} \cup \bar{B}. \tag{2-15}$$

These results are again checked by using a Venn diagram. Figure 2-11 shows (2-14) while Figure 2-12 shows (2-15).

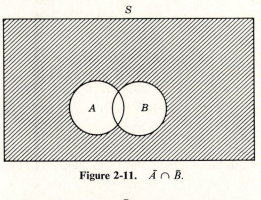

Figure 2-11. $\bar{A} \cap \bar{B}$.

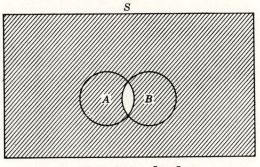

Figure 2-12. $\bar{A} \cup \bar{B}$.

15

Other results from this section are now summarized. These should be checked by using Venn diagrams and/or the more formal logic. Note that the Venn diagram is a useful aid and suffices for this book, but like all geometric pictures it is not a replacement for a formal proof when new results are to be proven.

$$\text{If} \quad A = B \quad \text{then} \quad \bar{A} = \bar{B}, \tag{2-16}$$

$$A \cup \bar{A} = S, \tag{2-17}$$

$$A \cap \bar{A} = \varnothing, \tag{2-18}$$

$$A \cup S = S, \tag{2-19}$$

$$A \cap S = A, \tag{2-20}$$

$$A \cup \varnothing = A, \tag{2-21}$$

$$A \cap \varnothing = \varnothing, \tag{2-22}$$

$$A \cup A = A, \tag{2-6}$$

$$A \cap A = A, \tag{2-7}$$

$$\overline{(\bar{A})} = A. \tag{2-23}$$

Example 2-11

$$(A \cap B) \cup (A - B) = ?$$

$$\begin{aligned}(A \cap B) \cup (A - B) &= (A \cap B) \cup (A \cap \bar{B}) \quad \text{by definition of difference} \\ &= A \cap (B \cup \bar{B}) \quad \text{using (2-12)} \\ &= A \cap S \quad \text{using (2-17)} \\ &= A \quad \text{using (2-20)}.\end{aligned}$$

Thus

$$(A \cap B) \cup (A - B) = A.$$

This result may also be seen from the Venn diagram in Figure 2-13.

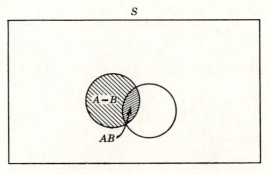

Figure 2-13. $AB \cup (A - B)$.

EXAMPLE 2-12

Express the set composed of the shaded region of Figure 2-14 in terms of the given sets.

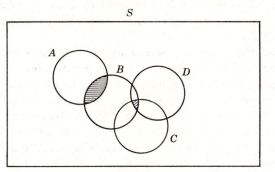

Figure 2-14

The set is

$$(A \cap B) \cup (B \cap C \cap D).$$

2-3 SAMPLE SPACE

In this section the concept of sets is specialized to the theory of probability. When talking about probability, the whole space will consist of elements that are *outcomes* of an *experiment*. In this text *an experiment is a sequence of actions which produces some properties of interest*. With this definition an experiment can be the measurement of a voltage, but it can also be the toss of a die or the turn of a roulette wheel. Although the latter two experiments are not commonly associated with science, they are experiments in the sense that the outcomes are not known in advance. This definition of experiment is broad enough to encompass the usual scientific experiment and other actions that are sometimes regarded as observations.

An outcome of the experiment is one of the possible properties of interest. The totality of all these outcomes is the sample space. Thus in applications of probability, outcomes correspond to elements and the sample space corresponds to S, the whole space. With these definitions an event may be defined as a collection of outcomes. Thus an event is a set or a subset of the sample space.

EXAMPLE 2-13

Consider the experiment of tossing a die. If the possible outcomes of this experiment are the possible up faces after the die stops, then the sample space

S consists of the events:

$$\{f_1\}, \{f_2\}, \{f_3\}, \{f_4\}, \{f_5\}, \{f_6\}$$

and all other events that can be generated by usual set operations (union, intersection, complementation) of these events. There are exactly 2^6 distinct events (see Problem 13).

To illustrate that a sample space is not fixed by the actions of performing the experiment, but by the interests of the person observing the outcome, consider the experiment of launching a satellite. To the public, a sample space for this experiment may consist of the two events, success and failure. To the designer of a TV camera used in the satellite the space may consist of three events: the camera performed satisfactorily, the camera failed, and the camera was not tried. Or the designer may wish to describe the space by a number of events which define levels of performance of the camera. A ballistics expert may choose as his space the set of possible trajectories of the satellite.

From the example it may be seen that the physical description of an experiment does not specify the events or the sample space. The events, which are groupings of outcomes, must be defined by the person working the problem.

Figure 2-15 shows a representation of a sample space with outcomes (points) and events (closed curves). More specific examples of sample spaces follow the definition of probability.

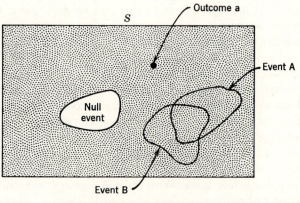

Figure 2-15. Sample space.

2-4 PROBABILITY MEASURE

Now the problem is to assign a "measure" to the events of the sample space in such a way that the measure corresponds to our intuitive notions of probability. That is, the measure should be defined so that results from applying the theory will be applicable no matter which of our earlier definitions of probability is adopted.

A set of axioms which result in a satisfactory theory is the following.*

1. $P(A) \geq 0$ for all events $A \subset S$. (A1)

2. $P(S) = 1$. (A2)

3. † If $A \cap B = \varnothing$,   then

$$P(A \cup B) = P(A) + P(B).$$ (A3)

Thus probability is a non-negative (axiom 1), normed (axiom 2), additive (axiom 3) set function (it is defined on a class of sets, i.e., its domain is a class of sets).

The following property can be deduced from the three axioms:

$$P(\varnothing) = 0.$$ (2-24)

This follows because from (2-21)

$$A \cup \varnothing = A.$$

Assigning probabilities to both sides

$$P(A \cup \varnothing) = P(A).$$

Using (A3) and the fact that $A \cap \varnothing = \varnothing$,

$$P(A) + P(\varnothing) = P(A)$$

or

$$P(\varnothing) = P(A) - P(A) = 0.$$

Another property is of interest here,

$$P(A) \leq 1.$$ (2-25)

This follows from

$$P(A \cup \bar{A}) = P(S) = 1,$$
$$P(A) + P(\bar{A}) = 1,$$
$$P(A) = 1 - P(\bar{A}) \leq 1,$$

where the last inequality is true because $P(\bar{A}) \geq 0$.

From the above development

$$P(\bar{A}) = 1 - P(A).$$ (2-26)

---

* The fact that the measure must be applied on a sigma algebra of sets rather than an arbitrary collection of sets is neglected in this elementary presentation. In the application of interest here, this is of no practical consequence.

† A later axiom (A3a) replaces (A3) if an infinite number of events is considered.

One other property is of interest at this time,

$$P(A \cap \bar{B}) = P(A) - P(A \cap B). \tag{2-27}$$

This follows from

$$A = (A \cap B) \cup (A \cap \bar{B}) \quad \text{and} \quad (A \cap B) \cap (A \cap \bar{B}) = \varnothing,$$
$$P(A) = P(A \cap B) + P(A \cap \bar{B}),$$
$$P(A \cap \bar{B}) = P(A) - P(A \cap B).$$

It should be observed that the axioms A1, A2, and A3 satisfy both the "equally likely" and the "relative frequency" definitions of probability. To check the former definition, let N_A be the number of equally likely outcomes of event A and N be the total number of outcomes. Certainly

$$\frac{N_A}{N} \geq 0, \quad \text{showing (A1)}.$$

$$\frac{N}{N} = 1, \quad \text{showing (A2)}.$$

If A and B are mutually exclusive

$$\frac{N_{(A \cup B)}}{N} = \frac{N_A + N_B}{N} = \frac{N_A}{N} + \frac{N_B}{N}, \quad \text{showing (A3)}.$$

Since properties (2-24) through (2-27) were derived from the axioms A1, A2, and A3, then automatically the "equally likely" definition satisfies properties (2-24) through (2-27). This demonstrates one of the advantages of an axiomatic system: after setting up the system we have less work in applications.

The relative frequency definition can be checked in exactly the same manner as shown above with small n's used in place of capital N's. The "personalistic" view will also fit the axioms as Savage* shows.

Examples

The application of probability to physical problems can now be attempted. Special emphasis will be placed on visualizing the probability space.

EXAMPLE 2-14

The ultimate goal is to apply the theory to engineering systems, but a game of chance will be used for this example because the experiment is easily described and probability can be assigned easily by using the equally likely definition of probability.

* See Ref. S2.

Let the experiment be the tossing of a true die. (By true, it is implied that each face is equally likely.) The sample space is described in Example 2-13 and is shown in Figure 2-16. What is the probability that f_1 or f_2 appears? Let $B = \{f_1 \text{ or } f_2\} = \{f_1\} \cup \{f_2\}$. Now $\{f_1\}$ is certainly mutually exclusive of $\{f_2\}$ (i.e., $\{f_1\} \cap \{f_2\} = \varnothing$) and since $P\{f_i\} = P\{f_j\}i, j = 1, \ldots, 6$, and

$$\sum_{i=1}^{6} P\{f_i\} = 1,$$

then

$$P\{f_i\} = \tfrac{1}{6}.$$

Thus

$$P(B) = P\{f_1\} + P\{f_2\} = \tfrac{1}{6} + \tfrac{1}{6} = \tfrac{1}{3}.$$

S

$\{f_1\}$	$\{f_2\}$
$\{f_3\}$	$\{f_4\}$
$\{f_5\}$	$\{f_6\}$

Figure 2-16

EXAMPLE 2-15

The experiment consists of drawing one card at random (implies each card is equally likely) from an ordinary deck of cards. The sample space is composed of 52 basic events, one for each of the equally likely and mutually exclusive outcomes and events generated from these 52 basic events (see Figure 2-17).

What is the probability that it is (a) a spade; (b) a heart, a diamond, or a club; (c) an ace; (d) the ace of clubs; (e) the ace of clubs *and* the ace of spades; (f) the four of spades or the ten of diamonds?

(a) Since there are four suits containing 13 outcomes the probability of a spade is $13/52 = 1/4$.

(b) Let A denote the event $\{$spade$\}$. Now let $\bar{A}$ denote the event $\{$not a spade$\}$. It follows $\{$a heart, a diamond, or a club$\} = \bar{A}$. Thus $P(\bar{A}) = 1 - P(A) = 3/4$.

(c) $P\{$ace$\} = 4/52 = 1/13$.

(d) $P\{$ace of clubs$\} = 1/52$.

(e) Let $\{$ace of clubs$\} = C$ and $\{$ace of spades$\} = B$;

$$B \cap C = \varnothing.$$

21

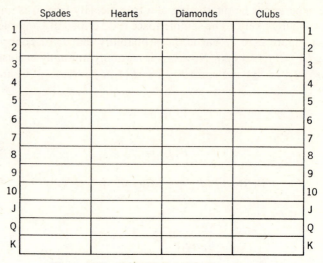

Figure 2-17. Sample space for drawing card.

Thus

$$P(B \cap C) = 0.$$

(f) $P(\{\text{four of spades}\} \cup \{\text{ten of diamonds}\}) = 1/52 + 1/52 = 1/26.$

EXAMPLE 2-16

In this example we consider an uncountable number of outcomes. However only a finite number of events are defined. A consideration of assignment of probability when more than a countable number of events are defined is postponed until random variables are defined in Chapter IV.

A plane is equally likely to arrive anytime between 10:00 p.m. and 11:00 p.m. Find the probability that the plane arrives (a) between 10:00 and 10:15; (b) between 10:15 and 11:00.

The probability space S can be considered to be the real line segment from 10 to 11. The event asked for in (a) will be called A and the event asked for in (b) is $\bar{A}$ (see Figure 2-18).

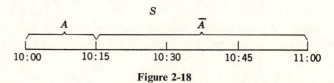

Figure 2-18

The probability of A, $P(A)$, is 15/60 or 1/4. Similarly

$$P(\bar{A}) = 1 - \tfrac{1}{4} = \tfrac{3}{4}.$$

22

EXAMPLE 2-17

In this example the outcomes are not precisely defined, and probability is defined as relative frequency. In the picture of the sample space, probability may be visualized as the area of the set.

It is known that the relative frequency of defectives fuses in a certain type of fuse is 1/100. What is the probability that a fuse is (a) good G, (b) not good $\bar{G}$? The sample space is shown in Figure 2-19.

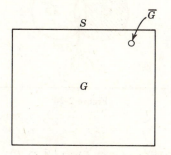

Figure 2-19

$$P(G) = .99$$
$$P(\bar{G}) = .01.$$

Probability of Unions

Consider now the probability of the union of more than two events. First, consider that all of the events are mutually exclusive, that is, $A_i \cap A_j = \varnothing$ for all $i \neq j$. Then if there are a finite number of events

$$P(A_1 \cup A_2 \cup A_3 \cup \cdots A_n) = P\left(\bigcup_{i=1}^{n} A_i\right) = \sum_{i=1}^{n} P(A_i). \tag{2-28}$$

This result can be shown by induction using (A3).

If the union is countably infinite the following axiom replaces A3.

If $A_i \cap A_j = \varnothing$ for all $i \neq j$, then

$$P\left(\bigcup_{i} A_i\right) = \sum_{i} P(A_i). \tag{A3a}$$

Now consider the union of two events when the events are not mutually exclusive. Given two arbitrary events A and B, then

$$P(A \cup B) = P(A) + P(B) - P(A \cap B). \tag{2-29}$$

23

This result can be shown by splitting $A \cup B$ into mutually exclusive sets and applying (A3). The Venn diagram shown in Figure 2-20 will be useful.

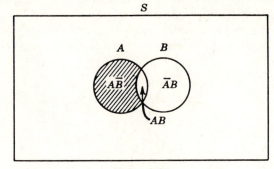

Figure 2-20

From Figure 2-20

$$A \cup B = (A \cap \bar{B}) \cup (A \cap B) \cup (\bar{A} \cap B) = A \cup (\bar{A} \cap B).$$

With the use of (A3)

$$P(A \cup (\bar{A} \cap B)) = P(A) + P(\bar{A} \cap B).$$

With the use of (2-27) on $P(\bar{A} \cap B)$

$$P(A \cup B) = P(A) + P(B) - P(A \cap B).$$

Consider the problem of three arbitrary sets A, B, C. Then

$$P(A \cup B \cup C) = P(A) + P(B) + P(C) - P(A \cap B) - P(A \cap C)$$
$$- P(B \cap C) + P(A \cap B \cap C). \qquad (2\text{-}30)$$

This may be shown in the same fashion as (2-29), but it is also instructive to think of the problem from the point of view of areas on the Venn diagram shown in Figure 2-21.

If one simply says

$$P(A \cup B \cup C) = P(A) + P(B) + P(C),$$

then the area $A \cap B$ (shaded) has been added twice. Similarly $A \cap C$ and $B \cap C$ have been added twice. Thus they should be subtracted, resulting in

$$P(A \cup B \cup C)$$
$$= P(A) + P(B) + P(C) - P(A \cap B) - P(A \cap C) - P(B \cap C).$$

This is correct except that the area $A \cap B \cap C$ (crosshatched) has been added three times and subtracted three times. To include this area it needs

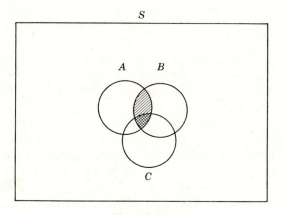

Figure 2-21

to be added. This produces result (2-30). The student should also show (2-30) by using the more formal method used in showing (2-29).

A general relation for the union of general events could be derived by an extension of the above reasoning. However, for our purposes, such a relation is not needed.

The following result is of quite some practical significance.

$$P(A_1 \cup A_2 \cup \cdots \cup A_n) \leq \sum_{i=1}^{n} P(A_i) \qquad (2\text{-}31)$$

where A_i, $i = 1, 2, \ldots, n$, are events.

This result is easily seen by using a Venn diagram. From this diagram it may also be seen that the equality sign holds only when all of the events are mutually exclusive. When the events are not mutually exclusive, some areas need to be subtracted, resulting in the inequality.

EXAMPLE 2-18

Suppose there is a group of 40 people of which 20 are engineers under 30 years of age, 10 are engineers over 30, four are nonengineers under 30, and 6 are nonengineers over 30. If one person is selected at random from the group, what is the probability that that person is "an engineer or a person over 30"?

Let E be the event {engineer} and F the event {over 30}. Then

$$P(E \cup F) = P(E) + P(F) - P(EF)$$
$$= \tfrac{30}{40} + \tfrac{16}{40} - \tfrac{10}{40} = \tfrac{36}{40}.$$

25

2-5 COMBINED EXPERIMENTS

The previous sections discussed the probability associated with one experiment. In most applications, however, the problems involve an experiment composed of more than one subexperiment. For example, the crapshooter is interested in the roll of two dice.

The roll of two dice is thought of as the combination of two experiments, each involving the roll of one die. In this book much attention is devoted to systems made up of subsystems or components. An experiment describing the system's behavior is thought of as a combination of experiments describing the subsystems or components.

The sample spaces of a total experiment, consisting of two subexperiments is, as before, composed of outcomes. Each outcome in S consists of the outcomes of the first and second subexperiment. If the events $A_1, A_2, \ldots, A_n$ are defined for the first subexperiment, and the events $B_1, B_2, \ldots, B_m$ are defined for the second subexperiment, then the event A_iB_j, $i = 1, \ldots, n$, $j = 1, \ldots, m$ is an event of the total experiment. An outcome in S determines an event A_iB_j which is an event A_i of the first subexperiment *and* an event B_j of the second subexperiment.

EXAMPLE 2-19

Consider the combined experiment of tossing a coin and rolling a die. Describe the total sample space.

$$S = \{h, f_1; h, f_2; h, f_3; h, f_4; h, f_5; h, f_6; t, f_1; t, f_2; t, f_3; t, f_4; t, f_5; t, f_6\}.$$

This is illustrated by the special Venn diagram or Karnaugh map shown in Figure 2-22. The outcome h, f_4 determines the event $\{h, f_4\}$ (shown XXX), the event $\{h\}$ (shown ////) which is associated with the tossing of the coin, and the event $\{f_4\}$ (shown \\\\) which is associated with rolling a die.

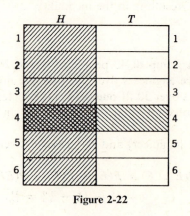

Figure 2-22

Joint Probability

The probability of an event which is the intersection of events from subexperiments is often called the *joint* probability of the event.

In example 2-19, if we assume the 12 outcomes are equally likely, then the joint probability of a head and f_4 is

$$P(\{h, f_4\}) = \tfrac{1}{12}.$$

Marginal Probability

If the events $A_1, A_2, \ldots$, associated with one subexperiment are mutually exclusive ($A_i A_j = \varnothing$, $i \neq j$) and exhaustive ($\cup_i A_i = S$), then

$$P(B_j) = P(B_j S) = P[B_j(\cup_i A_i)]$$
$$= P[\cup_i A_i B_j],$$
$$P(B_j) = \sum_i P(A_i B_j). \tag{2-32}$$

If the event B_j is associated with a second subexperiment, $P(B_j)$ is often called a marginal probability.

In Example 2-19 the probability of a head [using (2-32)] is

$$P(\text{head}) = P(\{h, f_1\} \cup \{h, f_2\} \cup \cdots \cup \{h, f_6\})$$
$$= \tfrac{1}{12} + \tfrac{1}{12} + \tfrac{1}{12} + \tfrac{1}{12} + \tfrac{1}{12} + \tfrac{1}{12} = \tfrac{1}{2}.$$

Note that the probability of a head and any outcome of the die is the same as the probability of a head, and the result, 1/2, is just as expected.

EXAMPLE 2-20

An examination of surveillance records on a specific transistor showed the following results when classified by manufacturer and class of defect:

| Manufacturer | Class of defect | | | | | |
	$B_1 =$ none	$B_2 =$ critical	$B_3 =$ serious	$B_4 =$ minor	$B_5 =$ incidental	Totals
M_1	124	6	3	1	6	140
M_2	145	2	4	0	9	160
M_3	115	1	2	1	1	120
M_4	101	2	0	5	2	110
Totals	485	11	9	7	18	530

What is the probability of a transistor selected at random from the 530 transistors (a) being from manufacturer M_2 and having no defects, (b) having a critical defect, (c) being from manufacturer M_1?

(a) This is a joint probability and is found by assuming that each transistor is equally likely to be selected. There are 145 transistors from M_2 having no defects out of a total of 530 transistors. Thus $P(M_2 B_1) = \frac{145}{530}$.

(b) This calls for a marginal probability.

$$P(B_2) = P(M_1 B_2) + P(M_2 B_2) + P(M_3 B_2) + P(M_4 B_2)$$
$$= \frac{6}{530} + \frac{2}{530} + \frac{1}{530} + \frac{2}{530} = \frac{11}{530}.$$

Note that $P(B_2)$ can also be found in the bottom margin of the table, that is, $P(B_2) = \frac{11}{530}$.

(c) Directly from the right margin
$$P(M_1) = \frac{140}{530}.$$

Conditional Probability

In most practical work joint probabilities are actually obtained from the probabilities associated with the subexperiments. To use these probabilities the concept of the conditional probability of an event A given an event B is needed.

The definition of the conditional probability of A given B is motivated by the following discussion where probability is defined by the equally likely definition. Let N be the total number of outcomes and let N_E be the number of outcomes corresponding to any event E. Then

$$P(AB) = \frac{N_{AB}}{N}$$

$$P(B) = \frac{N_B}{N}.$$

Given the event B has occurred we know the outcome is one of the outcomes in B. As far as we know each outcome in B is still equally likely. Thus the probability of A given B, $P(A \mid B)$ is the ratio of the outcomes in A and in B to the outcomes in B.

$$P(A \mid B) = \frac{N_{AB}}{N_B}.$$

Thus

$$P(AB) = \frac{N_{AB}}{N} = \frac{N_{AB} N_B}{N_B N} = P(A \mid B)P(B).$$

Based on this motivation we define

$$P(A \mid B) = \frac{P(AB)}{P(B)}. \tag{2-33}$$

This definition implies $P(B) \neq 0$.

We now show that $P(A \mid B)$ is indeed a probability. That is, we must show that

1. $P(A \mid B) \geq 0$;
2. $P(S \mid B) = 1$.
3. If $A \cap C = \varnothing$ then

$$P(A \cup C \mid B) = P(A \mid B) + P(C \mid B).$$

The first follows directly from the definition because the ratio of two non-negative numbers is non-negative. The second follows from the fact that $BS = B$. Thus

$$P(S \mid B) = \frac{P(BS)}{P(B)} = \frac{P(B)}{P(B)} = 1.$$

The third is shown as follows:

$$P(A \cup C \mid B) = \frac{P[(A \cup C)B]}{P(B)} = \frac{P(AB \cup CB)}{P(B)}$$

$$= \frac{P(AB) + P(CB)}{P(B)} = P(A \mid B) + P(C \mid B).$$

We now give a useful interpretation of conditional probability. Consider Figure 2-23 and interpret the area of the sets in Figure 2-23a as the probability of the events.

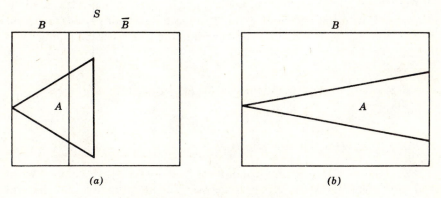

Figure 2-23. Conditional probability.

By the definition of conditional probability, (2-33), the probability of A given B is the area of AB divided by the area of B. That is, if B is taken as the whole space with an area of one as in Figure 2-23b, then the probability of A given B is the area of A (i.e. in B) when the area of B has been increased to one.

EXAMPLE 2-21

In Example 2-20 what is the probability of a transistor (a) having a critical defect given the transistor is from manufacturer M_2, (b) being from manufacturer M_1 given it has a critical defect?

(a) This conditional probability is found by the interpretation that given the transistor is from manufacturer M_2, there are 160 outcomes in the space, two of these have critical defects. Thus $P(B_2 \mid M_2) = \frac{2}{160}$. By the formal definition

$$P(B_2 \mid M_2) = \frac{P(B_2 M_2)}{P(M_2)} = \frac{2/530}{160/530} = \frac{2}{160}.$$

(b) $P(M_1 \mid B_2) = \frac{6}{11}$.

Note that joint probabilities can be computed from conditional probabilities. Indeed, from (2-33)

$$P(AB) = P(A \mid B)P(B) = P(B \mid A)P(A). \tag{2-34}$$

Marginal probabilities can also be computed from conditional probabilities. Using (2-32) and (2-34), where the A_i's are mutually exclusive and exhaustive, one obtains

$$P(B_j) = \sum_i P(B_j \mid A_i)P(A_i). \tag{2-35}$$

Thus the marginal probability of an event can be viewed as the weighted (by the probabilities of A_i) average of the conditional probabilities.

EXAMPLE 2-22

An experiment consists of tossing a coin. If the coin is heads one die is thrown and the result is recorded; if the coin is tails two dice are thrown and their sum is recorded. What is the probability that the recorded number will be 2?

Using (2-35)

$$P(2) = P(2 \mid 1 \text{ die})P(1 \text{ die thrown}) + P(2 \mid 2 \text{ dice})P(2 \text{ dice thrown})$$

$$P(2) = \tfrac{1}{6} \cdot \tfrac{1}{2} + \tfrac{1}{36} \cdot \tfrac{1}{2} = \tfrac{1}{12} + \tfrac{1}{72} = \tfrac{7}{72}.$$

More Than Two Subexperiments

This section will interpret conditional and marginal probability when there are more than two subexperiments. The interpretations are natural extensions of the definitions given for two subexperiments.

When the combined experiment consists of k subexperiments, then an outcome of S determines an event A_i in the first subexperiment and an event B_j in the second subexperiment, . . . , and an event Z_l in the kth subexperiment.

A set from the combined sample space will be denoted by

$$A_i B_j \cdots Z_l.$$

Using (2-34) with BC an event

$$P(ABC) = P(A \mid BC)P(BC). \tag{2-36}$$

Using (2-34) in (2-36) we obtain the product rule

$$P(ABC) = P(A \mid BC)P(B \mid C)P(C). \tag{2-37}$$

This may also be written as

$$P(ABC) = P(AB \mid C)P(C). \tag{2-38}$$

Note that (2-36) and (2-38) define the conditional probabilities $P(A \mid BC)$ and $P(AB \mid C)$. These definitions can be shown to agree with the relative frequency and equally likely definitions of probability.

The extension to a larger number of subexperiments should be obvious. We show the result for 5 subexperiments.

$$P(ABCDE) = P(A \mid BCDE)P(B \mid CDE)P(C \mid DE)P(D \mid E)P(E),$$

where

$$P(A \mid BCDE) = \frac{P(ABCDE)}{P(BCDE)},$$

$$P(B \mid CDE) = \frac{P(BCDE)}{P(CDE)}.$$

The idea of marginal probability can also be extended. Marginal probabilities are, as before, the probabilities assigned to the events in the individual subexperiments. Marginal and conditional probabilities are related by

$$P(A) = \sum_{i=1}^{n} \sum_{j=1}^{m} P(A \mid B_i C_j)P(B_i \mid C_j)P(C_j), \tag{2-39}$$

where $B_i B_j = \varnothing$, $i \neq j$ and $\bigcup_{i=1}^{n} B_i = S$; $C_i C_j = \varnothing$, $i \neq j$ and $\bigcup_{j=1}^{m} C_j = S$. This may be shown by a simple extension of the reasoning used to show (2-35).

2-6 INDEPENDENCE

In some cases it seems natural that the probabilities of events of one subexperiment would not be affected by the occurence of events of the other subexperiments. This idea has been formalized with the following definition. Two events A and B are statistically independent if

$$P(AB) = P(A)P(B). \tag{2-40}$$

This definition is exactly equivalent to either of the following statements:

$$P(A \mid B) = P(A) \tag{2-41a}$$

or

$$P(B \mid A) = P(B). \tag{2-41b}$$

With the probabilities for the subspaces and the joint probabilities given, it is a simple matter to check for independence. The real problem that must be faced in the application of interest in this book is to decide from reasoning, without the joint probabilities given, if events are statistically independent.

As will be illustrated more fully later, sometimes events are not statistically independent even though no direct cause-and-effect relation may be seen. For example consider the events that represent the failure (by opening) of two resistors. It is easily seen that if the two resistors are used in parallel, then the failure of one will change the current in the other, so one would expect these two events to be statistically dependent (not independent). However suppose they are not used in the same circuit but are used in the same piece of equipment. Then the failure of any one may change the ambient temperature, thereby changing the probability of failure of the other. A further removed type of dependence is illustrated by the thought that if one resistor failed, perhaps it failed because of poor design, and maybe the other resistor suffers from the same design philosophy.

The question of statistical independence is often a difficult one; the best one can do is to try to reason and collect data. It should be remembered that the defining equations are the criteria, not some vague words about cause and effect.

The concept of independence will now be extended to n events. The events $A_1, A_2, \ldots, A_n$ are statistically independent if for all choices of K integers, $i_1 < i_2 \cdots < i_K, 1 \leq i_1, i_K \leq K$, for $K = 2, \ldots, n$

$$P(A_{i_1} A_{i_2} \cdots A_{i_K}) = P(A_{i_1})P(A_{i_2}) \cdots P(A_{i_K}). \tag{2-42}$$

To illustrate this definition let $n = 3$. Then the definition requires

$$P(A_1 A_2) = P(A_1)P(A_2),$$
$$P(A_1 A_3) = P(A_1)P(A_3),$$
$$P(A_2 A_3) = P(A_2)P(A_3),$$
$$P(A_1 A_2 A_3) = P(A_1)P(A_2)P(A_3).$$

All of these must be satisfied. One might think that pairwise independence (the first three equations) would be sufficient. The following example shows otherwise. Let A be the event that the first of two coins is a head, let B be the event that the second coin is a head, and finally let C be the event that the two coins match (both heads or both tails). Assuming fair coins and independent coin tosses

$$P(A) = P(A \mid B) = P(A \mid C) = \tfrac{1}{2},$$
$$P(B) = P(B \mid A) = P(B \mid C) = \tfrac{1}{2},$$
$$P(C) = P(C \mid A) = P(C \mid B) = \tfrac{1}{2}.$$

Thus the events are pairwise independent. But they are not independent events, since

$$P(ABC) = \tfrac{1}{4} \neq P(A)P(B)P(C) = \tfrac{1}{8}.$$

In reasoning about independence often one will reason about the sub-experiments rather than about the events. Two subexperiments are said to be independent if any two events A_i and B_j, where A_i is selected from the first experiment and B_j is selected from the second experiment, are independent. More generally, n subexperiments are said to be independent if any set of n events A_i, B_j, ..., N_K, where A_i is selected from the first experiment, B_j from the second, ..., N_K is selected from the nth subexperiment are independent.

Conditional independence is also defined. Two events A and B are conditionally independent if

$$P(AB \mid C) = P(A \mid C)P(B \mid C). \qquad (2\text{-}43)$$

Note that conditional independence does not imply independence.

EXAMPLE 2-23

Consider two relay contacts activated by a single armature. Let A be the event that the first contact is closed, B be the event that the second contact is closed, and C be the event that the armature is energized. We are given

$$P(C) = .5,$$
$$P(A \mid C) = .9,$$
$$P(B \mid C) = .9,$$
$$P(A \mid \bar{C}) = .2,$$
$$P(B \mid \bar{C}) = .1.$$

The events A and B are conditionally independent. Find $P(AB \mid C)$ and $P(A)$, $P(B)$, $P(AB)$. Using (2-43)

$$P(AB \mid C) = P(A \mid C)P(B \mid C) = (.9)(.9) = .81.$$

With the use of (2-35)

$$P(A) = P(A \mid C)P(C) + P(A \mid \bar{C})P(\bar{C}) = (.9)(.5) + (.2)(.5) = .55,$$
$$P(B) = P(B \mid C)P(C) + P(B \mid \bar{C})P(\bar{C}) = (.9)(.5) + (.1)(.5) = .5,$$
$$P(AB) = P(AB \mid C)P(C) + P(AB \mid \bar{C})P(\bar{C}) = (.81)(.5) + (.02)(.5) = .415.$$

Note that $P(AB) = .415 \neq P(A)P(B) = .275$.

2-7 BAYES' RULE

From (2-34) if $P(B) \neq 0$, then

$$P(A \mid B) = \frac{P(B \mid A)P(A)}{P(B)}. \tag{2-44}$$

This is sometimes called Bayes' rule.

Also recall that if the events A_i, $i = 1, 2, \ldots, n$ are mutually exclusive $(A_i \cap A_j = \varnothing)$ and exhaustive $(\bigcup_{i=1}^{n} A_i = S)$, then

$$P(B) = \sum_{i=1}^{n} P(B \mid A_i)P(A_i). \tag{2-45}$$

Letting $A = A_j$ and using (2-44) and (2-45), we get the most common form of Bayes' rule:

$$P(A_j \mid B) = \frac{P(B \mid A_j)P(A_j)}{\sum_{i=1}^{n} P(B \mid A_i)P(A_i)}. \tag{2-46}$$

Equation 2-46 will be the basis for the estimation of probabilities from data discussed in some detail in Chapter VIII. Here we illustrate Bayes' theorem with two examples.

EXAMPLE 2-24

The probabilities are equal that any of three urns A_1, A_2, and A_3 will be selected. A_1 contains four white and one black ball. A_2 contains three white and two black balls. A_3 contains one white and four black balls. Given an urn is selected and a ball is drawn and observed to be black, what is the probability that the selected urn is A_3?

Using (2-46) we get

$P(A_3 \mid \text{black})$

$$= \frac{P(\text{black} \mid A_3)\, P(A_3)}{P(\text{black} \mid A_1)\, P(A_1) + P(\text{black} \mid A_2)\, P(A_2) + P(\text{black} \mid A_3)\, P(A_3)}$$

$$= \frac{\frac{4}{5}\frac{1}{3}}{\frac{1}{5}\frac{1}{3} + \frac{2}{5}\frac{1}{3} + \frac{4}{5}\frac{1}{3}} = \frac{4}{1 + 2 + 4} = \frac{4}{7}.$$

Thus we change from a probability of 1/3 to a probability of 4/7 by observing one black ball.

Now assume that three balls are drawn without replacement from the selected urn and that they are all black. Simple logic tells us we must have selected A_3. Bayes' rule also gives the same conclusion:

$P(A_3 \mid 3 \text{ black})$

$$= \frac{P(3 \text{ black} \mid A_3)\, P(A_3)}{P(3 \text{ black} \mid A_1)\, P(A_1) + P(3 \text{ black} \mid A_2)\, P(A_2) + P(3 \text{ black} \mid A_3)\, P(A_3)}$$

$$= \frac{(\frac{4}{5} \cdot \frac{3}{4} \cdot \frac{2}{3})\frac{1}{3}}{0\frac{1}{3} + 0\frac{1}{3} + (\frac{4}{5} \cdot \frac{3}{4} \cdot \frac{2}{3})\frac{1}{3}} = 1.$$

EXAMPLE 2-25

A binary communication channel is a system which carries data in the form of one of two types of signals, say, either zeros or ones. Because of noise a transmitted zero is sometimes received as a one and a transmitted one is sometimes received as a zero.

We assume that for a certain binary communication channel, the probability a transmitted zero is received as a zero is .95 and the probability that a transmitted one is received as a one is .90. We also assume the probability a zero is transmitted is .4. Find

1. Probability a one is received.
2. Probability a one was transmitted given a one was received.

Defining

$$A = \{\text{one transmitted}\},$$
$$\bar{A} = \{\text{zero transmitted}\},$$
$$B = \{\text{one received}\},$$
$$\bar{B} = \{\text{zero received}\}.$$

From the problem statement

$$P(A) = .6, \qquad P(B \mid A) = .90, \qquad P(B \mid \bar{A}) = .05.$$

1. With the use of (2-35)

$$P(B) = P(B \mid A)P(A) + P(B \mid \bar{A})P(\bar{A})$$
$$= .90\,(.6) + .05\,(.4)$$
$$= .56.$$

2. Using Bayes' rule, (2-44),

$$P(A \mid B) = \frac{P(B \mid A)P(A)}{P(B)} = \frac{(.90)(.6)}{.56} = \frac{27}{28}.$$

35

2-8 TWO USEFUL MODELS OF SEQUENTIAL EXPERIMENTS

Repeated independent trials will be considered. In this case the experiment consists of n subexperiments, each one of which describes the same situation. An important special case of repeated independent subexperiments are Bernoulli trials. In this special case each of the individual sample spaces consists only of the events $\{A, \bar{A}, \varnothing, S\}$. A common example of Bernoulli trials is a coin-tossing experiment.

This introduces questions such as: what is the probability that the event A will occur exactly 47 times in 100 trials.

To aid the reader in getting acquainted with problems of this type, consider the example of coin tossing.

1. A coin is tossed once: the results are either H or T.
2. A coin is tossed twice (or two coins are tossed simultaneously): the results are $H_1 H_2$; $H_1 T_2$; $T_1 H_2$; or $T_1 T_2$.
3. A coin is tossed three times (or three coins are tossed simultaneously): the results are $H_1 H_2 H_3$; $H_1 H_2 T_3$; $H_1 T_2 H_3$; $H_1 T_2 T_3$; $T_1 H_2 H_3$; $T_1 H_2 T_3$; $T_1 T_2 H_3$; or $T_1 T_2 T_3$. Assuming independent subexperiments

$$P(H_1 H_2 H_3) = P(H_1) \cdot P(H_2) \cdot P(H_3),$$

$$P(H_1 H_2 T_3) = P(H_1) \cdot P(H_2) \cdot P(T_3), \ldots ,$$

$$P(T_1 T_2 T_3) = P(T_1) \cdot P(T_2) \cdot P(T_3)$$

It is noted that if different orders are not distinguished, certain of the possibilities listed above are alike. That is, $H_1 H_2 T_3$, $H_1 T_2 H_3$, and $T_1 H_2 H_3$ are all equivalent (two heads and one tail), if order is not important. Then, the probabilities associated with certain types of events in which the order is unimportant are

$$P(3 \text{ heads}) = P(H_1 H_2 H_3) = [P(H)]^3$$

$$P(2 \text{ heads and 1 tail}) = P(H_1 H_2 T_3 \cup H_1 T_2 H_3 \cup T_1 H_2 H_3)$$

$$= [P(H)]^2 P(T) + [P(H)]^2 P(T) + [P(H)]^2 P(T)$$

$$= 3[P(H)]^2 P(T).$$

Similarly

$$P(2 \text{ tails and 1 head}) = 3[P(T)]^2 P(H)$$

$$P(3 \text{ tails}) = [P(T)]^3.$$

To generalize to n Bernoulli trials, a result from combinatorial analysis is eeded. If there are n objects to be divided into two groups, one group containing k of the objects and the other $n - k$ of the objects, then there are (n take k)

$$\binom{n}{k} = \frac{n!}{k!\,(n-k)!} = \frac{n(n-1)(n-2)\cdots 1}{[k(k-1)\cdots 1][(n-k)(n-k-1)\cdots 1]} \quad (2\text{-}47)$$

ways in which to divide the objects.

To see that there are $\binom{n}{k}$ ways in which k objects can be selected from a group of n objects, consider the following reasoning:

1. Select from the group of n objects one object for the first of the k objects. There are n possible choices.
2. Select from the $n - 1$ remaining objects, one object for the second of the k objects.
3. Continue until the kth object is selected. There are a total of $(n)(n-1)\cdots$ $(n - k + 1)$ ways of making the k selections. Note that

$$(n)(n-1)\cdots(n-k+1) = n!/(n-k)!.$$

4. $n!/(n - k)!$ is the number of ways in which k objects can be selected with the first element selected in the first position, the second for the second position, etc.
5. We do not wish to distinguish order in the group of k objects. Thus we have counted a group containing k objects too many times. In fact, we have counted the same group $k!$ times, because there are $(k)(k-1)\cdots 1$ ways in which a group of k objects can be ordered. Indeed, the first can be selected in k ways, etc.
6. The total number of ways of selecting a group of k where the order of selection is not important is

$$\frac{n(n-1)\cdots(n-k+1)}{k!} = \frac{n!}{(n-k)!\,k!}.$$

As an example, assume there are three objects, and they are to be divided into one group of two and one group of one. Then

$$\binom{3}{2} = \binom{3}{1} = \frac{3 \cdot 2 \cdot 1}{2 \cdot 1 \cdot 1} = 3.$$

This checks the coefficient that was obtained for the "two heads and one tail" problem worked above.

The number $\binom{n}{k}$ was probably (this is a personal probability) encountered by the student in the binomial expansion; that is

$$(p + q)^n = \sum_{r=0}^{n} \binom{n}{r} p^r q^{n-r}.$$

For this reason Bernoulli trials are sometimes referred to as a binomial experiment. We can now see that with Bernoulli trials the probability that an event with probability p occurs k times in n trials is

$$P(k \text{ occurrences in } n \text{ trials}) = \binom{n}{k} p^k (1 - p)^{n-k} \tag{2-48}$$

because p^k represents the probability of the event occurring k times, $(1 - p)^{n-k}$ represents the probability of the event not occurring $n - k$ times, and $\binom{n}{k}$ is the number of ways in which the n trials could be divided into two groups of size k and $n - k$.

EXAMPLE 2-26

What is the probability of exactly 47 heads in 100 flips of a fair coin?

$$P(47 \text{ heads}) = \binom{100}{47} P(\text{head})]^{47} [P(\text{tails})]^{53},$$

$$P(47 \text{ heads}) = \frac{100!}{47! \, 53!} \left(\frac{1}{2}\right)^{47} \left(\frac{1}{2}\right)^{53},$$

$$\simeq .0667.$$

Evaluation of binomial probabilities can be a tedious task, especially if several terms are involved. Tables are available* that give values not only of individual terms but also of cumulative sums. These tables can be used to good advantage when computation is necessary.

EXAMPLE 2-27

A manufacturing process produces parts which are one percent defective. Fifty of these parts selected at random are purchased. What is the probabilty that there are two or less defective parts?

$$P[\text{exactly 2 defective}] = \binom{50}{2} (.01)^2 (.99)^{48};$$

$$P[\text{exactly 1 defective}] = \binom{50}{1} (.01)(.99)^{49};$$

$$P[0 \text{ defective}] = (.99)^{50}.$$

* National Bureau of Standards, *Tables of the Binomial Probability Distribution*, Applied Mathematics Series 6, U.S. Government Printing Office, Washington, D.C. (1949).

$P[2 \text{ or less defective}] = P[2 \text{ defective}] + P[1 \text{ defective}] + P[0 \text{ defective}]$

$$= \binom{50}{2}(.01)^2(.99)^{48} + \binom{50}{1}(.01)(.99)^{49} + (.99)^{50}.$$

Further examples of Bernoulli trials are contained in problems.

Tree Diagrams

It is sometimes very helpful to give a graphical representation of an experiment that involves several subexperiments. This is especially helpful when the subexperiments are described by conditional probabilities. This can be done by what is called a tree, that is, a plane figure consisting of a finite number of line segments (branches), where each subexperiment is represented by a number of line segments in the same vertical column.

The events that occur in each subexperiment and their probabilities are usually shown on the tree. The event from the total experiment is often shown at the end of a branch. An example illustrates.

EXAMPLE 2-28

Urn A contains four red and two white balls. Urn B contains two white and three red balls. An urn is selected at random and two balls are drawn without replacing the first before drawing the second. Draw the tree diagram. What is the probability of drawing two white balls? See the tree diagram of Figure 2-24. The answer is

$$P(2 \text{ white}) = \tfrac{1}{2}\tfrac{2}{6}\tfrac{1}{5} + \tfrac{1}{2}\tfrac{2}{5}\tfrac{1}{4} = \tfrac{1}{30} + \tfrac{1}{20} = \tfrac{1}{12}.$$

Applications to engineering problems are given in the problems. We now consider an example where the subexperiments are independent. This is not a usual application of a tree diagram, but is useful to compare the results derived for Bernoulli trials to the results obtained from a tree diagram.

EXAMPLE 2-29

What is the probability of getting exactly two heads in four tosses of a coin where the probability of getting a head on each toss is p $(p + q = 1)$? See Figure 2-25. From this tree diagram it can be seen that the desired result is

$$P(\text{exactly 2 heads}) = ppqq + pqpq + pqqp + qppq + qpqp + qqpp$$
$$= 6p^2q^2.$$

Note that this problem could have been solved by using the result for Bernoulli trials, that is, $\binom{4}{2} = 6$.

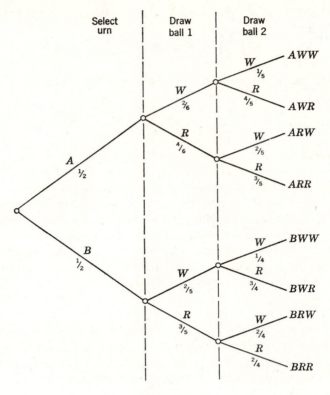

Figure 2-24

2-9 SUMMARY

The various definitions of probability were introduced, and the equally likely and relative frequency definitions were discussed. To avoid some of the problems associated with each of the definitions, a more general definition that included the others was adopted.

Some of the rules of set theory were introduced in order to use the more general definition of probability. Later these sets were identified with events that consisted of outcomes of an experiment.

Probability was defined as a non-negative, normed, additive function defined on these events. This definition was shown to include the intuitive definitions of probability discussed earlier. Various useful properties were derived from the axioms of probability.

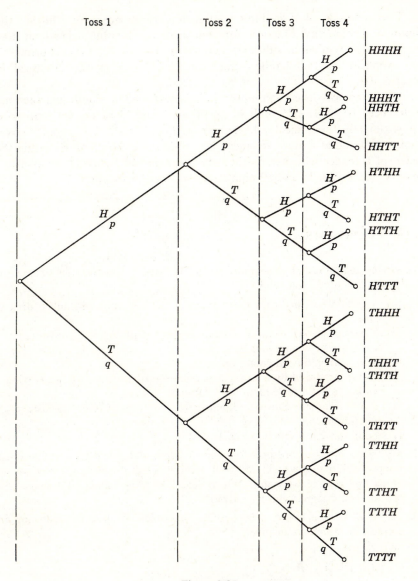

Figure 2-25

The concept of combined subexperiments was introduced, resulting in a definition of joint, conditional, and marginal probabilities. The concept of statistical independence was discussed, and Bayes' rule was introduced.

41

Two methods of describing combined experiments were studied: the binomial probability function for independent Bernoulli trials and tree diagrams for sequential subexperiments described by conditional probability. Applications to reliability and acceptance sampling are discussed in Chapter III.

The references listed at the end of this text contain additional reading. References B1, D3, H1, and R1 cover the main thrust of this chapter at approximately the same level. References D4, F1, P2, P3, and P5 cover similar material at a more advanced level, but these should all be within the reach of a reader who has finished this chapter. Reference D2 emphasizes history, while Ref. J1 and S2 emphasize personal probability.

2-10 PROBLEMS

1. Which definitions of probability would you use if you wanted to find the probability of the following:
   (a) A certain specified card is drawn from an ordinary well-shuffled deck of cards.
   (b) A resistor chosen at random from a group of $10 \, \Omega \pm 10\%$ resistors has a resistance between 9 and $10 \, \Omega$.
   (c) A coin when flipped comes up heads.
   (d) A coin produced by a stranger named Maverick comes up heads if he is willing to bet 4 to 1 that it comes up tails.
   (e) That a transistor chosen at random from a group of 10 \$1 transistors will function properly.
   (f) Your above answers being correct.

2. My friend wants to prospect for gold. In order to decide if I wish to go, I ask the probability that we will find gold. He replies it is obviously 1/2 since we will either find gold or we won't. Which definition of probability did he use, and why do you disagree with his answer?

3. Let A, B, C be arbitrary subsets of S. Determine which of the following are correct and which are incorrect.
   (a) $(A \cup B \cup C) - C = A \cup B$,
   (b) $(\overline{ABC}) = \bar{A} \cup \bar{B} \cup \bar{C}$,
   (c) $(A - AB)\bar{C} = A\overline{(B \cup C)}$,
   (d) $A(\overline{B \cup C}) = A\bar{B} \cup A\bar{C}$.

4. Find simple expressions for
   (a) $(A \cup B) \cap (A \cup \bar{B})$,
   (b) $(A \cup B) \cap (\bar{A} \cup B) \cap (A \cup \bar{B})$,
   (c) $(A \cup B) \cap (B \cup C)$.

5. The union of two sets, $A \cup B$, can be expressed as the union of two mutually exclusive sets, thus $A \cup B = A \cup (B\bar{A})$. Express in a similar way the union of three sets, A, B, C.

6. Find expressions for the occurrence of the following:
   (a) At least one of the three events A, B, C.
   (b) A and either B or C but not both B and C.
   (c) Two or more of A, B, C, D.

7. Let $S = \{1, 2, 3, 4, 5, 6, 7, 8, 9, 10\}$,
   $A = \{1, 2, 3, 4, 5\}$,
   $B = \{4, 5, 6, 7\}$,
   $C = \{6, 7, 8, 9, 10\}$.

   Write out the sets

   (a) $A \cup B$,
   (b) $A \cap C$,
   (c) $(A \cup B) - C$,
   (d) $A \cup C \cup \bar{B}$,
   (e) $\bar{C}$.

8. Show
   (a) $AB \subset B$,
   (b) $A = AB \cup A\bar{B}$,
   (c) If $A\bar{B} = \varnothing$, then $A \subset B$,
   (d) $(ABC) \cap (AB\bar{C}) = \varnothing$.

9. Set theory can be applied to switching circuits. In computer applications, components which perform basic operations are denoted as shown in Figure P2-1. Using intersection for "and" and union for "or," show that "Boolean" circuits in Figure P2-2 are equivalent. Note that the $+$ of "Boolean algebra" corresponds to $\cup$ in set theory and the $\cdot$ of "Boolean algebra" corresponds to $\cap$ in set theory. Thus such techniques as Karnaugh Maps and minterm and maxterm expansions can be used as aids in set theory.

10. An experiment consists of putting a known voltage across a resistor and measuring the current to the nearest milliampere. Describe the space if we want to determine:
    (a) The reliability, that is, the ratio of "good" resistors.
    (b) The probability of any resistance value when resistance is measured to the nearest .01 ohm and no resistance is above 10 ohms.

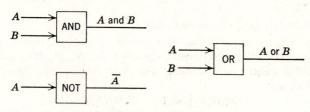

Figure P2-1

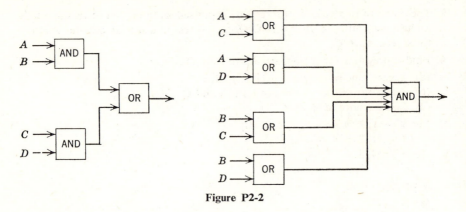

Figure P2-2

11. Show that the relative frequency definition of probability satisfies the axioms of probability.

12. Show that the disjoint and exhaustive events

$$\{f_1\}, \{f_2\}, \{f_3\}, \{f_4\}, \{f_5\}, \{f_6\}$$

generate 2^6 distinct events, where the events are the results of unions, intersections, and complementations.

13. If $A \subset B$, show that

$$P(A) \leq P(B).$$

14. A box contains 25 resistors and 15 capacitors. One is selected at random. Describe the outcomes and the sample space if one is interested in the events selecting a resistor or selecting a capacitor. What are the probabilities?

15. A circle is divided into 20 equal parts. A pointer is spun until it stops on one of the parts which are numbered from one to twenty. Describe the space and find
    (a) P(an even number),
    (b) P(the number 7),
    (c) $P(\{4\} \cup \{6\})$,
    (d) P(a number greater than 9).

16. If the probability of an event is $\mu/(\mu + 2)$, what is the probability that the event does not occur?

17. Show (2-30) by using the results of Problem 5 and the method used in the text to show (2-29).

18. Use an induction proof to show (2-31).

19. In a certain town there are two morning newspapers published, M_1 and M_2. There are three evening newspapers published, N_1, N_2, N_3. We assume that any person reads exactly one morning and one evening newspaper. The joint probabilities are

$$P(M_1 N_1) = .1 \qquad P(M_2 N_1) = .2$$
$$P(M_1 N_2) = .3 \qquad P(M_2 N_2) = .1$$
$$P(M_1 N_3) = .2 \qquad P(M_2 N_3) = .1$$

(a) What are the marginal probabilities of the following?

$$P(N_1) \quad P(N_2) \quad P(N_3) \quad P(M_1) \quad P(M_2)$$

(b) What are the conditional probabilities of the following?

$$P(N_1 \mid M_2) \quad P(M_1 \mid N_3) \quad P(N_1 \mid M_1)$$

(c) Are the events independent?

20. If two cards are drawn without replacement from an ordinary deck of cards, what is the probability?
   (a) They are both aces,
   (b) One is an ace and the other a king,
   (c) They are both spades,
   (d) One is an ace and the other a spade.

21. Verify (2-35). Hint: Look at (2-32).

22. Verify (2-39). Hint: Use same method as in Problem 21.

23. Two dice are thrown. Determine, the probability that the sum of the "up" faces $= K$ for $K = 2, 3, 4, \ldots, 12$.

24. Show $P(A \mid B) = P(A)$ implies $P(B \mid A) = P(B)$.

25. Given A_1, A_2, A_3, A_4, what equations must be satisfied in order for these four events to be independent? Hint: See (2-42).

26. If n coins are tossed, what is the probability they all show the same face?

27. Show that if the events A_1 and A_2, are independent, then

$$P(A_1 \bar{A}_2) = P(A_1)P(\bar{A}_2),$$
$$P(\bar{A}_1 \bar{A}_2) = P(\bar{A}_1)P(\bar{A}_2).$$

28. A simple filter consists of a resistor and a capacitor. The resistor can be either good R or short $\bar{R}$. The capacitor can be either good C or bad $\bar{C}$, and the probability of a good capacitor is less if the resistor has shorted. Tests have produced the following results:

$$P(R) = .9 \qquad P(C \mid R) = .95 \qquad P(C \mid \bar{R}) = .75.$$

Find
   (a) $P(RC)$,
   (b) $P(R \mid C)$,
   (c) $P(\bar{R} \mid \bar{C})$,
   (d) $P(C)$,

We might reason that the resistor is not caused to short by the capacitor failing; yet $P(R \mid C) \neq P(R)$. Explain.

29. If two events are mutually exclusive and statistically independent, what can be said about the events?

45

30. A single card is drawn at random from a deck of cards. Which of the following pairs of events are independent?

   (a) The card is a spade, the card is black.
   (b) Spade, Red.
   (c) Ace, Black.
   (d) Ace, Above 10.

31. Given a binary communication channel as in example 2-25. Let

$$P(A) = .5,$$
$$P(B \mid A) = .9 = P(\bar{B} \mid \bar{A}).$$

   Find $P(A \mid B)$ and $P(A \mid \bar{B})$.

32. The probability of passing a certain quiz given that a student studied is .90. The probability of passing the quiz given that a student did not study is .40. Assume the probability of studying is .8. Given a student passed the quiz, what is the probability he studied?

33. A certain type of relay will operate 98% of the time under certain specified conditions. In 10 trials under these conditions what are the probabilities of

   (a) all trials being successful;
   (b) one failure and nine successes;
   (c) the first nine trials being successful and the last a failure.

   Assume independent trials.

34. A certain type of welding machine produces defective joints .01% of the time. In welding four joints, what are the probabilities of getting zero, one, two, three, and four defective joints.

35. Use a tree diagram to determine in four draws without replacement from an ordinary deck of cards:

   (a) $P(3 \text{ hearts and } 1 \text{ spade})$;
   (b) $P(2 \text{ hearts, } 1 \text{ spade, and } 1 \text{ club})$;
   (c) $P(\text{at least } 2 \text{ hearts})$.

CHAPTER III

....

ENGINEERING APPLICATIONS OF PROBABILITY

The concepts introduced in Chapter II are sufficient to introduce two important engineering applications: reliability and acceptance sampling. Most engineers who work in design, manufacturing, or sales need the basic ideas of both these important areas.

3-1 RELIABILITY

A definition of reliability is now in order. The word reliability has come to mean many (related?) things. Frequently it is just another superlative in the jargon of a salesman. For our purposes there are two distinct usages of the word. First, it can be used as a concept and, second, as a measure. The use of reliability as a measure is the matter of present interest. In this sense the following definition will be adopted: Reliability is the probability of success in the use environment.

This definition is quite useful, as will be seen, although it is not as precise as one might desire. First, note that reliability is a probability, and probabilities will be calculated with the methods of the previous chapter. Success may be rather obvious at first glance, but a little more thought will reveal that success is rather subjective for a number of systems. For example, if a system is designed to put out a 10 V signal and it produces a 9 V signal, is this a success? Also different people may have different ideas about the purpose of a system, resulting in different definitions of success. For example, a manned space flight that safely returns to earth may be a success to most observers, but a failure to a scientist if the desired data were not obtained.

The last qualifying phrase in the definition, "in the use environment," often is overlooked and causes some confusion. From a technical point of view the experiment and the sample space are not defined until the environment is specified. For example a system may perform quite satisfactorily in a bench test, but when the system is mounted in a missile and used, heat, vibration, or radiation may cause a failure. Often the word quality is used to describe performance on a test while reliability is reserved for the use environment.

System Reliability Analysis

The first step in system reliability analysis is to analyze the system and decide whether the events under consideration are logically connected by "and" or by "or" relationships. That is, in the case of two events A, B decide whether event A *and* event B or event A *or* event B are required to produce the event of interest.

If the events are related by "and," then the set relation of interest is intersection of events from different subexperiments. That is, if the event of interest A is A_1 and A_2 and $\cdots$ and A_n, then any outcome that results in A results in A_1 and A_2 and $\cdots$ and A_n.

$$A = A_1 A_2 \cdots A_n = \bigcap_{i=1}^{n} A_i. \tag{3-1}$$

If the events are statistically independent

$$P(A) = P(A_1)\, P(A_2) \cdots P(A_n) = \prod_{i=1}^{n} P(A_i). \tag{3-2}$$

If the events are related by "or," then the set relation of interest is union, and the additive law of probability is needed. That is, if the event of interest A is A_1 or A_2 or A_3 or $\cdots$ or A_n, then any outcome that results in A_1 or A_2 or $\cdots$ or A_n results in A.

$$A = A_1 \cup A_2 \cup \cdots \cup A_n = \bigcup_{i=1}^{n} A_n. \tag{3-3}$$

Then if the events are mutually exclusive

$$P(A) = \sum_{i=1}^{n} P(A_i). \tag{3-4}$$

3-2 ELEMENTARY CALCULATION OF SYSTEM SUCCESS PROBABILITY

In this section a system is represented by a block diagram. Each block represents a component or part of the system which is either good (closed circuit) or bad (open circuit). The purpose of the system can be considered to be to pass current, and each one of the blocks can be considered to be a relay contact.

In this section all of the subexperiments associated with each block will be assumed to be statistically independent.

Series Circuits

First, consider a series circuit as shown in Figure 3-1. The following symbols are defined for $X = A, B, C, D, E$:

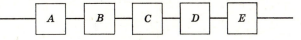

Figure 3-1

$p_X = $ probability device X is good (closed),

$q_X = $ probability device X is bad (open),

$$p_X + q_X = 1. \tag{3-5}$$

For the system shown in Figure 3-1, the probability, R of a closed circuit is (with the assumption of statistical independence)

$$R = p_A p_B p_C p_D p_E. \tag{3-6}$$

This follows from (3-2) and the fact that all parts must work.

The probability Q of open circuit of the same system is

$$Q = 1 - R = 1 - p_A p_B p_C p_D p_E. \tag{3-7}$$

An expression for Q may also be obtained by reasoning that there will be a system failure if any of the parts A or B or C or D or E fails. Thus

$$Q = P(F_A \cup F_B \cup F_C \cup F_D \cup F_E) \tag{3-8}$$

where F_X is the event representing an open circuit for X, $X = A, B, C, D, E$. This union of sets must be written as a disjoint union in order to utilize (3-4). (This is the technique for finding an equation for the probability of a general union of events.) This results in

$$Q = P(F_A \cup \bar{F}_A F_B \cup \bar{F}_A \bar{F}_B F_C \cup \bar{F}_A \bar{F}_B \bar{F}_C F_D \cup \bar{F}_A \bar{F}_B \bar{F}_C \bar{F}_D F_E) \tag{3-9}$$

where $\bar{F}_X$ is the event representing a closed circuit for X, $X = A, B, C, D,$ E. Now, using (3-4) and (3-9) and statistical independence,

$$Q = q_A + p_A q_B + p_A p_B q_C + p_A p_B p_C q_D + p_A p_B p_C p_D q_E. \tag{3-10}$$

49

Now (3-10) will be rearranged to show it is the same as (3-7). Using $p_X + q_X = 1$

$$Q = (1 - p_A) + p_A(1 - p_B) + p_A p_B(1 - p_C) \\ + p_A p_B p_C(1 - p_D) + p_A p_B p_C p_D(1 - p_E), \tag{3-11}$$

$$Q = 1 - p_A + p_A - p_A p_B + p_A p_B - p_A p_B p_C \\ + p_A p_B p_C - p_A p_B p_C p_D + p_A p_B p_C p_D - p_A p_B p_C p_D p_E.$$

This obviously telescopes to (3-7).

Some numerical values are now assigned to show the effect of putting parts in series and to explain an approximation often used in practice. Let

$$p_A = p_B = p_C = p_D = p_E = .99.$$

Then from (3-6),

$$R = (.99)^5 \simeq .951 \tag{3-12}$$

and

$$Q = 1 - R \simeq 1 - .951 = .049. \tag{3-13}$$

From (2-31),

$$Q \le \sum_{i=1}^{5} .01 = .05. \tag{3-14}$$

Thus if the inequality sign of (3-14) is replaced by an approximate equality, the approximation is quite good. This is often done in practice and results in the approximation (good for $Q \le .10$) for statistically independent events if

$$Q = P\left(\bigcup_{i=1}^{n} F_i\right),$$

$$Q \simeq \sum_{i=1}^{n} P(F_i). \tag{3-15}$$

Two examples will be given to illustrate the reliability of series circuits.

EXAMPLE 3-1

A system consists of 100 parts in series. Each part has a reliability of .99. What is the system reliability?

Since it is a series system

$$R = \prod_{i=1}^{n} P(\bar{F}_i)$$

$$R = (.99)^{100} \simeq .366.$$

Note that although each part is fairly good, on the average only about one out of three systems will perform satisfactorily.

50

EXAMPLE 3-2

A missile designed for manned space travel has 1000 components in series. If the mission is to have a reliability of .9 and if all parts have the same reliability, what is the required reliability of each part?

$$.9 = (p_X)^{1000}$$
$$p_X = (.9)^{.001} \simeq .99989.$$

Parallel Circuits

Now parallel circuits are considered. Consider the example given in Figure 3-2.

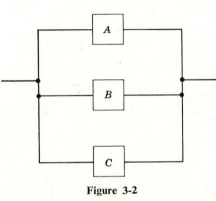

Figure 3-2

In this case the system will be a success (closed circuit) if A or B or C works (is closed), and the system will be a failure if A and B and C fail. Thus the probability R of success is

$$R = P(\bar{F}_A \cup \bar{F}_B \cup \bar{F}_C)$$
$$= P(\bar{F}_A \cup F_A \bar{F}_B \cup F_A F_B \bar{F}_C)$$

Assuming independence

$$R = p_A + q_A p_B + q_A q_B p_C. \tag{3-16}$$

The probability Q of failure is

$$Q = P(F_A F_B F_C)$$
$$Q = q_A q_B q_C. \tag{3-17}$$

It is left for the student to show that $R + Q = 1$.

The paralleling of components when only one is necessary may appear wasteful at first glance. However it is done in a number of practical systems. Such paralleling is called *redundancy*. Some examples illustrate the advantages of redundancy.

EXAMPLE 3-3

Consider the system shown in Figure 3-2. What is the reliability if each part has a reliability of .9?

$$R = 1 - Q = 1 - (.1)^3 = .999.$$

EXAMPLE 3-4

If a system has a required reliability of .99, how many components of reliability .5 must be paralleled?

$$Q = (q_X)^n = (.5)^n,$$
$$R \geq .99 \text{ implies } Q \leq .01,$$
$$(.5)^n \leq .01,$$
$$n \geq \log .01 / \log .5 \simeq 6.6.$$

Therefore set $n = 7$; then $Q = (.5)^7 = .0078$, which satisfies the requirement.

General Networks

Now more general networks of components will be considered. First, a combination of series and parallel components is studied. The procedure is to use the equations developed above for series and parallel components to reduce the system in a step by step procedure.

Because in practical systems, often there can be more than one type of failure (e.g., open circuit and short to ground), we emphasize equations for probability of failure Q. If there are no complications then $R + Q = 1$, so finding Q is equivalent to finding R.

Now consider the block diagram of Figure 3-3.

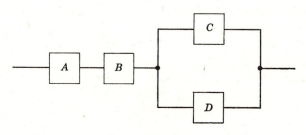

Figure 3-3

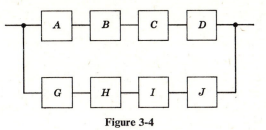

Figure 3-4

The following equation results from a consideration of Figure 3-3:

$$Q = P(F_A \cup F_B \cup F_C F_D).$$

With the assumption of statistical independence, the exact equation is

$$Q = q_A + q_B + q_C q_D - q_A q_C q_D - q_B q_C q_D - q_A q_B + q_A q_B q_C q_D$$

or

$$Q = q_A + p_A q_B + p_A p_B q_C q_D.$$

Also

$$Q \le q_A + q_B + q_C q_D$$

and if $Q \le .10$, then

$$Q \simeq q_A + q_B + q_C q_D.$$

For another example consider the block diagram of Figure 3-4. This can be reduced to the block diagram of Figure 3-5, where

$$\{F_K\} = \{F_A \cup F_B \cup F_C \cup F_D\}$$

and

$$\{F_L\} = \{F_G \cup F_H \cup F_I \cup F_J\}.$$

With the usual approximation

$$q_K \simeq q_A + q_B + q_C + q_D,$$
$$q_L \simeq q_G + q_H + q_I + q_J.$$

Now the block diagram of Figure 3-5 is a simple parallel circuit that has already been evaluated. So

$$Q = q_K q_L.$$

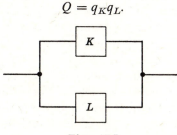

Figure 3-5

53

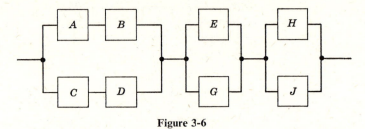

Figure 3-6

As a last example consider the block diagram of Figure 3-6. This may be reduced to the form shown in Figure 3-7, where

$$\{F_K\} = \{F_A \cup F_B\},$$
$$\{F_L\} = \{F_C \cup F_D\},$$
$$\{F_M\} = \{F_E F_G\},$$
$$\{F_N\} = \{F_H F_J\}.$$

Figure 3-7 may be further reduced to Figure 3-8, where $\{F_T\} = \{F_K F_L\}$. Now for this original system when Q is small

$$Q \simeq q_T + q_M + q_N = q_K q_L + q_M + q_N$$
$$\simeq (q_A + q_B)(q_C + q_D) + q_E q_G + q_H q_J. \qquad (3\text{-}18)$$

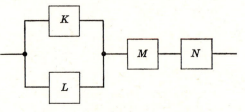

Figure 3-7

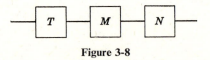

Figure 3-8

With some practice approximate equations such as (3-18) can be written directly without explicitly going through the intermediate steps.

The methods of series and parallel reduction will now be used in an example that illustrates how the probabilistic equations might be used in a simple design problem.

EXAMPLE 3-5

Compare the failure probabilities of the two systems shown in Figure 3-9.

Before doing the actual evaluation, note that if B and C fail or if A and D fail, then system (a) of Figure 3-9 will fail, while (b) will be successful.

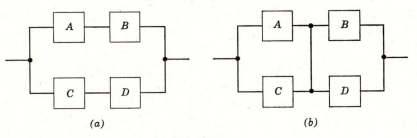

(a) (b)

Figure 3-9

Therefore (a) should have a higher failure probability.

$$Q_a \simeq (q_A + q_B)(q_C + q_D) = q_A q_C + q_A q_D + q_B q_C + q_B q_D,$$
$$Q_b \simeq q_A q_C + q_B q_D.$$

If

$$q_A = q_B = q_C = q_D = .01,$$

then

$$Q_a \simeq .0004,$$
$$Q_b \simeq .0002.$$

It might be illustrative to check the approximate result in this example by computing the exact expression.

$$Q_a = P\{(F_A \cup \bar{F}_A F_B)(F_C \cup \bar{F}_C F_D)\}$$
$$= [q_A + (1 - q_A)(q_B)][q_C + (1 - q_C)q_B]$$
$$= (.01 + .0099)^2 = .00039601$$
$$Q_b = P\{F_A F_C \cup [(\bar{F}_A \cup F_A \bar{F}_C)(F_B F_D)]\}$$
$$= q_A q_C + (1 - q_A q_C)q_B q_D$$
$$= .0001 + (.9999)(.0001) = .0001999.$$

These are both very good approximations, they are certainly good to the one significant digit that can be retained in such calculations.

One other type of system needs to be discussed. This is one which is not just a simple series and parallel combination. The methods of reasoning about such a system are the same (i.e., use "and" and "or" reasoning, transfer "and" to intersection and "or" to union, and write the probability equation from the set equation). An example of a bridge circuit as shown in Figure 3-10 illustrates the technique.

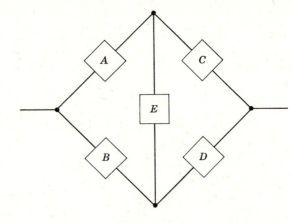

Figure 3-10

A tree diagram is used to describe this system. The events associated with each component occupy a certain vertical column of the diagram. A certain key component is chosen (key can be defined with experience; the beginner can be reassured by the fact that no choice ruins the method, some choices simply reduce the labor). All events associated with this component are shown emanating from the first node on the left. A next component is chosen. All events associated with this component are shown emanating from each termination of the first component. This process is continued until a result of the total experiment can be defined. Consider Figure 3-11. E is the first key component, then A, then C, then D, and finally B. Each branch is continued until system failure F or system success $\bar{F}$ is defined.

From Figure 3-11 the success and failure equations can be written directly as

$$R = p_E[p_A(p_C + q_C p_D) + q_A p_C(p_D p_B + q_D p_B) + q_A q_C p_D p_B]$$
$$+ q_E[p_A(p_C + q_C p_D p_B) + q_A p_D p_B].$$

The student should write the failure equation and check that $R + Q = 1$. The approximate failure equation is

$$Q = P(F_A F_B \cup F_C F_D \cup F_A F_E F_D \cup F_B F_E F_C)$$
$$\simeq q_A q_B + q_C q_D + q_A q_E q_D + q_B q_E q_C.$$

Although the approximate method may seem easier at this point, the tree diagram and concomitant exact method should be studied because in some cases exact equations may be needed, and because the tree diagram provides a general method that is especially useful when the subexperiments are not independent.

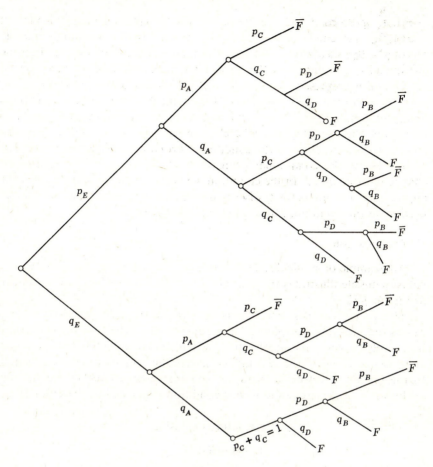

Figure 3-11

3-3 CONDITIONAL PROBABILITIES IN SYSTEM RELIABILITY ANALYSIS

If the systems analyzed in the previous section are considered as control systems, their purpose is to supply a signal at certain times and not supply a signal at other times. Thus there are two types of failures. The circuit can be open when it should be closed (this was the type of failure considered), and the circuit can be closed when it should be open (this is often called an early or premature type failure). It is not necessary to change the equations; it is only necessary that conditional probabilities be used and the conditional probabilities must be correctly interpreted.

Thus, if the condition of interest is when the circuit should be closed, R represents the conditional probability of success or reliability and Q represents the conditional probability of an open circuit failure. For the components, q_X represents the conditional probability of an open circuit failure and p_X represents the conditional probability of desired action.

If the condition of interest is when the circuit should be open, R represents the conditional probability of closed circuit or premature closure and Q represents the conditional probability of success or reliability. For the components, q_X represents the conditional probability of desired action and p_X represents the conditional probability of early closure. One practical change must be noted. In the condition when the circuit should be open, the approximate equations for Q discussed in the previous section are not useful because in this condition, for a practical system, $Q > .10$.

EXAMPLE 3-6

An example of an idealized system to control the firing of a warhead of a ballistic missile illustrates the concepts just introduced. The system is shown in Figure 3-12.

The system consists of a battery connected to an electric match which causes the warhead to explode. The path from the battery to the match will be completed when the acceleration switch senses enough g load to close, and one (or both) of the two radars detects the target and closes a switch. In the condition before acceleration and the radar target are present, the explosion is not desired, and in this condition (indicated by subscript 1)

$$R_1 = P\{\text{premature explosion}\},$$

$$Q_1 = P\{\text{no explosion}\},$$

$$1 - q_{B1} = p_{B1} = P\{\text{good battery}\},$$

$$1 - q_{A1} = p_{A1} = P\{\text{closed acceleration switch}\},$$

$$1 - q_{R11} = p_{R11} = P\{\text{closed radar 1 switch}\},$$

$$1 - q_{R21} = p_{R21} = P\{\text{closed radar 2 switch}\},$$

$$1 - q_{M1} = p_{M1} = P\{\text{good electric match}\},$$

$$R_1 = p_{B1}p_{M1}p_{A1}(p_{R11} + p_{R21} - p_{R11}p_{R21}).$$

Note that statistical independence was assumed. Also note that for a practical system, p_{B1} and p_{M1} should be very nearly 1, while p_{A1}, p_{R11} and p_{R21} should be small. Thus

$$R_1 \simeq p_{A1}(p_{R11} + p_{R21}).$$

58

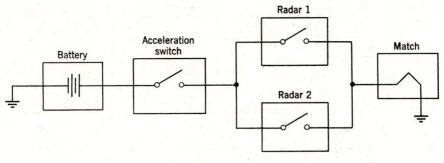

Figure 3-12

In the condition when acceleration has occurred and a radar target is present, the explosion is desired, and in this condition (indicated by subscript 2)

$$R_2 = P\{\text{success}\},$$

$$Q_2 = P\{\text{dud}\},$$

$$1 - q_{B2} = p_{B2} = P\{\text{good battery}\},$$

$$1 - q_{A2} = p_{A2} = P\{\text{closed acceleration switch}\},$$

$$1 - q_{R12} = p_{R12} = P\{\text{closed radar 1 switch}\},$$

$$1 - q_{R22} = p_{R22} = P\{\text{closed radar 2 switch}\},$$

$$1 - q_{M2} = p_{M2} = P\{\text{good electric match}\},$$

$$R_2 = p_{B2}p_{M2}p_{A2}(p_{R12} + p_{R22} - p_{R12}p_{R22}).$$

In this condition all of the p_X's should be close to one so that no approximations can be made.

The equations for R_1 and R_2 look very nearly the same, but the conditional component probabilities (at least for the acceleration switch and radars) are quite different from one condition to the other. Typical values might be

$$p_{R11} = p_{R21} = .01,$$

$$p_{A1} = .001,$$

$$p_{R12} = p_{R22} = p_{A2} = .99.$$

Component Conditional Probabilities

If the condition in which the system is to operate is specified, often there is still a need for conditional probabilities for the components. Consider Figure 3-13, which is an extremely simplified version of a system to control the power to system A. We are interested in the probability that system A will

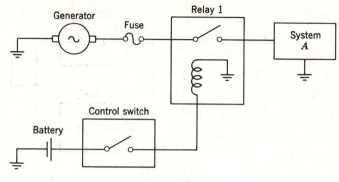

Figure 3-13

receive power given the control switch is turned to "on" and given system A does not draw an overload.

This example illustrates that with the system conditions defined, different conditional probabilities are needed for the relay. In addition we emphasize that a conditional probability of the fuse being good is needed. It is important that the final equation be in terms of component probabilities which can be estimated.

The probability of getting power to system A is the probability that the generator is good and the fuse and relay provide a closed circuit. Assuming statistical independence

$$R = p_G p_F p_C,$$

where R = probability of success,

p_G = probability of good generator,

p_F = probability of good fuse,

p_C = probability relay contact is closed.

We now examine p_F and p_C in more detail.

Recall that a fuse has two functions: to pass normal current and to open when abnormal current is present. Although a fuse is tested to determine the conditional probability of being good in conditions of both normal and abnormal current, by the stated conditions of this problem we are only interested in the probability of passing normal current. Therefore this conditional probability must be used for p_F.

The probability that the relay contact is closed can be computed in the following manner. The relay contact can be closed either because the relay receives power and works properly or because the relay does not receive power but the relay contact "fails" by closing. Thus [using (2-35)]

$$p_C = P(C_R \mid E)P(E) + P(C_R \mid \bar{E})P(\bar{E}),$$

where C_R represents closed relay contact and E represents power to relay. Note that the conditional probabilities in this equation can be determined from tests of the relay. (In most cases we would expect $P(C_R \mid E)$ to be close to 1 and $P(C_R \mid \bar{E})$ to be close to 0.)

$P(E)$, and thus $P(\bar{E})$, can be computed by analyzing the circuit consisting of B and the control switch. Under simplified assumptions

$$p(E) = p_B p_D,$$

where p_B = probability of a good battery, and p_D = probability of a good control switch given the switch is turned on. Thus

$$R = p_G p_F [p_B p_D P(C_R \mid E) + (1 - p_B p_D) P(C_R \mid \bar{E})],$$

where the terms are as defined above with the qualification that p_F is a conditional probability as discussed. Note that it is feasible to run tests for all of the terms in this final equation.

EXAMPLE 3-7

Consider the system shown in Figure 3-14. This example illustrates that more than two events may have to be defined for one component.

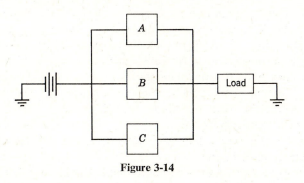

Figure 3-14

The probability of an open-circuit failure of the relays is

$$Q = q_A q_B q_C,$$

and if $q_A = q_B = q_C = .01$, then

$$Q = 10^{-6}.$$

In a real system the probability of a short to ground around one of the relay contacts is usually greater than 10^{-6}. This thought actually means that the model of the system is not complete. An event must be defined that describes the short to ground around each relay contact, so the model should look like Figure 3-15.

61

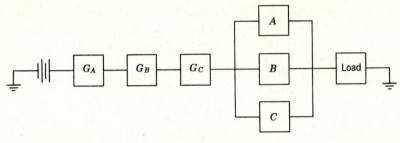

Figure 3-15

Note that although the contacts are in parallel, the events describing the shorts to ground are in series. This is because there will be system success if there is not a short to ground at *A and* there is not a short to ground at *B and* there is not a short to ground at *C*. Also, there will be a system failure if there is a short to ground at *A or* there is a short to ground at *B or* there is a short to ground at *C*.

3-4 SUMMARY OF RELIABILITY CALCULATIONS

The previous three sections have applied probability to the calculation of system reliability in terms of (conditional) probabilities of component behavior. In addition to exercising the concepts introduced in the last chapter, the purpose of these sections was to emphasize the idea of transforming a block diagram or a schematic diagram into a probabilistic model.

Such models are useful in preliminary system design to decide if the system is feasible, where redundancy might be needed, or which components must be replaced or further developed. Later in a development program such models can be useful in planning testing programs. After production, such models are useful in planning maintenance.

References S3 and B1 provide additional material on reliability.

3-5 ACCEPTANCE SAMPLING

When an engineer buys resistors for his radar, he will want to ensure that the resistors conform to his standards. More generally, when mass produced items are to be bought, the buyer wants to see if the items conform to standards. This inspection of the product may consist of every item being tested. However, in recent years there has been a trend to simply take a sample from each lot* and test the sample and then accept or reject the lot.

* A lot is a group of items produced with no significant changes in the manufacturing process.

Sampling rather than 100% inspection is used when (a) the testing is destructive (e.g., explosive devices); (b) the testing cost is great; and (c) the testing is so monotonous and boring that a high percentage of the defective items may not be detected by an inspector.

Furthermore, acceptance sampling may have an important psychological effect on the quality of production. With 100% sampling, workers and production managers may feel that no bad material will leave the plant because all bad items will be caught in the final inspection. However, with sampling inspection, total lots can be rejected and the cost of reworking rejected lots may place the emphasis on "make it right the first time."

At this point we consider acceptance sampling plans based upon Bernoulli trials. Such a model is useful when the results of tests are either pass or fail, and when the probability of occurrence of a defective item remains constant from one item to the next. In addition, with pass-fail testing, the Bernoulli trial model is a good approximation when there are a certain number of defective items in the lot and the lot size is much larger than the sample size. In this case the probability of *drawing* a defective sample remains approximately constant.

With the Bernoulli trial model, if p is the probability of a good unit and n is the sample size, then the probability of k defective units is [see (2-48)]

$$P(k \text{ defective units}) = \binom{n}{k}(q)^k(p)^{n-k},$$

where $q = 1 - p$ is the probability of a defective unit.

A sampling plan is usually described by the sample size n and the maximum allowable number c of defective items in the sample.

Given n and c, the probability of accepting a lot is

$$P(\text{acceptance}) = \sum_{k=0}^{c} \binom{n}{k}(q)^k(p)^{n-k}. \tag{3-19}$$

It is common practice to plot $P(\text{acceptance})$ versus q and call the resulting plot an O.C. curve, where O.C. stands for operating characteristic of the sampling plan. For instance if the sample size n is 40, and if the acceptance number c is 2, then

$$P(\text{acceptance}) = p^{40} + 40p^{39}q + \binom{40}{2}p^{38}q^2.$$

This is plotted versus q in Figure 3-16.

A sampling plan that may appear to be nearly the same as that plotted in Figure 3-16 is to test 20 samples and allow 1 failure. This plan, together with the earlier one that tests 40 samples and allows 2 failures, is plotted in Figure 3-17. From the figure it may be seen that the sampling plan with the smaller

63

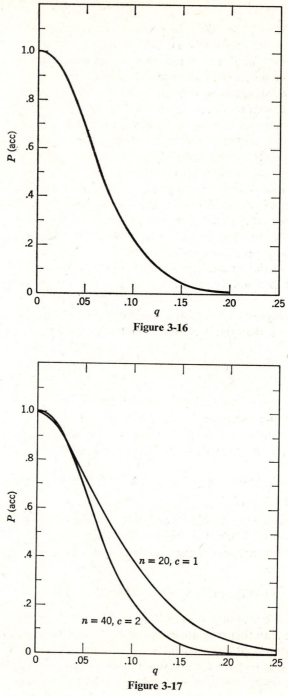

Figure 3-16

Figure 3-17

sample size results in a lower probability of accepting "good" lots (those with $q \leq .03$) and a higher probability of accepting "bad" lots.

A detailed analysis of operating characteristics of a sampling plan is a very useful way of displaying the protection offered to both buyer and seller.

EXAMPLE 3-8

A TV manufacturer wants to purchase resistors, and the supplier has suggested that from each lot (size 100,000), 100 resistors will be tested and any failures (resistances outside $\pm 10\%$) result in rejection of the lot. Evaluate the sampling plan by drawing an O.C. (operating characteristic) curve. What is the value of q (probability of defective) which will result in the lot being accepted 95% of the time? What is the value of q which will result in the lot being accepted 10% of the time?

See Figure 3-18, which was drawn from $P(\text{acceptance}) = (1 - q)^{100}$. The solution to the first question is

$$(.95) = (1 - q)^{100}.$$

This may be solved for q by using a table of logarithms, or approximately from Figure 3-18. The answer is $q \simeq .0005$.

The solution to $(.10) = (1 - q)^{100}$ is $q \simeq .0228$.

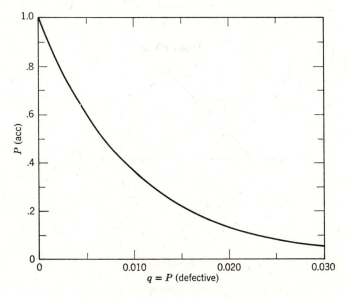

Figure 3-18

3-6 SUMMARY

Drawing a sample and testing it to decide if the lot is to be accepted or rejected is intended to provide a relatively inexpensive and efficient decision that is fair to both buyer and seller. An engineer who either buys or sells parts can judge a sampling plan from the O.C. curve. A more exact evaluation of sampling plans (or of any test) must be postponed until Chapter X. References B2, D5, and G1 provide additional reading.

3-7 PROBLEMS

1. Find expressions for the probability of success and the probability of failure of the following circuits:

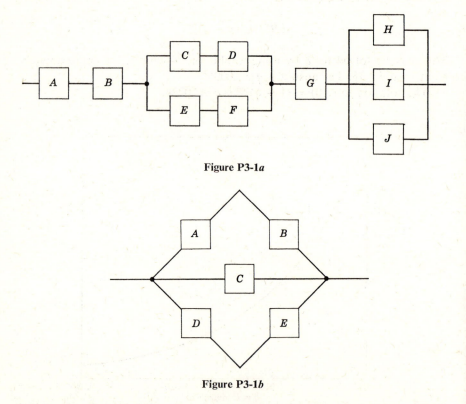

Figure P3-1*a*

Figure P3-1*b*

2. Assuming the probability of success of each component is .99, find the approximate probability of failure of the circuits shown in Figure P3-2.

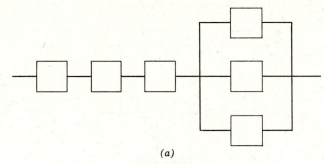

Figure P3-2a

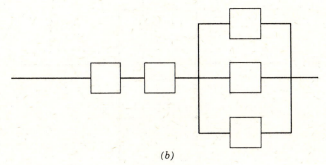

(b)

Figure P3-2b

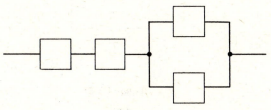

(c)

Figure P3-2c

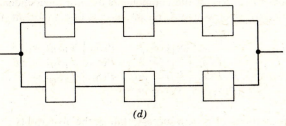

(d)

Figure P3-2d

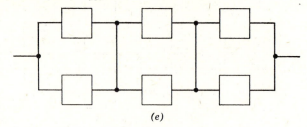

(e)

Figure P3-2e

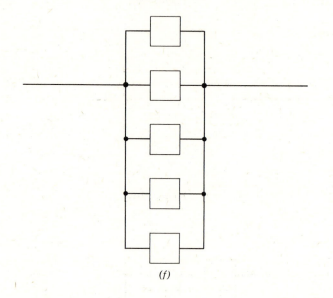

(f)

Figure P3-2f

3. There is a system composed of three components A, B, C. The system will function properly if any two of the three components function properly. The probabilities of success of the components are not independent. The following probabilities are given where S_x means success of component X:

$$
\begin{array}{ll}
P(S_A) = .7 & P(S_C \mid S_A \bar{S}_B) = .75 \\
P(S_B \mid S_A) = .8 & P(S_C \mid \bar{S}_A S_B) = .65 \\
P(S_B \mid \bar{S}_A) = .6 & P(S_C \mid \bar{S}_A \bar{S}_B) = .5 \\
P(S_C \mid S_A S_B) = .9 &
\end{array}
$$

Find the probability of system success, that is, the probability that two out of the three components are successful. (A tree diagram might help you.)

4. Given the circuit shown in Figure P3-3 the following probabilities are defined:

$B = P(\text{Good battery})$,
$A_b = P(\text{Acceleration switch is closed before actual acceleration})$,
$A_a = P(\text{Acceleration switch is closed after actual acceleration})$,
$\theta_1 = P(R_1 \text{ is open})$,
$\theta_2 = P(R_2 \text{ is open})$,
$S_1 = P(R_1 \text{ shorts to ground})$,
$S_2 = P(R_2 \text{ shorts to ground})$,
$R_1 = P(R_1 \text{ is good})$,
$R_2 = P(R_2 \text{ is good})$,
$L = P(\text{Good light})$.

Assume that

$$\theta_1 + S_1 + R_1 = 1,$$
$$\theta_2 + S_2 + R_2 = 1.$$

Assuming statistical independence of the subexperiments, find the probability of the light lighting in terms of the defined probabilities:

(a) after actual acceleration,
(b) before actual acceleration.

Hint: see Example 3-6

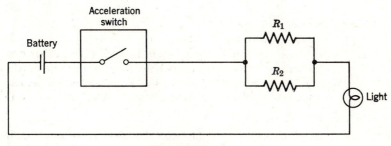

Figure P3-3

5. Given the circuit shown in Figure P3-4 probabilities of events are defined as follows:

$B = P(\text{Good battery})$,
$X_1 = P(\text{Timer } X \text{ is closed before set time})$,
$X_2 = P(\text{Timer } X \text{ is closed after set time})$,
$Y_1 = P(\text{Timer } Y \text{ is closed before set time})$,
$Y_2 = P(\text{Timer } Y \text{ is closed after set time})$,
$L = P(\text{Good light})$.

Assuming statistical independence, find the probability of the light lighting:

(a) before the set time occurs,
(b) after the set time occurs.

69

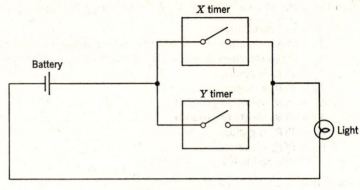

Figure P3-4

6. A system is composed of three subsystems A, B, C, all of which must work in order to have a successful system. Let the events that represent their proper functioning be a, b, and c, respectively. There are two possible environments E_1 and E_2 in which the systems can work. Given

$$P(abc \mid E_i) = P(a \mid E_i)P(b \mid E_i)P(c \mid E_i) \qquad i = 1, 2$$
$$P(a \mid E_1) = P(b \mid E_1) = P(c \mid E_1) = .9$$
$$P(a \mid E_2) = .9 \qquad P(b \mid E_2) = .8 \qquad P(c \mid E_2) = .7$$
$$P(E_1) = P(E_2) = \tfrac{1}{2}.$$

(a) Find $P(\text{systems success} \mid E_1)$.

(b) Find $P(\text{systems success})$.

(c) It is proposed that a redundant system be added (i.e., either one working results in a success). Find the probability that one or more of two identical systems functions properly. Assume conditional independence. Also assume independence. Why are the answers different? Which is the most reasonable assumption?

7. The designer of an inexpensive radio receiver needs to buy transistors. He decides to test 30 from each lot and allow two failures. Plot the O.C. curve.

His boss decides that it would be less expensive to test 10 and allow one failure. (The manufacturer has quoted a better price for transistors bought from this specification.) Plot the O.C. curve for this sampling plan on the same graph as the earlier plan was plotted.

The theory that determines which is the better plan is covered in a later chapter on decision theory. However, which plan seems best to you? You may have to assume something about cost.

CHAPTER IV

RANDOM VARIABLES

4-1 THE CONCEPT OF A RANDOM VARIABLE

It is often useful to describe the outcome of an experiment by a number, for example, the output voltage of a circuit, the time of failure of a piece of equipment, the number of cars in the queue at a traffic light.

The measured quantities associated with the outcomes of an experiment are often called variables, and since the outcome is not known in advance these quantities are loosely described as random variables. More formally a random variable is a function from the outcomes in the sample space to real numbers; that is, to every outcome s in the sample space S, a number $X(s)$ is assigned according to a rule X. X is called a random variable, and because it assigns one number to each outcome, X is a function.

As mentioned above the name random variable implies a variable that is random. However the presently accepted definition has attached the name random variable to X, the function which describes the mapping, while $X(s)$, a number which is a result of the mapping, is called a value of the random variable.

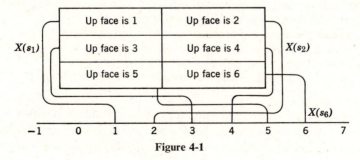

Figure 4-1

EXAMPLE 4-1

Consider the toss of one die. Let the random variable X represent the value of the up face. The mapping performed by X is shown in Figure 4-1. 1, 2, 3, 4, 5, 6, are called values of the random variable.

EXAMPLE 4-2

An experiment consists of building five radio receivers and counting the number of working receivers. We define a random variable Y to be the profit obtained. The random variable, profit, is defined by the following table:

Outcome	Value of Y
0 good	-100
1 good	-70
2 good	-40
3 good	-10
4 good	$+20$
5 good	$+50$

Probability

The purpose of a random variable is to carry the probability measure from the sets of the sample space to sets of points on the real line. Given the real numbers r_1 and r_2, we are interested typically in the probability that a random variable

(a) equals r_1;
(b) is greater than r_1;
(c) is between r_1 and r_2;
(d) is less than r_1; or
(e) is less than or equal to r_1.

For example, we are interested in

(a) the probability that the number of defects in a lot of components is 2;
(b) the probability that the "noise" voltage is greater than .01 V;
(c) the probability that a resistance is between 9 and 11 ohms;
(d) the probability the number of phone calls in a specified interval of time is less than 3000;
(e) the probability that the error of a position control system is less than or equal .1 in.

When we say the probability that a random variable X is within a certain set of real numbers (e.g., the sets given above in (a) through (e)), we mean the probability of the set of outcomes in the probability space which are mapped by X into the specified set of real numbers. For instance, by $P(X = r_1)$ we mean $P(\{s:X(s) = r_1\})$; by $P(X > r_1)$ we mean $P(\{s:X(s) > r_1\})$.

In Example 4-1, by $P(X = 2)$, we mean $P(\{s_2\})$ where s_2 is the outcome corresponding to the up face is 2 which is mapped by X into the value 2.

In Example 4-2, by $P(Y > 0)$ we mean $P(\{s:Y(s) > 0\}) = P(G_4 \cup G_5)$ where G_i is the event corresponding to i good.

A random variable transfers the probability assigned to events to the probability that the value of the random variable is within some specified set of real numbers. However there is one possible problem. Suppose the outcomes that correspond to the specified set of real numbers are not an event. Then we cannot talk about its probability. Although this can be a practical problem in some advanced applications, in this introductory book we assume that we only want to describe the probability that a random variable is within some interval (e.g., the intervals described in (a) through (e) above), and that the sets of outcomes that correspond to these intervals are events. In the practical problems of concern to us, this will suffice.

Discrete Random Variables

A random variable that can take on only a countable number of values is called a discrete random variable. Examples 4-1 and 4-2 describe discrete random variables. Other examples follow in later sections of this chapter.

A discrete random variable is often described by describing the probability that it takes on each of its possible values. That is, we describe a discrete random variable X by

$$P(X = x_i) = P(\{s:X(s) = x_i\}) \quad \text{for} \quad i = 0, 1, 2, \ldots$$

Continuous Random Variables

Often we wish to consider variables that can take on more than a countable number of values; for example, we often assume that a voltage can take on any value between -1 and $+1$. Such a random variable is called continuous. In this case the probability of a particular value within the interval is zero. Moreover if the probability of each value were greater than zero, the sum of an infinite number of positive quantities,* no matter how small, would not exist (would be infinite). Thus we describe such a continuous random variable not by the probability of taking on a certain value, but by the probability of being within a certain range of values. For example we might talk about the probability that a voltage is less than zero. Then we are left with the question: How is it possible that every value has probability zero, and yet the event that

---

* More exactly a positive greater lower bound is required.

the random variable is between, say, a and b has positive probability? A little reflection will reveal that this is very reasonable, and in fact any other assumption would be contradictory. For instance we talk about the charge density in a region; no point has nonzero charge, but the region does have. Also a falling body moves no distance at a point in time, but in a positive interval of time, a nonzero distance is traveled.

EXAMPLE 4-3

A telephone call is equally likely to occur any time between 12:00 noon and 1:00 p.m. Let X be a random variable that is 0 at 12:00 and is the time of call in fractions of an hour past 12:00 for any other outcome. Find

1. $P(0 \leq X \leq 1.0)$,
2. $P(X = .02)$,
3. $P(X \leq .5)$,
4. $P(.2 < X < .8)$.

The equally likely assumption implies that the probability of each interval is the length of the interval; thus

1. $P(0 \leq X \leq 1.0) = P(\{s : s \text{ between 12:00 noon and 1:00 p.m.}\})$
$$= P(S) = 1;$$
2. $P(X = .02) = P[(s : s = 12.02)] = 0$
   (The length of a point is zero.);
3. $P(X \leq .5) = .5;$
4. $P(.2 < X < .8) = .6.$

We now summarize the concept of a random variable with this definition.

Definition

A random variable is a function from the sample space to the real line such that

1. $P(X = -\infty) = P(X = +\infty) = 0.$
2. For a real number r, $\{s : X(s) \leq r\}$ is an event contained in the sample space.

Requirement 1 is only reasonable; it requires that a random variable always maps into a real number. Requirement 2 ensures that we can talk about the probability that a random variable is within any of the usual intervals.

4-2 DISTRIBUTIONS

Distribution Function

Given a random variable X, the distribution function F_X of the random variable X is

$$F_X(x) = P(X \leq x) = P(\{s : X(s) \leq x\}) \qquad (4\text{-}1)$$

where x is a real number.

Note that the subscript is a part of the identification of the distribution function, because it tells which random variable's distribution function is being described. The argument of the function is a possible value from the range of the random variable and thus is simply a real number.

EXAMPLE 4-4

Consider the toss of a fair die. Plot the distribution function of X where X is a random variable that equals the number on the up face. The solution is given in Figure 4-2.

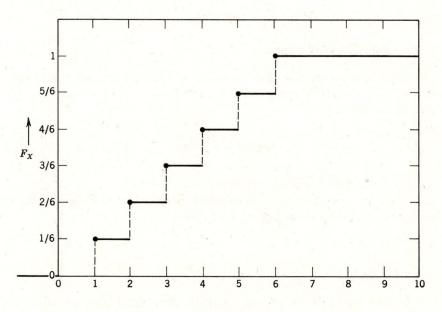

Figure 4-2

A distribution function of a random variable X has the following properties (see Problem 1).

1. If $x_1 < x_2$, then $F_X(x_1) \leq F_X(x_2)$ (i.e., F_X is nondecreasing).
2. $F_X(-\infty) = 0$, $F_X(\infty) = 1$.
3. $\lim\limits_{\substack{\epsilon \to 0 \\ \epsilon > 0}} F_X(x + \epsilon) = F_X(x)$ (i.e., F_X is continuous from the right).

Probability Mass Function

If the random variable is discrete, a probability mass function is defined as follows:

$$p_X(x) = P(X = x) = P(\{s : X(s) = x\}).\qquad(4\text{-}2)$$

In the example given in Example 4-4 the probability mass function is shown in Figure 4-3.

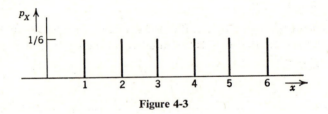

Figure 4-3

Probability Density Function

If the random variable X is continuous, then F_X is (absolutely) continuous and a derivative exists at all but a countable number of points. This derivative,

$$f_X(x) = \frac{dF_X(x)}{dx},\qquad(4\text{-}3)$$

is called a probability density function.

From the properties of a distribution function, a probability density function f must have the following properties:

1. $f(x) \geq 0$;

2. $\displaystyle\int_{-\infty}^{\infty} f(x)\, dx = 1$.

The first follows from the fact that the distribution function does not decrease, and the second follows from the fundamental theorem of integral calculus. See Problem 2. Also, from (4-3) and the fundamental theorem of integral calculus

$$F_X(x) = \int_{-\infty}^{x} f_X(\lambda)\, d\lambda. \tag{4-4}$$

Note that the dummy variable of integration has been changed from x to λ to avoid confusion with the x found in the upper limit of integration.

From the diagram

$$\{X \leq x_1\} \cup \{x_1 < X \leq x_2\} = \{X \leq x_2\}$$

and

$$\{X \leq x_1\} \cap \{x_1 < X \leq x_2\} = \varnothing.$$

Thus $\qquad P(x_1 < X \leq x_2) = P(X \leq x_2) - P(X \leq x_1)$

$$= F_X(x_2) - F_X(x_1) = \int_{x_1}^{x_2} f_X(x)\, dx. \tag{4-5}$$

Using (4-5) for Δx very small, we find

$$P(x - \Delta x/2 < X \leq x + \Delta x/2) \approx f_X(x)\, \Delta x.$$

Thus the density function can be thought of as a limit.

$$f_X(x) = \lim_{\Delta x \to 0} \frac{P(x - \Delta x/2 < X \leq x + \Delta x/2)}{\Delta x}. \tag{4-6}$$

Note that if X has a density function f_X, then

$$P(X = x) = \int_{x}^{x} f_X(\lambda)\, d\lambda = 0. \tag{4-7}$$

Mixed Distribution Function

It is possible for a random variable to have a distribution function as shown in Figure 4-4. In this case the distribution function is called mixed, because it consists of a part that has a density function and a part that has a probability mass function.

Four important concepts—random variable, distribution function, probability mass function, and probability density function—have been introduced. These concepts are illustrated by examples in the next section.

77

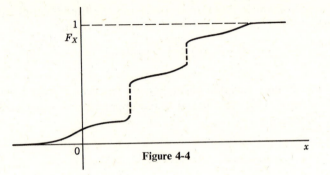

Figure 4-4

4-3 EXAMPLES OF DISTRIBUTION FUNCTIONS

In this section examples are given of distribution functions commonly used in engineering. In these examples, as is usual in practice, the distribution function will be specified by specifying either the probability mass function or the probability density function.

Binomial

The first distribution function to be discussed is the binomial distribution. The sample space for this distribution function is a space composed of independent Bernoulli trials. The random variable represents the number of times a certain event occurs. This random variable and its associated distribution function are useful in describing sequences of independent subexperiments where the probability of the events remains constant.

The binomial distribution of a random variable X is completely specified by n, the number of trials; p, the probability of the event of interest; and k, an integer value of the random variable X. From Chapter II, $P(X_n = k) = P(\text{the event of interest occurs } k \text{ times in } n \text{ trials}) = \binom{n}{k} p^k (1 - p)^{n-k}$.

EXAMPLE 4-5

A fair coin ($P\{\text{heads}\} = 1/2$) is tossed four times. What is the probability mass function for H, the number of heads?

$$p_H(0) = P(H = 0) = (1/2)^4 = 1/16,$$
$$p_H(1) = P(H = 1) = 4(1/2)^3(1/2) = 4/16,$$
$$p_H(2) = P(H = 2) = 6(1/2)^2(1/2)^2 = 6/16,$$
$$p_H(3) = P(H = 3) = 4(1/2)(1/2)^3 = 4/16,$$
$$p_H(4) = P(H = 4) = (1/2)^4 = 1/16.$$

The probability mass function p_H and the associated distribution function F_H are shown in Figure 4-5a and b, respectively.

Figure 4-5b is derived from 4-5a by using the relation

$$F_H(k) = \sum_{i=0}^{k} p_H(i), \quad k \text{ a non-negative integer,}$$

$$F_H(x) = F_H([x]),$$

where $[x]$ = greatest integer $\leq x$. From F_H one can answer questions such as, what is the probability of having three or fewer heads $[F_H(3) = P(H \leq 3)]$? The answer is 15/16.

EXAMPLE 4-6

Suppose there are five intermittent loads connected to a power supply and that each load demands either 2 w or no power. The probability of demanding 2 w is 1/4 for each load, and the demands are independent. What is the distribution function for W, a random variable representing power required?

$P(W = 0) = P(\text{no loads demand power}) = (3/4)^5 \approx .237,$
$P(W = 2) = P(1 \text{ load demands power}) = 5(1/4)(3/4)^4 \approx .395,$
$P(W = 4) = P(2 \text{ loads demand power}) = 10(1/4)^2(3/4)^3 \approx .264,$
$P(W = 6) = P(3 \text{ loads demand power}) = 10(1/4)^3(3/4)^2 \approx .088,$
$P(W = 8) = P(4 \text{ loads demand power}) = 5(1/4)^4(3/4) \approx .015,$
$P(W = 10) = P(5 \text{ loads demand power}) = (1/4)^5 \approx .001.$

The distribution function for W is shown in Figure 4-6.

If one had a power supply that could deliver 6 w, then the demanded load would be supplied with probability .984.

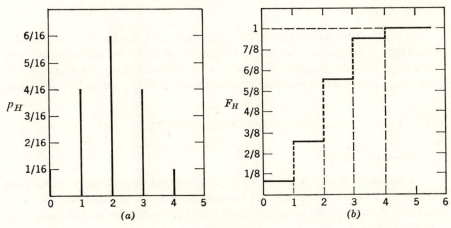

Figure 4-5

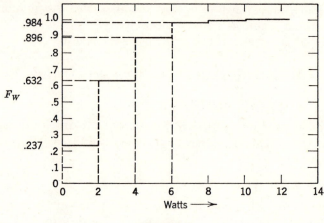

Figure 4-6

Poisson

Next, the Poisson distribution is considered. A random variable X is said to have a Poisson distribution or to have a Poisson probability mass function with parameter λ if

$$P(X = k) = e^{-\lambda} \frac{\lambda^k}{k!} \qquad k = 0, 1, \ldots . \qquad (4\text{-}8)$$

In Chapter XI (4-8) is derived from a differential and difference equation. Here we simply assume that certain experiments have been observed to have occurrences as described by a Poisson distribution.

EXAMPLE 4-7

A radioactive substance emits α particles, and the number of particles reaching a given portion of space during an interval of 1 sec has been observed to have a Poisson distribution with $\lambda = 10$. What is the probability that X, the number of particles reaching the given space during 1 sec, is 3? Using (4-8)

$$P(X = 3) = \exp [-10] \frac{10^3}{3 \cdot 2 \cdot 1}.$$

EXAMPLE 4-8

A machine produces sheet metal where the number of flaws X per yard follows a Poisson distribution. The average number of flaws per yard is 2 (λ in the Poisson probability function). Plot the probability mass function for X. See Figure 4-7.

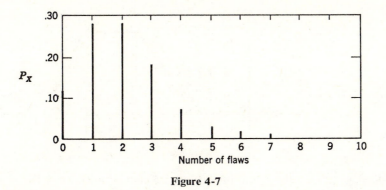

Figure 4-7

Uniform

The next distribution function to be studied is of the continuous type and has a density function. It is called the uniform density function. If a random variable X is equally likely to be anywhere between two points a and b where $b > a$, then it has a uniform density function. This is shown in Figure 4-8. The value of f_X is constant between a and b and can be determined from the fact that the area under a density function must be 1. Calling the constant value C, then

$$1 = \int_a^b C\,dx$$

implies

$$C = \frac{1}{b-a}.$$

The uniform distribution function is shown in Figure 4-9.

EXAMPLE 4-9

Resistors are produced that have a nominal value of 10 ohms and are $\pm 10\%$ resistors. Assume any possible value of resistance is equally likely. Find the density and distribution function of the random variable R, which represents resistance. Find the probability that a resistor, selected at random, is between 9.5 and 10.5 ohms.

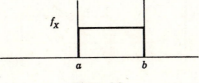

Figure 4-8

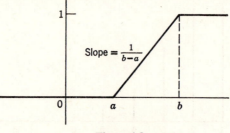

Figure 4-9

The density and distribution functions are shown in Figure 4-10.

$$P(9.5 < R \le 10.5) = F_R(10.5) - F_R(9.5) = \tfrac{3}{4} - \tfrac{1}{4} = \tfrac{1}{2}.$$

Or

$$P(9.5 < R \le 10.5) = \int_{9.5}^{10.5} \tfrac{1}{2}\, dr = \frac{10.5 - 9.5}{2} = \frac{1}{2}.$$

Normal

The next distribution to be studied is the normal or Gaussian distribution function. This distribution function is the most widely used distribution function in applications of probability and statistics. It is often said that everybody "believes" in the normal distribution; engineers believe that mathematicians have a theorem that says random variables are normally distributed, and mathematicians believe that engineers have observed the normal distribution in practice. Both statements are true with some reservations. The central limit theorem does imply that a random variable which is determined by a large number of independent causes tends to have a normal distribution function. Also many empirical (observed) distributions are nearly normal.

The normal density function is

$$f_X(x) = \frac{1}{\sqrt{2\pi\sigma^2}} \exp\left\{-\left[\frac{(x-\mu)^2}{2\sigma^2}\right]\right\}. \tag{4-9}$$

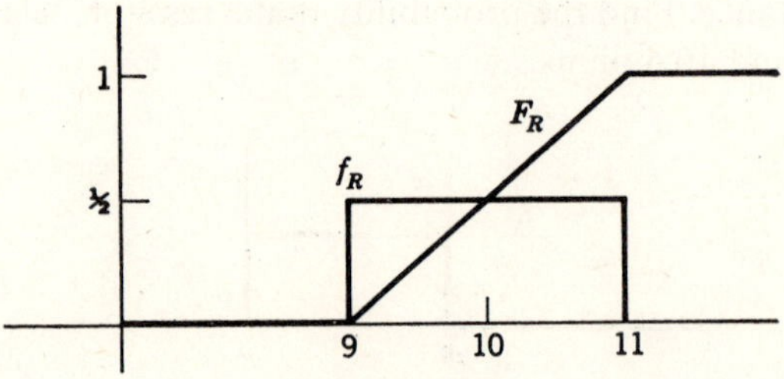

Figure 4-10

The normal distribution function cannot be expressed in closed form using elementary functions, but is

$$F_X(x) = \int_{-\infty}^{x} \frac{1}{\sqrt{2\pi\sigma^2}} \exp\left\{ -\left(\frac{(y-\mu)^2}{2\sigma^2}\right) \right\} dy. \tag{4-10}$$

In these expressions the parameter μ is the value at which the normal density function is a maximum, while σ is a parameter that characterizes the spread of the density function. The normal p.d.f. is shown in Figure 4-11 for several values of μ and σ.

We now show that the area under a normal p.d.f. is 1. Let

$$I = \int_{-\infty}^{\infty} \frac{1}{\sqrt{2\pi\sigma^2}} \exp\left(-\frac{(x-\mu)^2}{2\sigma^2} \right) dx.$$

Then

$$I^2 = \left[\int_{-\infty}^{\infty} \frac{1}{\sqrt{2\pi\sigma^2}} \exp\left(-\frac{(x-\mu)^2}{2\sigma^2} \right) dx \right]\left[\int_{-\infty}^{\infty} \frac{1}{\sqrt{2\pi\sigma^2}} \exp\left(-\frac{(y-\mu)^2}{2\sigma^2} \right) dy \right],$$

$$I^2 = \frac{1}{2\pi\sigma^2} \iint_{-\infty}^{\infty} \exp\left[-\frac{(x-\mu)^2}{2\sigma^2} - \frac{(y-\mu)^2}{2\sigma^2} \right] dx\, dy.$$

Letting $v = [(x-\mu)/\sigma]$ and $w = [(y-\mu)/\sigma]$

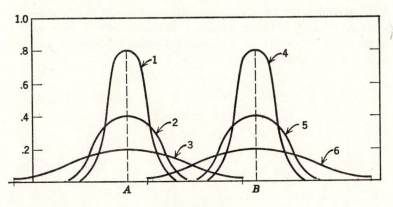

Figure 4-11

Graphs of the normal p.d.f. for several values of μ and σ:
(1) $\mu = A$, $\sigma = 0.5$; (2) $\mu = A$, $\sigma = 1$; (3) $\mu = A$, $\sigma = 2$;
(4) $\mu = B$, $\sigma = 0.5$; (5) $\mu = B$, $\sigma = 1$; (6) $\mu = B$, $\sigma = 2$.

$$I^2 = \frac{1}{2\pi\sigma^2} \int\!\!\int_{-\infty}^{\infty} \exp\left[-\frac{(v^2+w^2)}{2}\right] \sigma^2 \, dv \, dw.$$

Letting $r^2 = v^2 + w^2$, $\theta = \tan^{-1} w/v$

$$I^2 = \frac{1}{2\pi} \int_0^{2\pi} \int_0^{\infty} \exp\left[-\frac{r^2}{2}\right] r \, dr \, d\theta$$

$$= \frac{1}{2\pi} \left[2\pi \int_0^{\infty} r \exp\left[-\frac{r^2}{2}\right] dr\right] = -\exp\left[-\frac{r^2}{2}\right]\Big|_0^{\infty} = 1.$$

$I^2 = 1$ implies either $I = 1$ or $I = -1$. Since the density function is always positive, $I = 1$.

The transformation $v = (x - \mu)/\sigma$ used above is very useful in practice because it is often necessary to find the value of the distribution function, that is

$$F_X(a) = \int_{-\infty}^{a} \frac{1}{\sqrt{2\pi\sigma^2}} \exp\left[-\frac{(x-\mu)^2}{2\sigma^2}\right] dx.$$

Since this cannot be evaluated by elementary integration directly, the transformation $v = (x - \mu)/\sigma$ is used so that

$$F_X(a) = \int_{-\infty}^{(a-\mu)/\sigma} \frac{1}{\sqrt{2\pi}} \exp\left[-\frac{v^2}{2}\right] dv.$$

This transforms the density to one for which $\mu = 0$, $\sigma = 1$ and in this case it is called a standard normal distribution and is widely tabulated. See for example Ref. C3.

Various tables may tabulate any of the areas shown in Figure 4-12, so one must observe which is being tabulated. However any of the results can be obtained from the others by using the following relations for the standard ($\mu = 0$, $\sigma = 1$) normal random variable X:

$$P(X \leq x) = 1 - P(X > x),$$

$$P(-a \leq X \leq a) = 2P(-a \leq X \leq 0) = 2P(0 \leq X \leq a),$$

$$P(X \leq 0) = \tfrac{1}{2}.$$

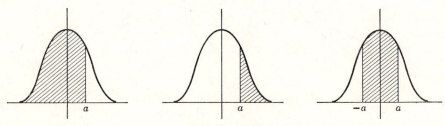

Figure 4-12

EXAMPLE 4-10

The voltage X at the output of a noise generator is a standard ($\mu = 0$, $\sigma = 1$) normal random variable. What is $P(1 \leq X \leq 2.3)$? What is $P(X > 2.3)$?

Using one of the tables of the standard normal distribution

$$P(1 \leq X \leq 2.3) = F_X(2.3) - F_X(1) \simeq .989 - .841 = .148,$$
$$P(X > 2.3) = 1 - P(X \leq 2.3) \simeq 1 - .989 = .011.$$

EXAMPLE 4-11

The velocity V of the wind at a certain location is a normal random variable with $\mu = 2$ and $\sigma = 5$. Determine

$$P(-3 \leq V \leq 8),$$

$$P(-3 \leq V \leq 8) = \int_{-3}^{8} \frac{1}{\sqrt{2\pi 25}} \exp\left[-\frac{(v-2)^2}{2(25)}\right] dv$$

$$= \int_{(-3-2)/5}^{(8-2)/5} \frac{1}{\sqrt{2\pi}} \exp\left[-\frac{x^2}{2}\right] dx = F_{SN}(1.2) - F_{SN}(-1)$$

where F_{SN} is the distribution function of the standard normal random variable.

$$F_{SN}(1.2) - F_{SN}(-1) \simeq .885 - .159 = .726.$$

Exponential

The last distribution function to be considered at this point is the exponential distribution function. The density function is

$$f(t) = ae^{-t/\lambda}, \qquad t \geq 0,$$
$$= 0, \qquad t < 0, \tag{4-11}$$

where λ is a positive parameter and a may be found from

$$1 = \int_0^\infty ae^{-t/\lambda}\, dt = -a\lambda e^{-t/\lambda}\Big|_0^\infty = a\lambda,$$

which implies $a = 1/\lambda$. Thus

$$F(x) = \int_0^x \frac{1}{\lambda} e^{-t/\lambda}\, dt = -e^{-t/\lambda}\Big|_0^x = 1 - e^{-x/\lambda}, \qquad x \geq 0,$$

$$F(x) = 0, \qquad x < 0.$$

85

EXAMPLE 4-12

Certain types of equipment have been observed to fail according to an exponential distribution. That is, the time to failure T has an exponential distribution function, and λ is called the mean time to failure. If the mean time to failure of a light bulb is 100 hr., find the probability that a light bulb will last for more than 150 hr. This is the same as the probability it will not fail in the first 150 hr.

$$P(T > 150) = 1 - F_T(150) = e^{-150/100} \simeq .224.$$

4-4 TWO RANDOM VARIABLES

Joint Distribution Function

We now consider the case where two random variables are defined on a sample space. For example both the voltage and current might be of interest in a certain experiment.

The probability of the joint occurrence of two events A, B was called the joint probability $P(A \cap B)$. If the event A is the event $\{X \le x\}$ and the event B is the event $\{Y \le y\}$, then the joint probability is called the joint distribution function of the random variables X and Y; that is

$$F_{X,Y}(x, y) = P\{(X \le x) \cap (Y \le y)\}. \tag{4-12}$$

From this definition it can be noted that

$$F_{X,Y}(-\infty, -\infty) = 0, \qquad F_{X,Y}(-\infty, y) = 0, \qquad F_{X,Y}(\infty, y) = F_Y(y),$$
$$F_{X,Y}(x, -\infty) = 0, \qquad F_{X,Y}(\infty, \infty) = 1, \qquad F_{X,Y}(x, \infty) = F_X(x).$$

Suppose that $x_1 < x_2$, (see Figure 4-13) then

$$P\{(X \le x_2) \cap (Y \le y_0)\}$$
$$= P\{(X \le x_1) \cap (Y \le y_0)\} + P\{(x_1 < X \le x_2) \cap (Y \le y_0)\}$$

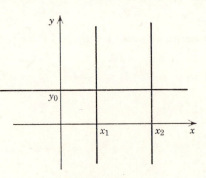

Figure 4-13

because the union of the two mutually exclusive events on the right side is the event on the left side. Thus

$$P\{(x_1 < X \le x_2) \cap (Y \le y_0)\} = F_{X,Y}(x_2, y_0) - F_{X,Y}(x_1, y_0) \ge 0. \quad (4\text{-}13)$$

Similarly for $y_1 < y_2$

$$P\{(X \le x) \cap (y_1 < Y \le y_2)\} = F_{X,Y}(x, y_2) - F_{X,Y}(x, y_1) \ge 0. \quad (4\text{-}14)$$

Both of the probabilities given above are non-negative; thus the right-hand sides are greater than zero, which shows that $F_{X,Y}$ is a nondecreasing function of both its arguments.

From

$$\{(x_1 < X \le x_2) \cap (Y \le y_2)\} = \{(x_1 < X \le x_2) \cap (Y \le y_1)\}$$
$$\cup \{(x_1 < X \le x_2) \cap (y_1 < Y \le y_2)\},$$

and from the fact that the two sets on the right are mutually exclusive,

$$P\{(x_1 < X \le x_2) \cap (y_1 < Y \le y_2)\} = P\{(x_1 < X \le x_2) \cap (Y < y_2)\}$$
$$- P\{(x_1 < X \le x_2) \cap (Y < y_1)\}.$$

Using (4-13)

$$P\{(x_1 < X \le x_2) \cap (y_1 < Y \le y_2)\}$$
$$= F_{X,Y}(x_2, y_2) - F_{X,Y}(x_1, y_2) - F_{X,Y}(x_2, y_1) + F_{X,Y}(x_1, y_1). \quad (4\text{-}15)$$

Note from the left-hand side of (4-15) that either side of (4-15) is non-negative.

Joint Probability Mass Function

If each of the random variables is discrete, then a joint probability mass function is defined as

$$p_{X,Y}(x_i, y_j) = P\{(X = x_i) \cap (Y = y_j)\}. \quad (4\text{-}16)$$

Also,

$$F_{X,Y}(x, y) = \sum_{x_i \le x} \sum_{y_j \le y} p_{X,Y}(x_i, y_j).$$

EXAMPLE 4-13

Find the joint probability mass function and joint distribution function of X, Y associated with the experiment of tossing two fair dice where X

represents the number appearing on the up face of one die and Y represents the number appearing on the up face of the other die.

$$p_{X,Y}(i, j) = \frac{1}{36}, \qquad i = 1, 2, \ldots, 6, \qquad j = 1, 2, \ldots, 6;$$

$$F_{X,Y}(x, y) = \sum_{i=1}^{x} \sum_{j=1}^{y} \frac{1}{36}, \qquad x = 1, 2, \ldots, 6, \qquad y = 1, 2, \ldots, 6$$

$$= \frac{xy}{36}.$$

If x and y are not integers and are between 0 and 7, $F_{X,Y}(x, y) = F_{X,Y}([x], [y])$ where $[x]$ is the greatest integer less than or equal to x. $F(x, y) = 0$ for $x < 1$ or $y < 1$. $F(x, y) = 1$ for $x \geq 6$ and $y \geq 6$.

EXAMPLE 4-14

In the same experiment as Example 4-13, find the joint probability mass function of X, Z where X is defined as above and Z is the sum of the two up faces. This answer is given in the following table:

X \ Z	2	3	4	5	6	7	8	9	10	11	12
1	$\frac{1}{36}$	$\frac{1}{36}$	$\frac{1}{36}$	$\frac{1}{36}$	$\frac{1}{36}$	$\frac{1}{36}$	0	0	0	0	0
2	0	$\frac{1}{36}$	$\frac{1}{36}$	$\frac{1}{36}$	$\frac{1}{36}$	$\frac{1}{36}$	$\frac{1}{36}$	0	0	0	0
3	0	0	$\frac{1}{36}$	$\frac{1}{36}$	$\frac{1}{36}$	$\frac{1}{36}$	$\frac{1}{36}$	$\frac{1}{36}$	0	0	0
4	0	0	0	$\frac{1}{36}$	$\frac{1}{36}$	$\frac{1}{36}$	$\frac{1}{36}$	$\frac{1}{36}$	$\frac{1}{36}$	0	0
5	0	0	0	0	$\frac{1}{36}$	$\frac{1}{36}$	$\frac{1}{36}$	$\frac{1}{36}$	$\frac{1}{36}$	$\frac{1}{36}$	0
6	0	0	0	0	0	$\frac{1}{36}$	$\frac{1}{36}$	$\frac{1}{36}$	$\frac{1}{36}$	$\frac{1}{36}$	$\frac{1}{36}$

Joint Probability Density Function

If $F_{X,Y}$ is continuous and has partial derivatives, then a joint density function is defined by

$$f_{X,Y}(x, y) = \frac{\partial^2 F_{X,Y}(x, y)}{\partial x \, \partial y}. \qquad (4\text{-}17)$$

Taking limits in (4-15) will show that

$$f_{X,Y} \geq 0.$$

From the fundamental theorem of integral calculus

$$F_{X,Y}(x, y) = \int_{-\infty}^{y} \int_{-\infty}^{x} f_{X,Y}(\mu, v) \, d\mu \, dv.$$

Since $F_{X,Y}(\infty, \infty) = 1$,

$$\int\!\!\int_{-\infty}^{\infty} f_{X,Y}(\mu, v)\, d\mu\, dv = 1. \tag{4-18}$$

A joint density function may be interpreted as

$$P\{(x < X \leq x + dx) \cap (y < Y \leq y + dy)\} = f_{X,Y}(x, y)\, dx\, dy.$$

The most common joint density function is the jointly normal density function. Two random variables, X_1 and X_2, are said to be jointly normal if

$$f_{X_1,X_2}(x_1, x_2) = \frac{1}{2\pi\sigma_1\sigma_2\sqrt{1 - r^2}}$$

$$\times \exp\left\{-\frac{1}{2(1 - r^2)}\left[\frac{(x_1 - \mu_1)^2}{\sigma_1^2} - \frac{2r(x_1 - \mu_1)(x_2 - \mu_2)}{\sigma_1\sigma_2} + \frac{(x_2 - \mu_2)^2}{\sigma_2^2}\right]\right\}. \tag{4-19}$$

Equation 4-19 describes a bell-shaped surface above the x_1, x_2 plane. The center of the bell occurs at $x_1 = \mu_1$ and $x_2 = \mu_2$, and $f_{X_1,X_2}(\mu_1, \mu_2) = (2\pi\sigma_1\sigma_2\sqrt{1 - r^2})^{-1}$, which is the peak height of the bell. When $r = 0$, if $\sigma_1 = \sigma_2$ the bell has a circular cross section; if $\sigma_1 > \sigma_2$ the cross section of the bell is elliptical with the axes of the ellipse parallel to the coordinate axes, and there is more spread in the x_1 direction.

EXAMPLE 4-15

The impact point of a projectile aimed at a point 100 ft north and 50 ft east of the reference has a jointly normal density with $\mu_1 = 50$, $\mu_2 = 100$, $\sigma_1 = 2 = \sigma_2$, and $r = 0$. Describe the density function.

Letting X_1 represent distance in the east direction and X_2 represent distance in the north direction, the density function is shown in Figure 4-14.

$$f_{X_1,X_2}(x_1, x_2) = \frac{1}{8\pi} \exp\left\{-\tfrac{1}{8}[(x_1 - 50)^2 + (x_2 - 100)^2]\right\}$$

Marginal Distributions

Since $\{A \cap S\} = \{A\}$,

$$P\{(X \leq x) \cap (Y \leq \infty)\} = P(X \leq x) = F_X(x).$$

Also, $\qquad P\{(X \leq x) \cap (Y \leq \infty)\} = F_{X,Y}(x, \infty).$

89

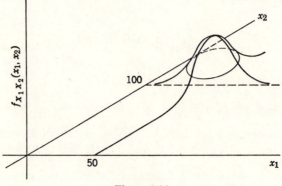

Figure 4-14

F_X is called the marginal distribution function of X and

$$F_{X,Y}(x, \infty) = F_X(x).$$

For discrete distribution functions

$$F_{X,Y}(x, \infty) = \sum_{x_i \leq x} \sum_{\text{all } j} p_{X,Y}(x_i, y_j) = \sum_{x_i \leq x} p_X(x_i).$$

Thus we have the basic relationship between joint probability mass functions and marginal probability mass functions:

$$p_X(x_i) = \sum_{\text{all } j} p_{X,Y}(x_i, y_j), \tag{4-20}$$

and p_X is called the marginal probability mass function of X, or simply the probability mass function of X.

Similarly, if a joint density function is defined,

$$F_{X,Y}(\infty, y) = \int_{-\infty}^{y} \int_{-\infty}^{\infty} f_{X,Y}(\mu, v) \, d\mu \, dv = \int_{-\infty}^{y} f_Y(v) \, dv. \tag{4-21}$$

From the last part of (4-21) we have the basic relationship between joint and marginal probability density functions:

$$f_Y(y) = \int_{-\infty}^{\infty} f_{X,Y}(\mu, y) \, d\mu, \tag{4-22}$$

and f_Y is called the marginal probability density function of Y, or simply the probability density function of Y.

EXAMPLE 4-16

The density function of X, Y is

$$f_{X,Y}(x, y) = axy, \qquad 1 \leq x \leq 3, 2 \leq y \leq 4,$$
$$= 0, \qquad\qquad \text{elsewhere.}$$

Find a, f_X, F_Y.

$$1 = \int_2^4 \int_1^3 axy \, dx \, dy = a \int_2^4 y \left[\frac{x^2}{2}\right]\Big|_1^3 dy$$

$$= a \int_2^4 4y \, dy = 4a \left[\frac{y^2}{2}\right]\Big|_2^4 = 24a,$$

$$a = \tfrac{1}{24} \, ;$$

$$f_X(x) = \frac{1}{24} \int_2^4 xy \, dy = \frac{x}{24} [8 - 2] = \frac{x}{4}, \qquad 1 \le x \le 3,$$

$$= 0, \qquad\qquad\qquad\qquad\qquad \text{elsewhere,}$$

$$F_Y(y) = 0, \qquad y \le 2,$$

$$= 1, \qquad y > 4,$$

$$= \frac{1}{24} \int_2^y \int_1^3 xv \, dx \, dv = \frac{1}{6} \int_2^y v \, dv$$

$$= \tfrac{1}{12} [y^2 - 4], \qquad 2 \le y \le 4.$$

EXAMPLE 4-17

If f_{X_1, X_2} is jointly normal as in (4-19), then f_{X_1} is normal.

$$f_{X_1}(x_1) = \int_{-\infty}^{\infty} \frac{1}{2\pi\sigma_1\sigma_2\sqrt{1 - r^2}}$$

$$\times \exp\left\{-\frac{1}{2(1 - r^2)}\left[\frac{(x_1 - \mu_1)^2}{\sigma_1^2} - \frac{2r(x_1 - \mu_1)(x_2 - \mu_2)}{\sigma_1\sigma_2} + \frac{(x_2 - \mu_2)^2}{\sigma_2^2}\right]\right\} dx_2.$$

It is left to the reader to perform the algebra to show

$$f_{X_1}(x_1) = \frac{1}{\sqrt{2\pi\sigma_1^2}} \exp\left[\frac{(x_1 - \mu_1)^2}{2\sigma_1^2}\right].$$

4-5 CONDITIONAL DISTRIBUTIONS

Because of the importance of conditions in engineering applications and in the interpretations of data covered in a latter part of this book, conditional distributions are given special attention here.

It is important to recall that conditional probabilities are probabilities defined on the probability space set up by the specified condition. Actually all probabilities are conditional upon the assumptions and model of the problem; certain of these are called conditional probabilities when it is desirable to display explicitly the given condition. It naturally follows that conditional distributions should have all of the properties of distributions. This will be shown below.

Conditional probabilities were defined by [see (2-33)]

$$P(A \mid B) = \frac{P(A \cap B)}{P(B)}.$$

Conditioning Based on an Event with Positive Probability

If A is the event $X = x_i$ and B is the event $Y = y_j$, where X and Y are both discrete, then the definition of a conditional probability mass function $p_{X|Y}(x_i \mid y_j)$ follows:

$$
\begin{aligned}
p_{X|Y}(x_i \mid y_j) &= P\{(X = x_i) \mid (Y = y_j)\} \\
&= \frac{P\{(X = x_i) \cap (Y = y_j)\}}{P(Y = y_j)} \\
&= \frac{p_{X,Y}(x_i, y_j)}{p_Y(y_j)}.
\end{aligned}
\tag{4-23}
$$

From (4-23), we see that $p_{X|Y}$ is a probability mass function. Indeed, using (4-20)

$$\sum_{\text{all } i} p_{X|Y}(x_i \mid y_j) = \frac{\sum\limits_{\text{all } i} p_{X,Y}(x_i, y_j)}{p_Y(y_j)} = \frac{p_Y(y_j)}{p_Y(y_j)} = 1.$$

Now consider the case where X is continuous. Let A be the event $\{X \leq x\}$; then

$$P(X \leq x \mid B) = \frac{P\{(X \leq x) \cap B\}}{P(B)}.$$

Following the convention for distribution functions, a conditional distribution function is defined:

$$F_{X|B}(x) = P(X \leq x \mid B) = \frac{P\{(X \leq x) \cap B\}}{P\{B\}}.
\tag{4-24}$$

If the derivative exists a conditional density function is defined by

$$f_{X|B}(x) = \frac{dF_{X|B}(x)}{dx}.
\tag{4-25}$$

Note that B may be an event defined by either a continuous or a discrete random variable.

One particular case is of some interest to most engineers. Let B be the event the random variable X is between two limits. This event is of interest because produced material can be described by some density function. It is then inspected and the material outside certain tolerance limits is rejected. The following example illustrates this.

EXAMPLE 4-18

Steel braces are designed to have a length X of 4 ± 0.2 in. The actual density function produced is called f_X and is shown in Figure 4-15.

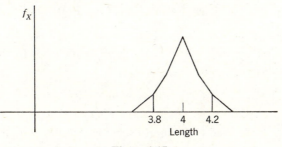

Figure 4-15

We are interested in the density after screening; that is, after all the braces above 4.2 in. and below 3.8 in. have been rejected. Using (4-24) and (4-25) with B the event that all braces are between 3.8 and 4.2 in.,

$$F_{X|B}(x) = \frac{P\{(X \le x) \cap (3.8 < X \le 4.2)\}}{P(3.8 < X \le 4.2)}.$$

Note that

$$\{(X \le x) \cap (3.8 < X \le 4.2)\} = \varnothing \qquad \text{if } x \le 3.8,$$
$$\{(X \le x) \cap (3.8 < X \le 4.2)\} = \{3.8 < X \le x\} \qquad \text{if } 3.8 < x \le 4.2,$$
$$\{(X \le x) \cap (3.8 < X \le 4.2)\} = \{3.8 < X \le 4.2\} \qquad \text{if } x > 4.2.$$

Thus

$$F_{X|B}(x) = 0, \qquad\qquad\qquad x \le 3.8,$$

$$= \frac{\displaystyle\int_{3.8}^{x} f_X(\lambda)\, d\lambda}{\displaystyle\int_{3.8}^{4.2} f_X(\theta)\, d\theta}, \qquad 3.8 < x \le 4.2,$$

$$= 1, \qquad\qquad\qquad x > 4.2.$$

From (4-25) and the above results

$$f_{X|B}(x) = 0, \qquad\qquad x \leq 3.8,$$
$$= C f_X(x), \qquad\quad 3.8 < x \leq 4.2,$$
$$= 0, \qquad\qquad\quad x > 4.2.$$

where
$$C = \frac{1}{\displaystyle\int_{3.8}^{4.2} f_X(\theta)\, d\theta} \geq 1.$$

Thus the conditional density function is zero outside the tolerance limits and is moved up by a constant factor inside the tolerance limits.

Conditioning Based on an Event With Zero Probability

The definitions given thus far would be sufficient except that we often want to condition by an event that has probability zero. In this case if the density functions exist

$$f_{X|Y}(x \mid y) = \frac{f_{X,Y}(x, y)}{f_Y(y)}. \tag{4-26}$$

This result can be justified if the limit exists by the following reasoning:

$$F_{X|Y}(x \mid y) = \lim_{\Delta y \to 0} F_{X|B}(x) \tag{4-27}$$

where
$$B = \{y < Y \leq y + \Delta y\}.$$

Using the previous definitions

$$F_{X|Y}(x \mid y) = \lim_{\Delta y \to 0} \frac{P\{(X \leq x) \cap (y < Y \leq y + \Delta y)\}}{P(y < Y \leq y + \Delta y)},$$

$$F_{X|Y}(x \mid y) = \lim_{\Delta y \to 0} \frac{\displaystyle\int_{-\infty}^{x} \int_{y}^{y+\Delta y} f_{X,Y}(\theta, \lambda)\, d\lambda\, d\theta}{\displaystyle\int_{y}^{y+\Delta y} f_Y(\lambda)\, d\lambda}$$

$$= \frac{\displaystyle\int_{-\infty}^{x} f_{X,Y}(\theta, y)\, d\theta\, dy}{f_Y(y)\, dy}. \tag{4-28}$$

Taking the derivative of both sides of (4-28) with respect to x produces (4-26). Such a limiting process will not be satisfactory in certain rare cases; however (4-26) is a useful concept.

94

From (4-28) and (4-26) we obtain the usual relation between density and distribution functions.

$$F_{X|Y}(x \mid y) = \int_{-\infty}^{x} f_{X|Y}(\lambda \mid y) \, d\lambda. \tag{4-29}$$

This distribution function is interpreted as

$$F_{X|Y}(x \mid y) = P[(X \le x) \mid Y = y].$$

From (4-22) and (4-26)

$$\int_{-\infty}^{\infty} f_{X|Y}(x \mid y) \, dx = \frac{\int_{-\infty}^{\infty} f_{X,Y}(x, y) \, dx}{f_Y(y)} = \frac{f_Y(y)}{f_Y(y)} = 1. \tag{4-30}$$

Note from (4-26) that $f_{X|Y} \ge 0$. Thus, as mentioned in the introduction to this section, $f_{X|Y}$ has both the required properties of a probability density function.

EXAMPLE 4-19

$$f_X(x) = \frac{1}{\sqrt{2\pi}\, a} \exp\left[-\frac{(x)^2}{2a^2}\right],$$

$$f_{Y|X}(y \mid x) = \frac{1}{\sqrt{2\pi}\, b(1 - r^2)^{1/2}} \exp -\left[\frac{\left(y - r\dfrac{b}{a}x\right)^2}{2b^2(1 - r^2)}\right].$$

Find $f_{X,Y}$.

Solution:

$$f_{X,Y}(x, y) = f_{Y|X}(y \mid x) f_X(x),$$

from (4-26). Thus

$$f_{X,Y}(x, y) = \frac{1}{2\pi ab(1 - r^2)^{1/2}} \exp\left\{-\left[\frac{1}{2(1 - r^2)}\right]\left[\frac{x^2}{a^2} - \frac{2rxy}{ab} + \frac{y^2}{b^2}\right]\right\}.$$

This is a jointly normal density function with $\mu_1 = \mu_2 = 0$, $\sigma_1 = a$, $\sigma_2 = b$. Compare with (4-19).

Marginal Distributions From Conditional Distributions

In Chapter II it was shown, (2-35), that

$$P(B_j) = \sum_i P(B_j \mid A_i) P(A_i) \tag{4-31}$$

where A_i, $i = 1, 2, \ldots,$ are mutually exclusive and exhaustive events. It follows directly that

$$p_X(x_i) = \sum_{\text{all } j} p_{X|Y}(x_i \mid y_j) p_Y(y_j). \tag{4-32}$$

95

The proof of this is left as a problem. An example will illustrate.

EXAMPLE 4-20

Let $Y = 1, 2, 3$ depending on the number of coins tossed. Let X be the number of heads. Assume $p_Y(1) = 1/4$, $p_Y(2) = 1/4$, $p_Y(3) = 1/2$. The coins are assumed to be fair coins. Find $p_X(i)$ for $i = 0, 1, 2, 3$.

From the problem statement

$$p_{X|Y}(0 \mid 1) = \tfrac{1}{2}, p_{X|Y}(1 \mid 1) = \tfrac{1}{2},$$
$$p_{X|Y}(0 \mid 2) = \tfrac{1}{4}, p_{X|Y}(1 \mid 2) = \tfrac{1}{2}, p_{X|Y}(2 \mid 2) = \tfrac{1}{4}$$
$$p_{X|Y}(0 \mid 3) = \tfrac{1}{8}, p_{X|Y}(1 \mid 3) = \tfrac{3}{8} = p_{X|Y}(2 \mid 3), p_{X|Y}(3 \mid 3) = \tfrac{1}{8}.$$

Now from (4-32)

$$p_X(0) = \sum_{j=1}^{3} p_{X|Y}(0 \mid j) p_Y(j)$$
$$= \tfrac{1}{2} \cdot \tfrac{1}{4} + \tfrac{1}{4} \cdot \tfrac{1}{4} + \tfrac{1}{8} \cdot \tfrac{1}{2} = \tfrac{1}{4}.$$

Similarly
$$p_X(1) = \sum_{j=1}^{3} p_{X|Y}(1 \mid j) p_Y(j) = \tfrac{7}{16},$$

$$p_X(2) = \sum_{j=2}^{3} p_{X|Y}(2 \mid j) p_Y(j) = \tfrac{4}{16},$$

$$p_X(3) = p_{X|Y}(3 \mid 3) p_Y(3) = \tfrac{1}{16}.$$

We now give a result similar to (4-32) for marginal and conditional density functions:

$$f_X(x) = \int_{-\infty}^{\infty} f_{X|Y}(x \mid y) f_Y(y) \, dy. \tag{4-33}$$

This result follows directly from (4-26) and a slightly different form of (4-22); that is

$$\int_{-\infty}^{\infty} f_{X|Y}(x \mid y) f_Y(y) \, dy = \int_{-\infty}^{\infty} f_{X,Y}(x, y) \, dy = f_X(x).$$

EXAMPLE 4-21

The position X of a temperature-sensitive element, given the temperature is described by

$$f_{X|T}(x \mid t) = \frac{1}{\sqrt{2\pi}} \exp\left[-\frac{\left(x - 10 - \dfrac{t}{100}\right)^2}{2} \right]$$

where the random variable T represents temperature and t is a value in the range of T. Find the probability density function of position if the probability density function of temperature is equally likely between 50 and 70.

The solution [using (4-33)] is

$$f_X(x) = \int_{50}^{70} \frac{1}{20} \frac{1}{\sqrt{2\pi}} \exp\left[-\frac{\left(x - 10 - \dfrac{t}{100}\right)^2}{2}\right] dt.$$

Letting $y = t/100 + 10 - x$, then

$$f_X(x) = \frac{1}{20} \int_{10.5-x}^{10.7-x} 100 \frac{1}{\sqrt{2\pi}} e^{-y^2/2} \, dy$$

$$= 5[F_{SN}(10.7 - x) - F_{SN}(10.5 - x)]$$

where F_{SN} is the standard normal distribution function.

4-6 RANDOM VECTORS (*n* RANDOM VARIABLES)

Consider an experiment where more than two random variables are defined. The definitions of distribution functions, density functions, and probability mass functions for random vectors are simply extensions of the definitions already given for one and two random variables. For instance, given n random variables, X_1, X_2, $\ldots$, X_n, the distribution function $F_{X_1,\ldots,X_n}$ is

$$F_{X_1,X_2,\ldots,X_n}(x_1, x_2, \ldots, x_n) = P\{(X_1 \leq x_1) \ldots (X_n \leq x_n)\}. \quad (4\text{-}34)$$

Density functions and probability functions are defined in analogous fashions as for two random variables, and

$$1 = \sum_{\text{all } i_1} \cdots \sum_{\text{all } i_n} p_{X_1,X_2,\ldots,X_n}(i_1, i_2, \ldots, i_n), \quad (4\text{-}35)$$

$$1 = \int_{-\infty}^{\infty} \cdots \int_{-\infty}^{\infty} f_{X_1,X_2,\ldots,X_n}(x_1, \ldots, x_n) \, dx_1 \cdots dx_n. \quad (4\text{-}36)$$

The marginal density concept is extended to include such ideas as a marginal joint density. For instance, if $n = 5$,

$$f_{X_1,X_2}(x_1, x_2) = \int\!\!\int\!\!\int_{-\infty}^{\infty} f_{X_1,X_2,X_3,X_4,X_5}(x_1, x_2, x_3, x_4, x_5) \, dx_3 \, dx_4 \, dx_5. \quad (4\text{-}37)$$

Similarly the definition of conditional density functions is extended to include

$$f_{X_1,\ldots,X_K|X_{K+1},\ldots,X_n} = \frac{f_{X_1,\ldots,X_n}}{f_{X_{K+1},\ldots,X_n}}. \quad (4\text{-}38)$$

"Removing" Random Variables

From (4-37) and (4-38) follow very useful rules for removing various random variables. We employ the shortened notation which is often used:

$$f_{X_1,\ldots,X_i|X_k,\,\ldots,\,X_n}(x_1, \ldots, x_i \mid x_k, \ldots, x_n) = f(x_1, \ldots, x_i \mid x_k, \ldots, x_n).$$

That is, the subscript identifying the density function is deleted, letting the argument identify the function.

To remove conditioning (right side of vertical bar) random variables multiply by their conditional density function and integrate with respect to the random variables being removed. For example,

$$f(x_1, x_2 \mid x_5) = \int\int_{-\infty}^{\infty} f(x_1, x_2 \mid x_3, x_4, x_5) f(x_3, x_4 \mid x_5) \, dx_3 \, dx_4. \qquad (4\text{-}39)$$

The proof of (4-39) follows:

$$f_{X_1,X_2|X_3,X_4,X_5} = \frac{f_{X_1,X_2,X_3,X_4,X_5}}{f_{X_3,X_4,X_5}}$$

and

$$f_{X_3,X_4|X_5} = \frac{f_{X_3,X_4,X_5}}{f_{X_5}},$$

by definition (4-38). From the above,

$$f_{X_1,X_2 \mid X_3,X_4,X_5} f_{X_3,X_4|X_5} = \frac{f_{X_1,X_2,X_3,X_4,X_5}}{f_{X_5}}.$$

Then the right side of (4-39) becomes

$$\int\int_{-\infty}^{\infty} \frac{f(x_1, x_2, x_3, x_4, x_5)}{f(x_5)} \, dx_3 \, dx_4 = \frac{f(x_1, x_2, x_5)}{f(x_5)} = f(x_1, x_2 \mid x_5).$$

To remove random variables which are conditioned (left side of vertical bar), simply integrate with respect to the random variables which are to be removed. For example,

$$f(x_1 \mid x_4, x_5) = \int\int_{-\infty}^{\infty} f(x_1, x_2, x_3 \mid x_4, x_5) \, dx_2 \, dx_3. \qquad (4\text{-}40)$$

The proof of (4-40) is given as a problem.

Jointly Normal Random Vector

The random variables X_1, X_2, $\ldots$, X_n are said to be jointly normally distributed if

$$f_{X_1, \ldots, X_n}(x_1, \ldots, x_n)$$

$$= \frac{1}{|C|^{1/2}(2\pi)^{n/2}} \exp\left\{-\frac{1}{2} \sum_{i=1}^{n} \sum_{j=1}^{n} D_{ij}(x_i - \mu_i)(x_j - \mu_j)\right\} \quad (4\text{-}41)$$

where C is called the covariance matrix (a matrix of known constants),

$$C_{ij} = E[(X_i - \mu_i)(X_j - \mu_j)], \qquad D = C^{-1}, \text{ that is, inverse of } C,$$

$$|C| = \text{determinant of } C, \qquad D_{ij} = ij\text{th element of } D.$$

For example, when $n = 2$

$$C_{11} = \sigma_1^2$$

$$C_{12} = C_{21} = r\sigma_1\sigma_2,$$

$$C_{22} = \sigma_2^2,$$

$$C = \begin{bmatrix} \sigma_1^2 & r\sigma_1\sigma_2 \\ r\sigma_1\sigma_2 & \sigma_2^2 \end{bmatrix},$$

$$D = \frac{1}{\sigma_1^2\sigma_2^2(1 - r^2)} \begin{bmatrix} \sigma_2^2 & -r\sigma_1\sigma_2 \\ -r\sigma_1\sigma_2 & \sigma_1^2 \end{bmatrix}.$$

Thus

$$f_{X_1 X_2}(x_1, x_2) = \frac{1}{2\pi\sigma_1\sigma_2\sqrt{1 - r^2}} \exp\left\{-\frac{1}{2\sigma_1^2\sigma_2^2(1 - r^2)} [\sigma_2^2(x_1 - \mu_1)^2 \right.$$

$$\left. - 2r\sigma_1\sigma_2(x_1 - \mu_1)(x_2 - \mu_2) + \sigma_1^2(x_2 - \mu_2)^2]\right\}$$

which is the same as (4-19).

4-7 INDEPENDENT RANDOM VARIABLES

Two events A, B are defined as independent if

$$P(AB) = P(A)P(B).$$

Let the event A be $\{X \leq x\}$ and the event B be $\{Y \leq y\}$, then independence of A and B implies

$$F_{X,Y}(x, y) = F_X(x)F_Y(y). \quad (4\text{-}42)$$

If (4-42) is true for all values x and y, then it is taken as the definition of statistical independence of the random variables, X and Y.

If independent random variables are discrete, then for all x and y

$$p_{X,Y}(x, y) = p_X(x)p_Y(y); \tag{4-43}$$

while if the densities exist, independence is defined by

$$f_{X,Y}(x, y) = f_X(x)f_Y(y) \text{ for all } x \text{ and } y. \tag{4-44}$$

EXAMPLE 4-22

$$f_{X,Y}(x, y) = \frac{1}{2\pi} \exp\left[-\left(\frac{x^2}{2} + \frac{y^2}{2}\right)\right].$$

Are X and Y independent?

Because $\quad f_{X,Y}(x, y) = \dfrac{1}{\sqrt{2\pi}} \exp\left[-\dfrac{x^2}{2}\right] \dfrac{1}{\sqrt{2\pi}} \exp\left[-\dfrac{y^2}{2}\right]$

can be factored into the product of two density functions, one containing only x and the other only y, then X and Y are independent.

Random Vectors

Independence of n random variables follows directly from the definition of independence of events. The random variables $X_1, \ldots, X_n$ are independent if

$$F_{X_1,\ldots,X_n} = F_{X_1}F_{X_2}\ldots F_{X_n}. \tag{4-45}$$

Note that independence of n random variables implies any set of less than n is also independent. As with events, the converse is not true.

If density functions exist then (4-45) implies

$$f_{X_1,\ldots,X_n} = f_{X_1}f_{X_2}\ldots f_{X_n}. \tag{4-46}$$

Conditional Independence

The random variables X_1 and X_2 are said to be conditionally independent given X_3 if

$$F_{X_1,X_2|X_3} = F_{X_1|X_3}F_{X_2|X_3}. \tag{4-47}$$

Reasoning About Independence

If joint probability distribution functions are given, then the definition of independence can be applied to determine if the random variables are statistically independent. However in many engineering applications, the procedure is different. The marginal distributions are available from data and the various random variables must be assumed to be independent. Thus we briefly consider when it is reasonable to assume statistical independence. We consider groups of components made from the same specifications.

First a rather obvious dependence is mentioned. If X_1 and X_2 are two different "parameters" of the same component, then from direct cause-and-effect reasoning, one would not usually assume independence. For instance the diameter and weight of metal rods would not be independent, the gain and storage time of transistors would not be independent, etc.

Consider now the case of parameters taken from different components. At first thought the resistance R_1 of, say, a 1-ohm resistor should be independent of the value of resistance R_2 of, say, a 2-ohm resistor. Let us restate this more precisely. The nominal values and tolerance of R_1 and R_2 have been set. The question is

$$f_{R_1|R_2} \overset{?}{=} f_{R_1}.$$

At first thought, the resistor chosen for R_2 in a particular circuit does not affect the resistor chosen for R_1. Therefore

$$f_{R_1|R_2} = f_{R_1}.$$

However, although there is no direct cause-and-effect relation, one must look further. For example, let X be the number of beer parlors in a city and let Y be the number of ministers in a city. Data show that larger values of X tend to go with larger values of Y or that X and Y are not independent. Although some might argue differently, most people will not attribute this lack of independence to direct cause and effect; rather to the fact that in larger cities there are more of both. That is, a third factor (size of city) was causing the dependence. Similarly temperature may cause R_1 not to be independent of R_2.

Such factors must be considered in engineering problems. It is possible that some environmental factor such as temperature may influence the values of random variables. Such dependence can be treated by considering conditional distributions. That is, the random variables are considered with the environmental factor fixed. Later, averaging over the environmental variable can be considered. We point out that in testing of parts to be used over a range of environments, tests at various environments (e.g., high and low temperature) should be and actually usually are run. Thus data for conditional distributions are often available.

4-8 SUMMARY

A random variable was defined as a function from outcomes in the sample space to the real line. Then the concept of distribution function was introduced. Discrete random variables are those that have a probability mass function, while continuous random variables are those whose distribution functions have derivatives, and the derivative of the distribution function is called a probability density function.

The concepts of random variables, distribution functions, probability mass functions, and density functions were extended to more than one random variable, or to random vectors. The concept of marginal distributions was introduced and particular emphasis was placed on conditional distributions. The point of view adopted was that all random variables are conditional on assumptions, and it is often necessary to make these assumptions explicit in engineering applications.

Independence and conditional independence of random variables were defined.

Random variables are discussed in almost all books on probability or statistics. Readings at about the same level as this text can be found in Ref. B1, D3, G2, and H1. More sophisticated descriptions that should be within the range of students who have read this chapter are contained in Ref. D4, P2, P3, and P5.

4-9 PROBLEMS

1. Show that the three properties of a distribution function follow from the definition of a distribution function.

2. Show that if F_X has a derivative everywhere and f_X is as defined by (4-3), then the two properties of a density function follow.

3. What are the properties of a probability mass function?

In the next four problems find the probability mass function and the distribution function of the random variables defined.

4. Four cards are dealt from an ordinary deck without replacement. Let X be the number of spades.

5. A coin is tossed until a tail appears. Let N be the number of tosses.

6. Two dice are tossed. X_1 is the up face of the first die. X_2 is the up face of the second die. $Y = X_1 + X_2$.

7. A machine makes fuses with an average of 1% defective. Let X be the number of defective fuses in the sample of ten.

8. Assume that the production rules that apply to Problem 7 are that if one or more defectives are found then production is stopped. What is the probability production is stopped?

9. If three fair dice are rolled, what is the probability function of X, where X is the largest of the numbers appearing on the up face.

10. Which of the following functions, $f(x)$, are probability density functions?

a. $f(x) = e^{-x}, \quad x \geq 0,$
$\quad = 0, \qquad x < 0.$

b. $f(x) = 2e^{-2x}.$

c. $f(x) = \dfrac{1}{\pi}\dfrac{1}{1 + x^2}.$

d. $f(x) = \frac{2}{3}(x - 1), \quad 0 \leq x \leq 3,$
$\quad = 0, \qquad\qquad \text{elsewhere.}$

11. Which of the following are distribution functions?

a. $F(x) = 0, \qquad x < 0,$
$\quad = x^2, \qquad 0 \leq x \leq 1,$
$\quad = 1, \qquad x > 1.$

b. $F(x) = 0, \qquad x \leq 4,$
$\quad = 1, \qquad x > 4.$

c. $F(x) = 0, \qquad x < 0,$
$\quad = \frac{1}{2}, \qquad 0 \leq x < 2,$
$\quad = (x - \frac{3}{2}), \qquad 2 \leq x < 2\frac{1}{2},$
$\quad = 1, \qquad x \geq 2\frac{1}{2}.$

d. $F(x) = 1 - e^{-10x}.$

12. Find the distribution function for any of the parts of 10 which are density functions.

13. A random variable Y has a Poisson distribution with $\lambda = 10$.

Find $P(2 < Y \leq 4)$.

14. A random variable X has the following density function:

$$f_X(\theta) = \frac{\sin \theta}{2}, \qquad 0 \leq \theta \leq \pi,$$

$$= 0, \qquad \text{elsewhere.}$$

a. Find $F_X(\theta)$.
b. Find $F_X(2\pi)$.

15. Given the Poisson distribution

$$P(X = k) = e^{-\lambda}\frac{\lambda^k}{k!}.$$

Show
$$\sum_{k=0}^{\infty} P(X = k) = 1.$$

103

16. The dielectric strength of material A has a mean (μ) of 542 V/mil and a standard deviation (σ) of 32 V/mil. Material B has a mean (μ) of 575 V/mil and a standard deviation of 75 V/mil. Assuming normal distributions, find which material has the higher probability of being below 450 V/mil.

17. The circuit shown in Figure P4-1 has switches which are operated independently.

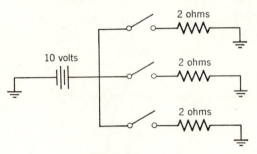

Figure P4-1

For each switch the probability that the switch is closed is .7. I is a random variable that represents the current from the battery. Give the probability function and the distribution function of I. (Hint: this is a binomial.)

18. The joint density function of two random variables is

$$f_{X,Y}(x, y) = e^{-x}e^{-y}, \qquad x \geq 0, y \geq 0,$$
$$= 0, \qquad\qquad \text{elsewhere.}$$

a. Find the marginal density function of X. (Give the answer for all values of the random variable.)

b. Find the marginal distribution function of Y. (Give the answer for all values of the random variable.)

19. Are X and Y independent if

a. $f_{X,Y}(x, y) = ae^{-x}e^{-y}, \qquad x \geq 0, y \geq 0,$
$\qquad\quad = 0, \qquad\qquad\qquad \text{elsewhere.}$

b. $f_{X,Y}(x, y) = bxy, \qquad\qquad 0 \leq x \leq y, 0 \leq y \leq 2,$
$\qquad\quad = 0, \qquad\qquad\qquad\qquad \text{elsewhere.}$

20. The joint probability density function of two random variables is

$$f_{X,Y}(x, y) = c(1 + xy), \qquad 0 \leq x \leq 1, \qquad 0 \leq y \leq 2,$$
$$= 0, \qquad\qquad\qquad\qquad\qquad \text{elsewhere.}$$

a. Find $F_{X,Y}(.5, 1.0)$.
b. Find $f_{X,Y}(x, 1)$.
c. Find $f_{X|Y}(x \mid 1)$.

21. The joint p.d.f. of X and Y is

$$f_{X,Y}(x, y) = K(x + y) \qquad 0 \le x \le 1, \qquad 0 \le y \le 2,$$
$$= 0, \qquad\qquad\qquad\qquad \text{elsewhere.}$$

a. Find $F_{X,Y}(x, y)$.
b. Find $f_{X|Y}(x \mid y)$.
c. Find $F_{X|Y}(x \mid y)$.

22. Prove

$$f_{X_1, X_2 | X_4}(x_1, x_2 \mid x_4) = \int_{-\infty}^{\infty} f_{X_1, X_2, X_3 | X_4}(x_1, x_2, x_3 \mid x_4) \, dx_3.$$

23. Shafts are machined such that the density function of diameter D is

$$f_D(x) = \frac{1}{\sqrt{2\pi}\, 2} \exp\left\{-\frac{(x - 100)^2}{8}\right\}.$$

a. Find the probability that the diameter D is between 95 and 105 [i.e., find $P(95 < D \le 105)$].
b. The shafts are screened at 95 and 105, that is, all shafts with diameters below 95 or above 105 are thrown away. Given this situation, find the probability the diameter is between 99 and 101. [i.e., find

$$P(99 < D \le 101 \mid 95 < D \le 105)].$$

24. The time to failure T of a radar set is exponentially distributed

$$f_T(t) = 0, \qquad t < 0,$$
$$= ae^{-bt}, \qquad t \ge 0.$$

Find conditions on a and b such that f_T is a probability density function.

25. Show that (4-32) is correct.

26. R is a random variable that is equally likely to be any value between 80 and 100.

a. Find $P(90 \le R \le 95)$.
b. Find $P(90 \le R \le 95 \mid 85 \le R \le 95)$.

27. Show that $F_X(x) = \sum_{i=1}^{n} F_{X|B_i}(x) P(B_i)$

if $\bigcup_{i=1}^{n} B_i = S$,

and $B_i \cap B_j = \emptyset$ for every $i \ne j$.

CHAPTER V

....

EXPECTED VALUES

5-1 MEAN VALUE

One may wish to describe a random variable by one number or by just a few numbers rather than a complete distribution. A natural choice of one number is the average $\bar{X}$ or mean of the sample values of the random variable. Usually to find the average, one sums all of the values and divides by the number of values n. That is

$$\bar{X} = \frac{X_1 + X_2 + \cdots + X_n}{n}$$

$$= X_1 \frac{1}{n} + X_2 \frac{1}{n} + \cdots + X_n \frac{1}{n}.$$

If a certain value occurs twice, say $X_K = X_{K+1}$, then

$$\bar{X} = X_1 \frac{1}{n} + X_2 \frac{1}{n} + \cdots + X_{K-1} \frac{1}{n} + X_K \frac{2}{n} + X_{K+2} \frac{1}{n} + \cdots + X_n \frac{1}{n}.$$

Thus each value is weighed by its relative frequency of occurrence and summed.

This idea motivates the concept of the expected value $E[X]$ of a random variable X. The expected value is also called the mean.

$$\mu_X = E[X] = \sum_{\text{all } i} x_i p_X(x_i), \tag{5-1}$$

or

$$\mu_X = E[X] = \int_{-\infty}^{\infty} x f_X(x)\, dx. \tag{5-2}$$

That is if X is discrete, the expected value is the weighted (by probability of occurrence) average of the possible values. If X is continuous, again each possible value of X is weighted by its probability density and averaged.

106

Note that (5-1) and (5-2) are like center of gravity calculations except the usual denominator [which in this case is either $\sum p_X(x_i)$ or $\int f_X(x)\,dx$] is omitted. It should be obvious that in the case of probabilities, the denominator is always one.

EXAMPLE 5-1

What is the expected value of X where X is the value on the up face of a die?

$$\mu = E[X] = \sum_{i=1}^{6} p_X(i)i$$

$$= 1(\tfrac{1}{6}) + 2(\tfrac{1}{6}) + 3(\tfrac{1}{6}) + 4(\tfrac{1}{6}) + 5(\tfrac{1}{6}) + 6(\tfrac{1}{6}) = \tfrac{21}{6} = 3\tfrac{1}{2}.$$

EXAMPLE 5-2

What is the mean value of X where X is a continuous random variable uniformly distributed between 2 and 4?

$$f_X(x) = \tfrac{1}{2}, \qquad 2 \le x \le 4,$$

$$= 0, \qquad \text{otherwise.}$$

$$\mu = E[X] = \int_2^4 x\tfrac{1}{2}\,dx = \frac{x^2}{4}\Big|_2^4 = 4 - 1 = 3.$$

5-2 VARIANCE

While the expected value or mean is used more than any other one number to describe a random variable, usually one also desires to know how much "spread" there is. While various measurements of spread could be used, the mathematically most convenient measure and the one most frequently used is the variance

$$E[(X - \mu_X)^2] = E[(X - E[X])^2].$$

The variance is denoted by σ^2 and its positive square root σ is called the standard deviation. The variance is computed by

$$\sigma^2 = \sum_{\text{all } i} (x_i - \mu)^2 p(x_i) \tag{5-3a}$$

or by

$$\sigma^2 = \int_{-\infty}^{\infty} (x - \mu)^2 f_X(x)\,dx \tag{5-3b}$$

where μ is the mean as defined by (5-1) and (5-2).

Just as the mean is analogous to the center of gravity of a density function, the variance is analogous to the radius of gyration.

We now give an inequality which shows that the variance provides an upper limit on the deviation of a random variable from its mean.

Chebycheff's Inequality

Chebycheff's inequality is

$$P\{|X - \mu| > K\} \leq \frac{\sigma^2}{K^2}, \tag{5-4}$$

where $\sigma^2 = E[(X - \mu)^2]$.

In the case in which X has a density function, this can be shown as follows:

$$\sigma^2 = \int_{-\infty}^{\infty} (x - \mu)^2 f_X(x)\, dx$$

$$= \int_{-\infty}^{\mu-K} (x - \mu)^2 f_X(x)\, dx + \int_{\mu-K}^{\mu+K} (x - \mu)^2 f_X(x)\, dx$$

$$+ \int_{\mu+K}^{\infty} (x - \mu)^2 f_X(x)\, dx.$$

Because $(x - \mu)^2$ and $f_X(x)$ are always non-negative

$$\sigma^2 \geq \int_{-\infty}^{\mu-K} (x - \mu)^2 f_X(x)\, dx + \int_{\mu+K}^{\infty} (x - \mu)^2 f_X(x)\, dx.$$

In both integrals $(x - \mu)^2 \geq K^2$. Thus

$$\sigma^2 \geq K^2 \left[\int_{-\infty}^{\mu-K} f_X(x)\, dx + \int_{\mu+K}^{\infty} f_X(x)\, dx \right].$$

From the definition of f_X

$$\int_{-\infty}^{\mu-K} f_X(x)\, dx + \int_{\mu+K}^{\infty} f_X(x)\, dx = P\{|X - \mu| > K\}.$$

Thus
$$\sigma^2 \geq K^2 P\{|X - \mu| > K\}.$$

The above result can also be shown without assuming that X is continuous.

EXAMPLE 5-3

If $\sigma_X^2 = \frac{1}{4}$, find an upper bound on the probability that X deviates by more than 2 from its mean μ. Also find a lower bound on the probability that X is within 2 of its mean.

Using Chebycheff's inequality, (5-4),

$$P\{|X - \mu| > 2\} \leq \frac{\frac{1}{4}}{4} = \frac{1}{16},$$

$$P\{|X - \mu| \leq 2\} = 1 - P\{|X - \mu| > 2\}.$$

Thus $\qquad P\{|X - \mu| \leq 2\} \geq 1 - \frac{1}{16} = \frac{15}{16}$

5-3 EXPECTED VALUE OF FUNCTIONS OF RANDOM VARIABLES

By a function of a random variable we mean the following. For a fixed point in the sample space X takes on the value x; then if $Y = g(X)$ the random variable* Y takes on the value $y = g(x)$. Y is thus defined for every outcome. For example, if $Y = X^2$ and an outcome is mapped by X into the number 3, that outcome is mapped by Y into the number 9.

Similarly if $Y = h(X_1, X_2, \ldots, X_n)$, then for a given outcome if $X_1 = x_1, \ldots, X_n = x_n$ then $y = h(x_1, x_2, \ldots, x_n)$. For example, if $Y = X_1^2 + X_2^2$ then for any values, x_1 and x_2 of X_1 and X_2, $y = x_1^2 + x_2^2$.

Definition of Expected Value

Based on the above, the more general concept of the expected value of a function g of a random variable X is defined as follows. If X has a probability mass function

$$E[g(X)] = \sum_{\text{all } i} g(x_i)p_X(x_i). \tag{5-5}$$

If X has a probability density function

$$E[g(X)] = \int_{-\infty}^{\infty} g(x)f_X(x)\, dx. \tag{5-6}$$

The mean and variance are included as special cases of (5-5) and (5-6). Indeed, if $g(X) = X$, then (5-5) and (5-6) become equations for the mean, and if $g(X) = (X - \mu)^2$, then (5-5) and (5-6) become expressions for the variance.

---

* It is beyond the scope of this book to prove that Y is a random variable. However, if g is any of the usual analytic functions, then $Y = g(X)$ is a random variable.

Moments

The mean is sometimes called the first moment, and the variance is called the second central moment of a random variable. Other moments can also be defined. The Kth moment about the origin is $E[X^K]$ while the Kth central moment is $E[(X - \mu)^K]$.

Random Vectors

In the following it is assumed that density functions exist. If there are two random variables X, Y, defined on a sample space, then the expected value of a function g of X and Y is

$$E[g(X, Y)] = \int\!\!\!\int_{-\infty}^{\infty} g(x, y) f_{X,Y}(x, y)\, dx\, dy. \tag{5-7}$$

Similarly for n random variables

$$E[g(X_1, X_2, \ldots, X_n)] = \int_{-\infty}^{\infty} \cdots \int_{-\infty}^{\infty} g(x_1, x_2, \ldots, x_n)$$

$$\times f_{X_1 \ldots X_n}(x_1, \ldots, x_n)\, dx_1 \cdots dx_n. \tag{5-8}$$

5-4 PROPERTIES OF EXPECTED VALUES

We now state some important properties of expected values. In the proof of these properties it is assumed that density functions exist. The same ideas are involved in proving the results for discrete random variables.

$P1$: $E[C] = C$ where C is a constant.

Proof: $\qquad E[C] = \int_{-\infty}^{\infty} C f_X(x)\, dx = C \int_{-\infty}^{\infty} f_X(x)\, dx = C.$

$P2$: $E[CX] = CE[X]$.

Proof: $\qquad E[CX] = \int_{-\infty}^{\infty} C x f_X(x)\, dx = C \int_{-\infty}^{\infty} x f_X(x)\, dx = CE[X].$

$P3$: $E[X_1 + X_2] = E[X_1] + E[X_2]$.

110

Proof:

$$E[X_1 + X_2] = \int\limits_{-\infty}^{\infty}\!\!\int (x_1 + x_2)f_{X_1,X_2}(x_1, x_2)\,dx_1\,dx_2$$

$$= \int\limits_{-\infty}^{\infty}\!\!\int x_1 f_{X_1,X_2}(x_1, x_2)\,dx_1\,dx_2 + \int\limits_{-\infty}^{\infty}\!\!\int x_2 f_{X_1,X_2}(x_1, x_2)\,dx_1\,dx_2$$

$$= \int_{-\infty}^{\infty} x_1 f_{X_1}(x_1)\,dx_1 + \int_{-\infty}^{\infty} x_2 f_{X_2}(x_2)\,dx_2$$

$$= E[X_1] + E[X_2].$$

P4: $E\left[\sum\limits_{i=1}^{n} a_i X_i\right] = \sum\limits_{i=1}^{n} a_i E[X_i].$

Proof: The proof involves using *P2* and *P3* in an induction proof.

P5: If X and Y are independent

$$E[XY] = E[X]E[Y].$$

Proof:

$$E[XY] = \int\limits_{-\infty}^{\infty}\!\!\int xy f_{X,Y}(x, y)\,dx\,dy$$

$$= \int\limits_{-\infty}^{\infty}\!\!\int xy f_X(x)f_Y(y)\,dx\,dy \quad \text{(independence)}$$

$$= \left[\int_{-\infty}^{\infty} x f_X(x)\,dx\right]\left[\int_{-\infty}^{\infty} y f_Y(y)\,dy\right] = E[X]E[Y].$$

Some results are now given.

$$\sigma^2 = E[X^2] - \mu^2. \tag{5-9}$$

Proof: $\quad \sigma^2 = E[(X - \mu)^2] = E[X^2 - 2\mu X + \mu^2]$

$\qquad\qquad = E[X^2] - 2\mu E[X] + E[\mu^2], \quad \text{using } P4 \text{ and } P2$

$\qquad\qquad = E[X^2] - 2\mu^2 + \mu^2$

$\qquad\qquad = E[X^2] - \mu^2.$

If $X_1, X_2, \ldots, X_n$ are independent and $E[X_i] = 0$, for all i, then

$$E\left[\left(\sum_{i=1}^{n} X_i\right)^2\right] = \sum_{i=1}^{n} E[X_i^2]. \tag{5-10}$$

111

Proof:

$$E\left[\left(\sum_{i=1}^{n} X_i\right)^2\right] = E\left[X_1^2 + X_2^2 + \cdots + X_n^2 + X_1\left(\sum_{i=2}^{n} X_i\right)\right.$$

$$\left. + X_2\left(\sum_{\substack{i=1 \\ i \neq 2}}^{n} X_i\right) + \cdots + X_n\left(\sum_{i=1}^{n-1} X_i\right)\right]$$

$$= E[X_1^2] + E[X_2^2] + \cdots + E[X_n^2] + E[X_1]\left\{\sum_{i=2}^{n} E[X_i]\right\}$$

$$+ \cdots + E[X_n]\left\{\sum_{i=1}^{n-1} E[X_i]\right\} = \sum_{i=1}^{n} E[X_i^2],$$

where the last step follows from $E[X_i\,X_j] = E[X_i]E[X_j] = 0$, $i \neq j$, because of the assumptions of independence and $E[X_i] = 0$.

5-5 EXAMPLES OF MEAN AND VARIANCE

EXAMPLE 5-4

The mean and variance of a binomial random variable is considered.

$$P(X = K) = \binom{n}{K} p^K q^{n-K}; \qquad q = 1 - p, \qquad K = 0, 1, \ldots, n$$

$$E[X] = \sum_{K=0}^{n} KP(X = K) = \sum_{K=0}^{n} K\binom{n}{K} p^K q^{n-K}.$$

$$E[X] = \sum_{K=1}^{n} K \frac{n!}{K!\,(n-K)!} p^K q^{n-K}$$

$$= \sum_{K=1}^{n} \frac{n(n-1)!}{(K-1)!\,(n-K)!} pp^{K-1} q^{n-K}$$

$$= np \sum_{K=1}^{n} \frac{(n-1)!}{(K-1)!\,(n-K)!} p^{K-1} q^{n-K} = np(p+q)^{n-1}$$

$$= np.$$

$$E[X^2] = \sum_{K=0}^{n} K^2 \binom{n}{K} p^K q^{n-K}$$

$$= \sum_{K=0}^{n} K(K-1)\binom{n}{K} p^K q^{n-K} + \sum_{K=0}^{n} K\binom{n}{K} p^K q^{n-K}$$

$$= \sum_{K=2}^{n} \frac{n(n-1)(n-2)!}{(K-2)!(n-K)!} p^2 p^{K-2} q^{n-K} + np$$

$$= p^2 n(n-1) \sum_{K=2}^{n} \binom{n-2}{K-2} p^{K-2} q^{n-K} + np$$

$$= p^2 n(n-1) + np = n^2 p^2 + np - np^2.$$

With the use of (5-9)

$$\sigma^2 = E[X^2] - \{E[X]\}^2 = n^2 p^2 + np - np^2 - n^2 p^2 = np(1-p) = npq.$$

EXAMPLE 5-5

The mean and variance of a Poisson random variable are found.

$$P(X = K) = e^{-\lambda} \frac{\lambda^K}{K!}, \qquad K = 0, 1, \ldots$$

$$E[X] = \sum_{K=0}^{\infty} K e^{-\lambda} \frac{\lambda^K}{K!} = \lambda e^{-\lambda} \sum_{K=1}^{\infty} \frac{\lambda^{(K-1)}}{(K-1)!}$$

$$= \lambda e^{-\lambda}\left\{1 + \lambda + \frac{\lambda^2}{2!} + \frac{\lambda^3}{3!} + \cdots\right\} = \lambda e^{-\lambda} e^{\lambda} = \lambda.$$

$$E[X^2] = \sum_{K=0}^{\infty} K^2 e^{-\lambda} \frac{\lambda^K}{K!} = \sum_{K=0}^{\infty} K(K-1) e^{-\lambda} \frac{\lambda^K}{K!} + E[X],$$

$$E[X^2] = \lambda^2 \sum_{K=2}^{\infty} e^{-\lambda} \frac{\lambda^{K-2}}{(K-2)!} + \lambda = \lambda^2 + \lambda,$$

$$\sigma^2 = \lambda^2 + \lambda - \lambda^2 = \lambda.$$

EXAMPLE 5-6

The mean and variance of a continuous random variable uniformly distributed between a and b, $a < b$, as shown in Figure 5-1 are found. The mean is

$$E[X] = \int_a^b x \frac{1}{b-a} dx = \frac{a+b}{2}.$$

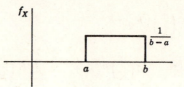

Figure 5-1

The variance is

$$\sigma^2 = E[(X - \mu)^2] = E\left[\left(X - \frac{a+b}{2}\right)^2\right],$$

$$\sigma^2 = \int_a^b \left(x - \frac{a+b}{2}\right)^2 \frac{1}{b-a}\, dx.$$

Let $\quad y = x - \dfrac{a+b}{2}$,

$$\sigma^2 = \int_{(a-b)/2}^{(b-a)/2} (y)^2 \frac{1}{b-a}\, dy$$

$$= \frac{1}{b-a} \frac{y^3}{3}\Big|_{\frac{a-b}{2}}^{\frac{b-a}{2}} = \frac{(b-a)^2}{(8)3} + \frac{(a-b)^3}{3(a-b)8} = \frac{1}{12}(b-a)^2.$$

EXAMPLE 5-7

The mean and variance of an exponential random variable are considered.

$$E[X] = \int_0^\infty x \frac{1}{\lambda} e^{-\frac{x}{\lambda}}\, dx = \lambda\left[e^{\frac{-x}{\lambda}}\left(-\frac{x}{\lambda}-1\right)\right]_0^\infty = \lambda,$$

$$E[X^2] = \int_0^\infty x^2 \frac{1}{\lambda} e^{-\frac{x}{\lambda}}\, dx = -x^2 e^{-\frac{x}{\lambda}}\Big|_0^\infty + 2\lambda E[X] = 2\lambda^2,$$

$$\sigma^2 = 2\lambda^2 - \lambda^2 = \lambda^2.$$

EXAMPLE 5-8

To illustrate that an expected value may not exist, consider the game where a fair coin is tossed. The payoff M of the game is $2 if the first head appears on the first toss, $4 if the first head appears on the second toss, and in general 2^n if the first head appears on the nth toss. What is the expected value of M?

$$P(M = 2^m) = (\tfrac{1}{2})^m,$$

$$E[M] = \sum_{i=1}^\infty (2)^i(\tfrac{1}{2})^i = \sum_{i=1}^\infty 1$$

and this series does not converge; therefore the expected value of M does not exist. The reader might ask himself how much he would pay to play this game.

5-6 CHARACTERISTIC FUNCTION

One other function of a random variable is of particular interest. If $g(X) = e^{jX\omega}$, then $\varphi_X(\omega) = E[e^{jX\omega}]$ is called the characteristic function.* If X is continuous

$$\varphi_X(\omega) = E[e^{jX\omega}] = \int_{-\infty}^{\infty} e^{jx\omega} f_X(x)\, dx$$

Note that except for a change in sign in the exponent, φ_X is simply the Fourier transform of f_X. Thus f_X can be obtained from φ_X by using a table of Fourier transforms, and if f_X is continuous the transformation is unique.

If X is discrete

$$\varphi_X(\omega) = E[e^{jX\omega}] = \sum_{\text{all } i} e^{j\omega x_i} p_X(x_i).$$

Moments From the Characteristic Function

One of the important uses of the characteristic function is to find moments. We now show how this may be accomplished. If $\varphi_X(\omega)$ is expanded in a Maclaurin's series

$$\varphi_X(\omega) = \alpha_0 + \alpha_1\omega + \alpha_2\omega^2 + \alpha_3\omega^3 + \cdots \tag{5-11}$$

then

$$\varphi_X(0) = \alpha_0$$

$$\left.\frac{d\varphi_X(\omega)}{d\omega}\right|_{\omega=0} = \alpha_1,$$

$$\left.\frac{d^2\varphi_X(\omega)}{d\omega^2}\right|_{\omega=0} = 2\alpha_2,$$

$$\left.\frac{d^k\varphi_X(\omega)}{d\omega^k}\right|_{\omega=0} = k!\,\alpha_k.$$

* For the first time a complex random variable, $e^{jX\omega}$, is considered. The theory of complex random variables is not discussed; rather we rely on knowledge of Fourier series and transforms.

115

We now identify the coefficients with moments as follows. Expanding $e^{j\omega X}$ in a series

$$e^{jX\omega} = 1 + j\omega X + \frac{(j\omega X)^2}{2!} + \cdots + \frac{(j\omega X)^k}{k!} + \cdots.$$

Thus

$$\varphi_X(\omega) = E[e^{jX\omega}] = \int_{-\infty}^{\infty}\left[1 + j\omega x + \frac{(j\omega x)^2}{2!} + \cdots\right]f_X(x)\,dx.$$

Interchanging the order of integration and summation

$$\varphi_X(\omega) = \int_{-\infty}^{\infty} f_X(x)\,dx + \int_{-\infty}^{\infty} j\omega x f_X(x)\,dx + \int_{-\infty}^{\infty} \frac{(j\omega x)^2}{2} f_X(x)\,dx + \cdots$$

$$\varphi_X(\omega) = 1 + j\omega E[X] + \frac{(j\omega)^2}{2} E[X^2] + \cdots + \frac{(j\omega)^k}{k!} E[X^k] + \cdots. \quad (5\text{-}12)$$

Equating like powers of ω in (5-11) and (5-12)

$$\alpha_0 = 1,$$

$$\alpha_1 = jE[X] \qquad \text{or} \qquad E[X] = \frac{\alpha_1}{j},$$

$$\alpha_2 = \frac{j^2}{2} E[X^2] \qquad \text{or} \qquad E[X^2] = 2\frac{\alpha_2}{j^2},$$

$$\alpha_k = \frac{(j)^k}{k!} E[X^k] \qquad \text{or} \qquad E[X^k] = \frac{k!}{(j)^k}\alpha_k. \quad (5\text{-}13)$$

Thus if $\varphi_X(\omega)$ is expanded in a series, the coefficients of the Maclaurin expansion can be converted to moments. Alternatively $\varphi_X(\omega)$ can be differentiated K times and ω set equal to zero. In this case

$$\varphi_X(0) = \alpha_0 = 1,$$

$$\frac{d\varphi_X(\omega)}{d\omega}\bigg|_{\omega=0} = \alpha_1 = jE[X],$$

$$\frac{d^2\varphi_X(\omega)}{d\omega^2}\bigg|_{\omega=0} = 2\alpha_2 = j^2 E[X^2],$$

$$\frac{d^k\varphi_X(\omega)}{d\omega^k}\bigg|_{\omega=0} = j^k E[X^k]. \quad (5\text{-}14)$$

116

EXAMPLE 5-9

Find the characteristic function of an exponential random variable X and from this find $E[X]$, $E[X^2]$, and $E[X^k]$.

$$\varphi_X(\omega) = E[e^{j\omega X}] = \int_0^\infty \frac{1}{\lambda} e^{j\omega x} e^{-x/\lambda} \, dx$$

$$= \frac{1}{\lambda} \frac{1}{j\omega - \frac{1}{\lambda}} e^{(j\omega - 1/\lambda)x} \Big|_0^\infty = \frac{1}{1 - j\omega\lambda},$$

$$\frac{1}{1 - j\omega\lambda} = 1 + j\omega\lambda + (j\omega\lambda)^2 + \cdots + (j\omega\lambda)^k + \cdots.$$

Thus
$$E[X] = \lambda,$$

$$\tfrac{1}{2}E[X^2] = \lambda^2 \qquad \text{or} \qquad E[X^2] = 2\lambda^2,$$

$$\frac{1}{k!} E[X^k] = \lambda^k \qquad \text{or} \qquad E[X^k] = k! \, \lambda^k.$$

EXAMPLE 5-10

The characteristic function of a normal random variable X is found and expanded in a series to find the mean $E[X]$ and variance Var $[X]$. It will be shown that the mean is the parameter μ and the variance is the parameter σ^2 of the normal density function.

$$\varphi_X(\omega) = \int_{-\infty}^\infty e^{j\omega x} \frac{1}{\sqrt{2\pi\sigma^2}} \exp\left[-\frac{(x-\mu)^2}{2\sigma^2}\right] dx$$

$$= \int_{-\infty}^\infty \frac{1}{\sqrt{2\pi\sigma^2}} \exp\left[-\frac{(x^2 - 2\mu x - 2j\omega\sigma^2 x + \mu^2)}{2\sigma^2}\right] dx$$

$$= \int_{-\infty}^\infty \frac{1}{\sqrt{2\pi\sigma^2}} \exp\left[\frac{\begin{array}{c}-(x^2 - 2\mu x - 2j\omega\sigma^2 x + \mu^2 - \sigma^4\omega^2 \\ + 2j\mu\sigma^2\omega) + 2j\mu\sigma^2\omega - \sigma^4\omega^2\end{array}}{2\sigma^2}\right] dx$$

$$= \exp\left[+j\omega\mu - \frac{\sigma^2\omega^2}{2}\right] \int_{-\infty}^\infty \frac{1}{\sqrt{2\pi\sigma^2}} \exp\left[-\frac{(x - \mu - j\omega\sigma^2)^2}{2\sigma^2}\right] dx.$$

The integrand appears to be a normal density function with a complex mean. Except for the complex mean, the integral would be one, as has been shown. By using other techniques it may be shown that in fact the integral is one.

117

Thus
$$\varphi_X(\omega) = \exp\left[j\mu\omega - \frac{\sigma^2\omega^2}{2}\right], \tag{5-15}$$

$$\frac{d\varphi_X(\omega)}{d\omega} = (j\mu - \sigma^2\omega)\varphi_X(\omega), \qquad \frac{d\varphi_X(\omega)}{d\omega}\bigg|_0 = j\mu,$$

$$\frac{d^2\varphi_X(\omega)}{d\omega^2} = (j\mu - \sigma^2\omega)^2\varphi_X(\omega) - \sigma^2\varphi_X(\omega),$$

$$\frac{d^2\varphi_X(\omega)}{d\omega^2}\bigg|_0 = -\mu^2 - \sigma^2.$$

With the use of (5-14)

$$E[X] = \mu,$$
$$E[X^2] = \sigma^2 + \mu^2,$$
$$\text{Var } [X] = E[X^2] - \{E[X]\}^2 = \sigma^2 + \mu^2 - \mu^2 = \sigma^2.$$

5-7 COVARIANCE AND CORRELATION

The covariance and correlation functions are convenient measures of linear dependence of two random variables. The covariance σ_{XY} of two random variables X and Y is

$$\sigma_{XY} = E[(X - \mu_X)(Y - \mu_Y)]. \tag{5-16}$$

This can also be expressed as

$$\sigma_{XY} = E[XY - X\mu_Y - \mu_X Y + \mu_X\mu_Y]$$
$$= E[XY] - \mu_Y E[X] - \mu_X E[Y] + \mu_X\mu_Y,$$
$$\sigma_{XY} = E[XY] - \mu_X\mu_Y. \tag{5-17}$$

The correlation coefficient ρ_{XY} is defined by

$$\rho_{XY} = \frac{\sigma_{XY}}{\sigma_X\sigma_Y}. \tag{5-18}$$

We now show

$$-1 \leq \rho_{XY} \leq 1.$$

For all real λ

$$E\{[\lambda(X - \mu_X) - (Y - \mu_Y)]^2\} \geq 0$$

because it is the expected value of a non-negative quantity.

$$E\{[\lambda(X - \mu_X) - (Y - \mu_Y)]^2\} = \lambda^2\sigma_X{}^2 - 2\lambda\sigma_{XY} + \sigma_Y{}^2 \geq 0. \tag{5-19}$$

Equality in (5-19) occurs when

$$\lambda = \frac{\sigma_{XY} \pm \sqrt{\sigma_{XY}{}^2 - \sigma_X{}^2\sigma_Y{}^2}}{\sigma_X{}^2}. \tag{5-20}$$

If there are two distinct real roots to (5-20), this implies there exist values of λ that make the quadratic in λ both positive and negative, contradicting (5-19). Therefore there must not be two distinct real roots, or

$$\sigma_{XY}{}^2 - \sigma_X{}^2\sigma_Y{}^2 \leq 0.$$

This is equivalent to

$$\sigma_{XY}{}^2 \leq \sigma_X{}^2\sigma_Y{}^2$$

or

$$|\rho_{XY}| = \frac{|\sigma_{XY}|}{\sigma_X\sigma_Y} \leq 1$$

which gives the result that was to be shown.

The significance of the correlation coefficient is explained by considering some extreme cases. When X and Y are independent

$$E[XY] = E[X]E[Y] = \mu_X\mu_Y.$$

Thus

$$\rho_{XY} = \frac{E[XY] - \mu_X\mu_Y}{\sigma_X\sigma_Y} = 0.$$

That is, independent random variables are uncorrelated. However, if two random variables are uncorrelated, this does not imply they are independent. An example is given in Chapter IX. If $Y = aX$ and $\mu_X = 0$, then

$$E[XY] = E[X(aX)] = a\sigma_X{}^2,$$

$$\rho = \frac{a\sigma_X{}^2}{\sigma_X\sqrt{a^2\sigma_X{}^2}} = \frac{a}{\sqrt{a^2}} = \pm 1;$$

that is, if Y is a linear function of X and $a > 0$, then $\rho = 1$, and, if $a < 0$, $\rho = -1$.

These extreme examples give an idea of the significance of the correlation coefficient. Figure 5-2 shows typical data that would result from various values of ρ. Linear estimation, which is discussed in Chapter IX, provides further interpretation of ρ.

Jointly Normal Random Variables

We show first that if X_1 and X_2 are jointly normal random variables as described by (4-19), then (this is shown with less algebra in the next section)

$$\sigma_{X_1X_2} = r\sigma_1\sigma_2.$$

119

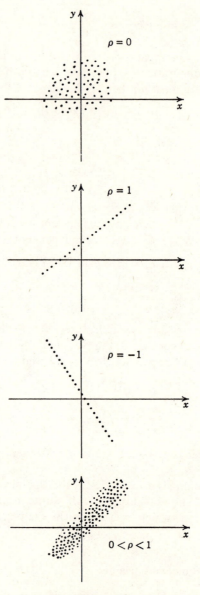

Figure 5-2

With the use of (5-16)

$$\sigma_{X_1 X_2} = \int\int_{-\infty}^{\infty} \frac{(x_1 - \mu_1)(x_2 - \mu_2)}{2\pi\sigma_1\sigma_2\sqrt{1-r^2}} \exp\left\{-\frac{1}{2(1-r^2)}\left[\frac{(x_1-\mu_1)^2}{\sigma_1^2}\right.\right.$$

$$\left.\left. -\frac{2r(x_1-\mu_1)(x_2-\mu_2)}{\sigma_1\sigma_2} + \frac{(x_2-\mu_2)^2}{\sigma_2^2}\right]\right\} dx_1\, dx_2.$$

Making the substitution $y_1 = x_1 - \mu_1$ and $y_2 = x_2 - \mu_2$

$$\sigma_{X_1 X_2} = \int\int_{-\infty}^{\infty} \frac{y_1 y_2}{2\pi\sigma_1\sigma_2\sqrt{1-r^2}} \exp\left[-\frac{1}{2(1-r)^2}\left(\frac{y_1^2}{\sigma_1^2} - \frac{2ry_1 y_2}{\sigma_1\sigma_2} + \frac{y_2^2}{\sigma_2^2}\right)\right] dy_1\, dy_2$$

$$= \int_{-\infty}^{\infty} \frac{y_2}{\sqrt{2\pi}\sigma_2} \exp\left[-\frac{1}{2(1-r^2)}\left(\frac{y_2^2}{\sigma_2^2} - \frac{r^2 y_2^2}{\sigma_2^2}\right)\right]$$

$$\times \left\{\left(\int_{-\infty}^{\infty} \frac{y_1}{\sqrt{2\pi}\sigma_1\sqrt{1-r^2}} \exp\left[-\frac{\left(y_1 - r\frac{\sigma_1}{\sigma_2}y_2\right)^2}{2(1-r^2)\sigma_1^2}\right] dy_1\right)\right\} dy_2.$$

The last integral is simply the mean of a normal random variable; hence

$$\sigma_{X_1 X_2} = \int_{-\infty}^{\infty} \frac{y_2}{\sqrt{2\pi}\sigma_2} \exp\left[-\frac{1}{2(1-r^2)}\frac{y_2^2(1-r^2)}{\sigma_2^2}\right]\frac{r\sigma_1 y_2}{\sigma_2} dy_2$$

$$= \frac{r\sigma_1}{\sigma_2} \int_{-\infty}^{\infty} \frac{y_2^2}{\sqrt{2\pi}\sigma_2} \exp\left\{-\frac{y_2^2}{2\sigma_2^2}\right\} dy_2$$

$$= \frac{r\sigma_1}{\sigma_2}\sigma_2^2 = r\sigma_1\sigma_2.$$

Thus the parameter r in the joint normal density function is the correlation coefficient.

For jointly normal random variables if the correlation coefficient is zero then they are independent (see Example 4-22). Thus in this important special case, independence and uncorrelatedness are equivalent.

5-8 CONDITIONAL EXPECTATION

Given the event defined by $X = x$ and the conditional probability mass function $p_{Y|X}$, the conditional expectation of $g(Y)$ is

$$E[g(Y)\,|\,X = x] = \sum_{\text{all } i} g(y_i)p_{Y|X}(y_i\,|\,x). \tag{5-21}$$

121

Similarly given the event B and a conditional density function $f_{X|B}$

$$E[g(X) \mid B] = \int_{-\infty}^{\infty} g(x) f_{X|B}(x)\, dx. \tag{5-22}$$

Letting the event B be defined in terms of a random variable and letting g be a function of two variables

$$E[g(X, Y) \mid X = x] = \int_{-\infty}^{\infty} g(x, y) f_{Y|X}(y \mid x)\, dy. \tag{5-23}$$

The function $E[g(X, Y) \mid X = x]$ is a function of the parameter x. $E[g(X, Y) \mid X]$ denotes the above function when x is allowed to vary over the range of the random variable X. Thus $E[g(X, Y) \mid X]$ is a function of the random variable X and thus is a random variable. Returning to fundamentals, $E[g(X, Y) \mid X]$ varies depending on the outcome in the sample space; thus $E[g(X, Y) \mid X]$ is a random variable, and its expectation may be taken as

$$E\{E[g(X, Y) \mid X]\} = \int_{-\infty}^{\infty} E[g(X, Y) \mid X = x] f_X(x)\, dx.$$

This gives a result that is important in applications:

$$E[g(X, Y)] = E\{E[g(X, Y) \mid X]\}. \tag{5-24}$$

To prove this result in the case where density functions exist, note that

$$E\{E[g(X, Y) \mid X]\} = \int_{-\infty}^{\infty} \left[\int_{-\infty}^{\infty} g(x, y) f_{Y|X}(y \mid x)\, dy \right] f_X(x)\, dx$$

$$= \iint_{-\infty}^{\infty} g(x, y) f_{X,Y}(x, y)\, dx\, dy$$

$$= E[g(X, Y)].$$

A special case of (5-24) when $g(X, Y) = Y$ is

$$E[Y] = E[E[Y \mid X]]. \tag{5-25}$$

EXAMPLE 5-11

A series of missiles is to be fired at a target. Each missile has a probability of p of hitting the target, and the trials are independent. A missile is fired; if it hits, no further missiles are fired. If it misses, the next is fired, and so on until a hit is achieved. Define a random variable N as the number of missiles fired at a target. What is $E[N]$? With $q = 1 - p$

$$E[N] = 1p + 2qp + 3q^2p + 4q^3p + \cdots$$
$$= p[1 + 2q + 3q^2 + 4q^3 + \cdots]$$
$$= p\frac{1}{p^2} = \frac{1}{p}.$$

This method of solution involves summing a series. We now work the problem by conditional expectation. Define the random variable H to be 1 if there is a hit on the first shot and $H = 0$ if no hit. Then from (5-25)

$$E[N] = E[E[N \mid H]],$$
$$E[N] = pE[N \mid H = 1] + qE[N \mid H = 0],$$
$$E[N] = p1 + q(1 + E[N]),$$
$$E[N] - qE[N] = 1(p + q) = 1,$$
$$E[N] = \frac{1}{1 - q} = \frac{1}{p}.$$

Normal Random Variables

We find the conditional expected value of a random variable that is jointly normally distributed. This is then used to find $E[X_1X_2]$ and the result is compared to the calculation of $\sigma_{X_1X_2}$ in the last section.

Given f_{X_1,X_2} as defined in (4-19) and that f_{X_2} is normal,

$$f_{X_1\mid X_2}(x_1 \mid x_2) = \frac{\dfrac{1}{2\pi\sigma_1\sigma_2\sqrt{1 - r^2}} \exp\left\{-\dfrac{1}{2(1 - r^2)}\times\left[\dfrac{(x_1 - \mu_1)^2}{\sigma_1^2} - \dfrac{2r(x_1 - \mu_1)(x_2 - \mu_2)}{\sigma_1\sigma_2} + \dfrac{(x_2 - \mu_2)^2}{\sigma_2^2}\right]\right\}}{\dfrac{1}{\sqrt{2\pi}\sigma_2}\exp\left\{-\dfrac{(x_2 - \mu_2)^2}{2\sigma_2^2}\right\}}$$

$$= \frac{1}{\sqrt{2\pi}\sigma_1\sqrt{1 - r^2}} \exp\left\{-\frac{1}{2(1 - r^2)}\times\left[\frac{(x_1 - \mu_1)^2}{\sigma_1^2} - \frac{2r(x_1 - \mu_1)(x_2 - \mu_2)}{\sigma_1\sigma_2} + \frac{r^2(x_2 - \mu_2)^2}{\sigma_2^2}\right]\right\},$$

$$f_{X_1\mid X_2}(x_1 \mid x_2) = \frac{1}{\sqrt{2\pi}\sigma_1\sqrt{1 - r^2}} \exp\left\{-\frac{\left[x_1 - \mu_1 - r\dfrac{\sigma_1}{\sigma_2}(x_2 - \mu_2)\right]^2}{2\sigma_1^2(1 - r^2)}\right\}.$$

$$(5\text{-}26)$$

123

With the use of (5-23)

$$E[X_1 \mid X_2 = x_2] = \int_{-\infty}^{\infty} \frac{x_1}{\sqrt{2\pi}\sigma_1\sqrt{1-r^2}}$$

$$\times \exp\left\{-\frac{\left[x_1 - \mu_1 - r\dfrac{\sigma_1}{\sigma_2}(x_2 - \mu_2)\right]^2}{2\sigma_1^2(1-r^2)}\right\} dx_1.$$

This can be integrated directly, but it is easier to note that

$$\mu_1 + r(\sigma_1/\sigma_2)(x_2 - \mu_2)$$

is the mean of this normal random variable. Thus

$$E[X_1 \mid X_2] = \mu_1 + r\frac{\sigma_1}{\sigma_2}(X_2 - \mu_2). \tag{5-27}$$

To find $E[X_1X_2]$, (5-24) is used;

$$E[X_1X_2] = E\{E[X_1X_2 \mid X_2]\}.$$

We now show

$$E\{E[X_1X_2 \mid X_2]\} = E\{X_2E[X_1 \mid X_2]\}, \tag{5-28}$$

$$E[X_1X_2] = E\{E[X_1X_2 \mid X_2] = \int_{-\infty}^{\infty}\left[\int_{-\infty}^{\infty} x_2x_1 f_{X_1\mid X_2}(x_1 \mid x_2)\, dx_1\right] f_{X_2}(x_2)\, dx_2$$

$$= \int_{-\infty}^{\infty} x_2\left[\int_{-\infty}^{\infty} x_1 f_{X_1\mid X_2}(x_1 \mid x_2)\, dx_1\right] f_{X_2}(x_2)\, dx_2.$$

Thus

$$E[X_1X_2] = E\{X_2E[X_1 \mid X_2]\}.$$

With the use of (5-27)

$$E[X_1X_2] = E\left\{X_2\left[\mu_1 + r\frac{\sigma_1}{\sigma_2}(X_2 - \mu_2)\right]\right\}$$

$$= \mu_1\mu_2 + r\frac{\sigma_1}{\sigma_2}[\sigma_2^2 + \mu_2^2 - \mu_2^2],$$

$$E[X_1X_2] = \mu_1\mu_2 + r\sigma_1\sigma_2 \tag{5-29}$$

or

$$\sigma_{12} = E[X_1X_2] - \mu_1\mu_2 = r\sigma_1\sigma_2.$$

Note the similarity of this development and the one in the preceding section.

5-9 SUMMARY

The concept of expected value was introduced and particular attention was directed toward the mean, variance, and characteristic function.

Important properties of expected value were given, and the mean and variance of common random variables were found in the examples.

Chebycheff's inequality was introduced, and this inequality, with the variance, gives a upper bound on the probability that a random variable exceeds any given deviation from the mean.

The characteristic function is very nearly (sign change in exponent) the Fourier transform of the probability density function. It was shown that the characteristic function can be used to find central moments.

Emphasis was placed upon conditional expectation. It was noted and illustrated by examples that expected values may be found by first finding a conditional expectation and then averaging the conditional expected values.

Additional reading may be found in Ref. B1, D3, D4, F1, H1, P2, P3, and P5.

5-10 PROBLEMS

Compute the mean and variance of the random variable X when

1. $p_X(0) = \frac{1}{3}, p_X(2) = \frac{2}{3}$

2. $p_X(x) = \dfrac{1}{x}, \qquad x = 6, 3, 2,$

$\qquad\quad = 0, \qquad$ elsewhere.

3. $p_X(x) = \dbinom{6}{x}\left(\dfrac{2}{3}\right)^x\left(\dfrac{1}{3}\right)^{6-x}, \qquad x = 0, 1, 2, \ldots, 6,$

$\qquad\quad = 0, \qquad\qquad\qquad\qquad$ elsewhere.

4. $f_X(x) = 2x, \qquad 0 \le x \le 1,$

$\qquad\quad = 0, \qquad$ elsewhere.

5. $f_X(x) = x, \qquad\quad 0 \le x \le 1,$

$\qquad\quad = 2 - x, \qquad 1 \le x \le 2,$

$\qquad\quad = 0, \qquad\qquad$ elsewhere.

6. $F_X(x) = 0, \qquad\quad x < 0,$

$\qquad\quad = x^{1/2}, \qquad 0 \le x \le 1,$

$\qquad\quad = 1, \qquad\quad x > 1.$

7. $F_X(x) = 0, \qquad\qquad x < 0,$

$\qquad\quad = 1 - e^{-x/\lambda}, \qquad x \ge 0.$

8. If the time to failure of electronic tubes is exponentially distributed with parameter $\lambda = 100$ hr, what is the mean time to failure?

125

9. Show, using Chebycheff's inequality, that

$$P\{|X - \mu| \le K\} \ge 1 - \frac{\sigma^2}{K^2}.$$

10. Let Y be a non-negative random variable, that is, $f_Y(y) = 0$ for $y < 0$. Using the same steps in the proof of Chebycheff's inequality, show that

$$P\{Y > K\} \le \frac{E[Y]}{K}.$$

11. a. Using Chebycheff's inequality, if $\sigma^2 = 4$, find the upper bound on the probability that a random variable deviates from its mean by 3, that is, find an upper bound on

$$P\{|X - \mu| > 3\}.$$

b. If the random variable is normal, find

$$P\{|X - \mu| > 3\}.$$

12. Find the characteristic function of the following density function.

$$f_X(x) = 1/2, \quad 3 \le x \le 5,$$
$$= 0, \quad \text{elsewhere.}$$

13. Show that the characteristic function of a binomial random variable is

$$\varphi(\omega) = (q + e^{j\omega}p)^n.$$

From this find $E[X]$, $E[X^2]$, and σ^2.

14. If $\varphi_X(\omega) = e^{j\omega - \omega^2/2}$, find f_X.

15. If $\varphi_X(\omega) = (\frac{1}{2} + e^{j\omega}\frac{1}{2})^{10}$, find $E[X]$.

16. If $f_{X,Y}$ is uniform in the region shown in Figure P5-1 that is,

$$f_{X,Y}(x, y) = \frac{1}{2}, \quad 0 \le x \le 1, 0 \le y \le 1,$$
$$= \frac{1}{2}, \quad -1 \le x \le 0, -1 \le y \le 0,$$
$$= 0, \quad \text{elsewhere,}$$

find $E[XY]$ and ρ.

17. If $f_{X,Y}$ is uniform in the region shown in Figure P5-2 find $E[XY]$ and ρ.

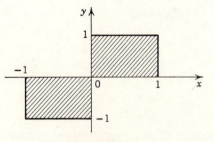

Figure P5-1

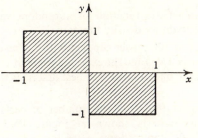

Figure P5-2

18. The joint density function of two random variables X and Y is

$$f_{X,Y}(x, y) = \frac{1}{\pi\sqrt{3}} \exp\{-\tfrac{2}{3}(x^2 - xy + y^2)\}.$$

The marginal density function of x is

$$f_X(x) = \frac{1}{\sqrt{2\pi}} e^{-x^2/2}.$$

The marginal density function of Y is

$$f_Y(y) = \frac{1}{\sqrt{2\pi}} e^{-y^2/2}.$$

(a) Are X and Y independent?
(b) What is the conditional density function of X given $Y = 1$? [i.e., find $f_{X|Y}(x \mid 1)$].
(c) Simplify the expression from (b) into the form: $(1/a)e^{-b(x-c)^2}$, and identify the conditional mean of X given $Y = 1$.
(d) Identify the conditional variance, that is

$$E\{[X - E(X \mid Y = 1)]^2 \mid Y = 1\}.$$

(e) Find $E[X \mid Y]$.
(f) Find $E[XY]$.
(g) Find ρ.

19. Show that

$$E[Y \mid X] = E[Y]$$

if X and Y are independent.

20. The thief of Bagdad has been placed in a prison which has three doors. One of the doors leads him on a one-day trip, after which he is dumped on his head (which destroys his memory as to which door he chose). Another door is similar except he takes a three-day trip before being dumped on his head. The third door leads to freedom. Assume he chooses a door immediately and with probability 1/3 when he has a chance. Find his expected number of days to freedom. (Hint: use conditional expectation.)

127

21. Compare the characteristic function of a random variable to the Laplace transform of the probability density function.

22. We read the voltage of 100 power supplies with the result that 99 readings are 10 V while 1 reading is 0 V. Find the expected value of voltage. Do you consider this expected value the one number that best describes the voltage of the power supplies?

23. The conditional expectation E of the number of parts we can sell given the number i of competitors is shown in Figure P5-3. If we consider $P(0$ competitors$) = 1/2$, $P(i$ competitors$) = 1/6$ for $i = 1, 2, 3$, find the expected number of parts sold.

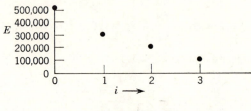

Figure P5-3

CHAPTER VI

....

DISTRIBUTIONS OF FUNCTIONS OF RANDOM VARIABLES

The problem of finding the distribution of a function of random variables is emphasized because of its importance in engineering. Often we do have mathematical models of the system of interest, that is, the output as a function of the input and system components, and we know something about the distribution of the input and the parts or components. We naturally desire to know the distribution of the output. Examples and problems illustrate a few of the many practical applications of this theory.

6-1 FUNCTIONS OF ONE RANDOM VARIABLE

In this section the problem is to find the distribution function of Y when $Y = g(X)$ where g is some known function and the distribution function of the random variable X is known. As mentioned in Chapter V, if g is any of the usual functions, Y will be a random variable. The proof is beyond the scope of this text.

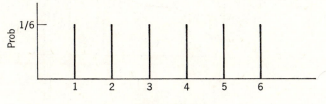

Figure 6-1

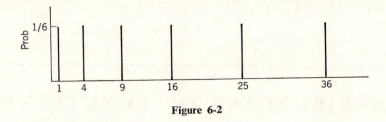

Figure 6-2

Discrete Random Variable

The discrete case is easy. One example will illustrate the technique. Suppose X represents the up face on a thrown die, that is, X has the probability mass function shown in Figure 6-1.

Now suppose $Y = X^2$; what is the probability mass function for Y? The outcome in the sample space that maps into $Y = 1$ is exactly the same outcome that maps into $X = 1$; the outcome that maps into $Y = 4$ is exactly the same outcome that maps into $X = 2$; etc. Therefore the probability function for Y is as shown in Figure 6-2.

Continuous Random Variable

The same basic idea is involved when X has a density function, but the computations are a bit more involved.

First it is assumed that g is an increasing function. Such a function is shown in Figure 6-3. In this situation

$$P(X \leq x_0) = P[g(X) \leq g(x_0)] = P(Y \leq y_0).$$

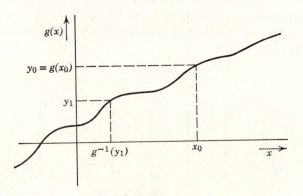

Figure 6-3

This is true for all x_0 and all $y_0 = g(x_0)$. Thus

$$F_X(x_0) = F_Y[g(x_0)]$$

or

$$\int_{-\infty}^{x_0} f_X(x)\, dx = \int_{-\infty}^{y_0} f_Y(y)\, dy.$$

In the first integral let $y = g(x)$, and since g is strictly increasing, one can uniquely find $x = g^{-1}(y)$ (see Figure 6-3). This results in $x = g^{-1}(y)$ and $(dx/dy) = (dg^{-1}(y)/dy)$. Thus

$$\int_{-\infty}^{y_0} f_X[g^{-1}(y)]\frac{dg^{-1}(y)}{dy}\, dy = \int_{-\infty}^{y_0} f_Y(y)\, dy.$$

Because the above must be true for all y_0, the integrands must be equal and this implies the basic result

$$f_Y(y) = f_X[g^{-1}(y)]\frac{dg^{-1}(y)}{dy}. \tag{6-1}$$

EXAMPLE 6-1

An example will illustrate the use of (6-1). Assume that

$$f_X(x) = e^{-x}, \qquad x \geq 0,$$
$$= 0, \qquad \text{otherwise.}$$
$$Y = 2X + 3.$$

What is f_Y?

$$g^{-1}(y) = \frac{y-3}{2},$$

$$\frac{dg^{-1}(y)}{dy} = \frac{1}{2},$$

$$f_Y(y) = e^{-\frac{y-3}{2}}\left(\tfrac{1}{2}\right), \qquad y \geq 3,$$

$$= 0, \qquad \text{otherwise.}$$

Now suppose g is a decreasing function. Such a function is shown in Figure 6-4. In this case

$$P(X \geq x_0) = P[g(X) \leq g(x_0)] = P(Y \leq y_0).$$

Thus

$$\int_{x_0}^{\infty} f_X(x)\, dx = \int_{-\infty}^{y_0} f_Y(y)\, dy.$$

Letting $y = g(x)$, or $x = g^{-1}(y)$, in the first integral

$$\int_{y_0}^{-\infty} f_X[g^{-1}(y)]\frac{dg^{-1}(y)}{dy}\, dy = \int_{-\infty}^{y_0} f_Y(y)\, dy.$$

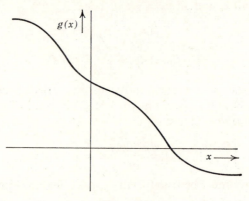

Figure 6-4

But because g is a decreasing function, $(dg^{-1}(y)/dy)$ is negative, or

$$\frac{dg^{-1}(y)}{dy} = -\left|\frac{dg^{-1}(y)}{dy}\right|.$$

Therefore

$$\int_{-\infty}^{y_0} f_X[g^{-1}(y)]\left|\frac{dg^{-1}(y)}{dy}\right|dy = \int_{-\infty}^{y_0} f_Y(y)\,dy.$$

This results in

$$f_Y(y) = f_X[g^{-1}(y)]\left|\frac{dg^{-1}(y)}{dy}\right|. \tag{6-2}$$

Note that in (6-1), $(dg^{-1}(y)/dy)$ is positive since g is increasing; therefore in (6-1)

$$\frac{dg^{-1}(y)}{dy} = \left|\frac{dg^{-1}(y)}{dy}\right|.$$

Thus (6-2) may be used for either an increasing or a decreasing function.

Now consider a function $Y = g(X)$ composed of both increasing and decreasing segments such as the example shown in Figure 6-5. Now

$$P(a \leq Y \leq b) = P(x_1 \leq X \leq x_2) + P(x_3 \leq X \leq x_4) + P(x_5 \leq X \leq x_6)$$

Thus for y_0 between a and b, (6-2) is applied three times, resulting in

$$f_Y(y_0) = f_X[g_1^{-1}(y_0)]\left|\frac{dg_1^{-1}(y_0)}{dy_0}\right| + f_X[g_2^{-1}(y_0)]\left|\frac{dg_2^{-1}(y_0)}{dy_0}\right|$$

$$+ f_X[g_3^{-1}(y_0)]\left|\frac{dg_3^{-1}(y_0)}{dy_0}\right|$$

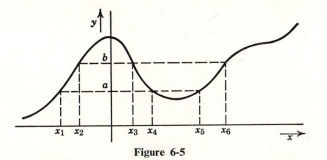

Figure 6-5

where $g_1^{-1}(y_0)$ is the inverse of $g(x)$ that applies between x_1 and x_2, $g_2^{-1}(y_0)$ is the inverse of $g(x)$ that applies between x_3 and x_4, and $g_3^{-1}(y_0)$ is the inverse of $g(x)$ that applies between x_5 and x_6.

There are two remaining possibilities to be considered. If g has a jump, such as shown in Figure 6-6, then $P(c < Y < d) = 0$ and $f_Y(y) = 0$, $c < y < d$.

If g is constant as shown in Figure 6-7 then $P(Y = c) = P(a < X < b) = \int_a^b f_X(x)\, dx$. We now illustrate with examples.

EXAMPLE 6-2 (see Figure 6-8)

$$f_X(x) = \tfrac{2}{15}x, \qquad 1 \le x \le 4,$$
$$= 0, \qquad \text{otherwise.}$$
$$Y = (X - 2)^2.$$

Find f_Y.

$$X^2 - 4X + 4 - Y = 0,$$

$$g^{-1}(y) = \frac{+4 \pm \sqrt{4^2 - 4(4 - y)}}{2} = 2 \pm \sqrt{4 - 4 + y},$$

$$g_1^{-1}(y) = 2 - \sqrt{y}, \qquad g_2^{-1}(y) = 2 + \sqrt{y},$$

$$\left| \frac{dg_i^{-1}(y)}{dy} \right| = \frac{1}{2}\frac{1}{\sqrt{y}} \qquad \text{for} \qquad i = 1 \text{ or } 2.$$

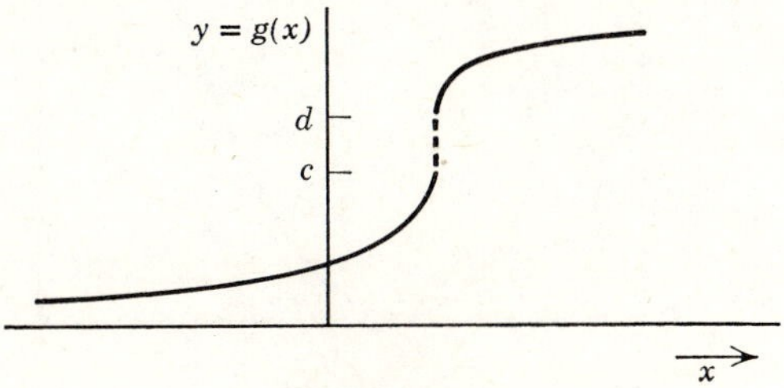

Figure 6-6

133

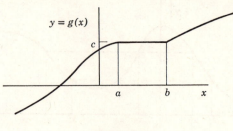

Figure 6-7

For $1 \leq y \leq 4$, $x = g_2^{-1}(y)$, thus

$$f_Y(y) = \tfrac{2}{15}(2 + \sqrt{y})\left|\frac{1}{2}\frac{1}{\sqrt{y}}\right|, \qquad 1 \leq y \leq 4,$$

$$= \frac{1}{15}\left(\frac{2}{\sqrt{y}} + 1\right), \qquad 1 \leq y \leq 4.$$

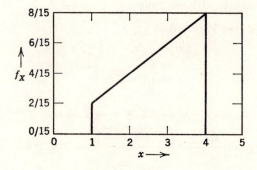

Figure 6-8a

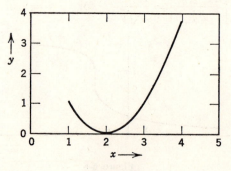

Figure 6-8b

For $0 \le y < 1$, two values of x, $g_2^{-1}(y)$ and $g_1^{-1}(y)$, exist; thus

$$f_Y(y) = \tfrac{2}{15}(2 + \sqrt{y}) \left| \frac{1}{2} \frac{1}{\sqrt{y}} \right| + \tfrac{2}{15}(2 - \sqrt{y}) \left| \frac{1}{2} \frac{1}{\sqrt{y}} \right|, \qquad 0 \le y < 1,$$

$$f_Y(y) = \frac{4}{15} \frac{1}{\sqrt{y}}, \qquad 0 \le y < 1.$$

Finally

$$f_Y(y) = 0, \qquad y < 0 \text{ or } y > 4.$$

EXAMPLE 6-3

A nominally 10-ohm resistor has a resistance uniformly distributed between 9 and 11 ohms (see Figure 6-9). What is the density function of conductance

$$G = g(R) = \frac{1}{R},$$

Letting c be a value of G

$$g^{-1}(c) = \frac{1}{c},$$

$$\frac{dg^{-1}(c)}{dc} = -\frac{1}{c^2},$$

$$f_G(c) = \frac{1}{2} \frac{1}{c^2}, \qquad \tfrac{1}{11} \le c \le \tfrac{1}{9},$$

$$= 0, \qquad \text{otherwise.}$$

This is shown in Figure 6-10.

EXAMPLE 6-4

A sinusoidal voltage generator produces a voltage of the form

$$V = \sin \omega t.$$

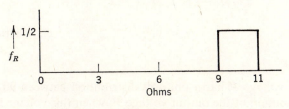

Figure 6-9

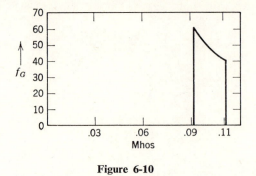

Figure 6-10

If the voltage is randomly sampled, what is the density function of the sampled voltage?

Since each cycle is the same, this problem is the same as determining the voltage when time is a random variable uniformly distributed between $-\pi/2\omega$ and $+\pi/2\omega$. Thus

$$f_t(\lambda) = \frac{\omega}{\pi}, \qquad -\frac{\pi}{2\omega} \le \lambda \le \frac{\pi}{2\omega},$$

$$= 0, \qquad\qquad \text{elsewhere.}$$

$$V = g(t) = \sin \omega t.$$

$$g^{-1}(v) = \frac{1}{\omega} \sin^{-1} v,$$

$$\frac{dg^{-1}(v)}{dv} = \frac{1}{\omega} \frac{1}{\sqrt{1 - v^2}},$$

$$f_V(v) = \frac{\omega}{\pi} \frac{1}{\omega} \frac{1}{\sqrt{1 - v^2}} = \frac{1}{\pi\sqrt{1 - v^2}}, \qquad -1 < v < 1,$$

$$= 0, \qquad\qquad \text{elsewhere.}$$

EXAMPLE 6-5

A random voltage V has a uniform distribution between 90 and 100 V. Find the distribution of the output voltage W when this voltage is put into a

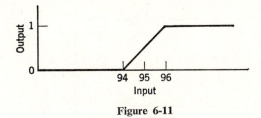

Figure 6-11

nonlinear device, a limiter, with the characteristic shown in Figure 6-11.

$$P(W = 0) = P(V \le 94) = \tfrac{4}{10},$$

$$P(W = 1) = P(V \ge 96) = \tfrac{4}{10},$$

$$W = \tfrac{1}{2}(V - 94), \qquad 94 \le V \le 96,$$

$$V = g^{-1}(W) = 2(W + 47), \qquad 0 \le W \le 1,$$

$$\frac{dg^{-1}(w)}{dw} = 2.$$

$$f_W(\lambda) = \tfrac{1}{10}(2) = \tfrac{1}{5}, \qquad 0 \le \lambda \le 1.$$

The plot of F_W is shown in Figure 6-12. Note that the distribution function is a combination of jumps and a part that has a density function. Thus W has a mixed distribution.

EXAMPLE 6-6

In the circuit shown in Figure 6-13 the input voltage is a constant 10 V; R_0 is a constant 1 ohm, R is a random variable with a density f_R. Find the

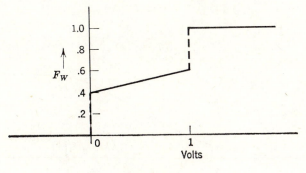

Figure 6-12

137

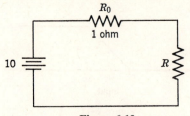

Figure 6-13

density of V_R, the voltage across R.

$$V_R = 10\,\frac{R}{R+1}\,,$$

$$R = g^{-1}(V_R) = \frac{V_R}{10 - V_R}\,, \qquad \frac{\partial R(V_R)}{\partial V_R} = \frac{10}{(10 - V_R)^2}\,,$$

$$f_{V_R}(\lambda) = f_R\!\left(\frac{\lambda}{10 - \lambda}\right)\!\left[\frac{10}{(10 - \lambda)^2}\right].$$

If
$$f_R(r) = \tfrac{2}{3}r, \qquad 1 \le r \le 2,$$
$$= 0, \qquad\quad \text{otherwise.}$$

Then
$$f_{V_R}(\lambda) = \frac{2}{3}\!\left(\frac{\lambda}{10 - \lambda}\right)\!\left[\frac{10}{(10 - \lambda)^2}\right], \qquad 5 \le \lambda \le \tfrac{20}{3},$$
$$= 0, \qquad\qquad\qquad\qquad\qquad\qquad \text{otherwise.}$$

EXAMPLE 6-7

Let X have a distribution function F and a density function f. Then if

$$Y = F(X),$$

show that Y is uniformly distributed $(0, 1)$ (see Figure 6-14).

$$y = F(x), \qquad x = F^{-1}(y),$$

$$\frac{dy}{dx} = f(x), \qquad \frac{dx}{dy} = \frac{1}{f(x)} = \frac{1}{f\{F^{-1}(y)\}}\,.$$

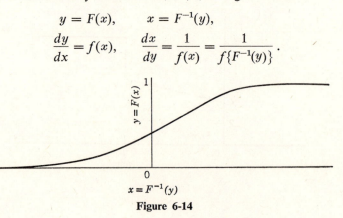

Figure 6-14

With the use of (6-2)

$$f_Y(y) = f\{F^{-1}(y)\} \frac{1}{f\{F^{-1}(y)\}} = 1, \qquad 0 \le y \le 1,$$

$$= 0, \qquad\qquad\qquad\qquad \text{elsewhere.}$$

6-2 SUM OF TWO RANDOM VARIABLES

Consider the sum of two independent discrete random variables. To be specific, consider $Y = X_1 + X_2$ where X_1 is the number appearing on one die and X_2 is the number appearing on the second die. The probability mass function of Y is described by

$$P(Y = 2) = P(X_1 = 1)P(X_2 = 1) = 1/36,$$
$$P(Y = 3) = P(X_1 = 1)P(X_2 = 2) + P(X_1 = 2)P(X_2 = 1) = 2/36;$$
$$P(Y = 4) = P(X_1 = 1)P(X_2 = 3) + P(X_1 = 2)P(X_2 = 2)$$
$$+ P(X_1 = 3)P(X_2 = 1) = 3/36$$

$$\cdot$$
$$\cdot$$
$$\cdot$$

$$P(Y = i) = \sum_{j=1}^{6} P(X_1 = j)P(X_2 = i - j) \tag{6-3}$$

$$\cdot$$
$$\cdot$$
$$\cdot$$

$$P(Y = 12) = P(X_1 = 6)P(X_2 = 6) = 1/36.$$

The complete probability mass function is shown in Figure 6-15. An expression similar to (6-3) can be found for the sum of any two discrete independent

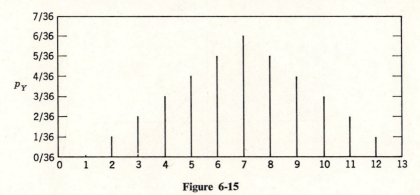

Figure 6-15

139

random variables. In the case of independent random variables that have density functions, then by analogy with (6-3)

$$f_Y(y) = \int_{-\infty}^{\infty} f_{X_1}(x) f_{X_2}(y - x)\, dx. \qquad (6\text{-}4)$$

Equation (6-4) may be derived as follows:

$$P(Y \le y) = F_Y(y) = P(X_1 + X_2 \le y) = \iint\limits_{x_1 + x_2 \le y} f_{X_1, X_2}(x_1, x_2)\, dx_1\, dx_2.$$

The region $x_1 + x_2 \le y$ can be seen in Figure 6-16. Thus

$$F_Y(y) = \int_{-\infty}^{\infty} \int_{-\infty}^{y - x_1} f_{X_1, X_2}(x_1, x_2)\, dx_2\, dx_1,$$

$$f_Y(y) = \frac{dF_Y(y)}{dy} = \int_{-\infty}^{\infty} f_{X_1, X_2}(x_1, y - x_1)\, dx_1. \qquad (6\text{-}5)$$

Equation 6-5 is valid for random variables that are dependent. If X_1 and X_2 are independent

$$f_{X_1, X_2}(x_1, y - x_1) = f_{X_1}(x_1) f_{X_2}(y - x_1).$$

Thus
$$f_Y(y) = \int_{-\infty}^{\infty} f_{X_1}(x_1) f_{X_2}(y - x_1)\, dx_1$$

which is the same as (6-4).

The integral shown in (6-4) has been given the name convolution and is often written as

$$f_Y = f_{X_1} * f_{X_2}.$$

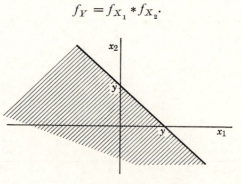

Figure 6-16

EXAMPLE 6-8

$Y = X_1 + X_2$ where X_1 and X_2 are independent and

$$f_{X_1}(x) = \frac{1}{\sqrt{2\pi}\sigma} \exp\left[-\frac{(x-\mu)^2}{2\sigma^2}\right],$$

$$f_{X_2}(x) = \frac{1}{\sqrt{2\pi}\sigma} \exp\left[-\frac{(x-\eta)^2}{2\sigma^2}\right].$$

With the use of (6-4)

$$f_Y(y) = [f_{X_1} * f_{X_2}](y) = \int_{-\infty}^{\infty} f_{X_1}(x_1) f_{X_2}(y-x_1)\, dx_1$$

$$= \int_{-\infty}^{\infty} \frac{1}{\sqrt{2\pi}\sigma} \exp\left[-\frac{(x-\mu)^2}{2\sigma^2}\right] \frac{1}{\sqrt{2\pi}\sigma} \exp\left[-\frac{(y-x-\eta)^2}{2\sigma^2}\right] dx.$$

At this point convolution becomes simply the problem of evaluating the integral. The steps that follow are the rather tedious algebra necessary to evaluate the integral. This same problem is worked in a more general case with less effort in Section 6-4.

$$f_Y(y) = \int_{-\infty}^{\infty} \frac{1}{2\pi\sigma^2} \exp\left[\frac{-2x^2 + 2x(\mu - \eta + y) - \mu^2 - y^2 - \eta^2 + 2\eta y}{2\sigma^2}\right] dx$$

$$= \int_{-\infty}^{\infty} \frac{1}{2\pi\sigma^2} \exp\left[\{-x^2 + 2x\tfrac{1}{2}(\mu - \eta + y) - \tfrac{1}{4}[\mu^2 + \eta^2 + y^2\right.$$

$$\left. - 2(\mu\eta + \eta y - \mu y)] - \tfrac{1}{4}(\mu^2 + y^2 + \eta^2) + \tfrac{1}{2}(\mu y + \eta y - \mu\eta)\} \frac{1}{\sigma^2}\right] dx$$

$$= \frac{1}{2\sigma\sqrt{\pi}} \exp\left\{-\frac{[y^2 - 2y(\mu + \eta) + (\mu + \eta)^2]}{4\sigma^2}\right\} \int_{-\infty}^{\infty} \frac{1}{\sqrt{\pi}\sigma}$$

$$\times \exp\left\{-\frac{[x - \tfrac{1}{2}(\mu - \eta + y)]^2}{\sigma^2}\right\} dx$$

$$= \frac{1}{2\sigma\sqrt{\pi}} \exp\left\{\frac{-[y - (\mu + \eta)]^2}{4\sigma^2}\right\}.$$

Thus Y is normal with mean $\mu + \eta$ and variance $2\sigma^2$.

The reader familiar with convolution can skip the next section.

6-3 CONVOLUTION

First we show that convolution is commutative, or

$$f_{X_1} * f_{X_2} = f_{X_2} * f_{X_1}.$$

$$[f_{X_1} * f_{X_2}](y) = \int_{-\infty}^{\infty} f_{X_1}(x) f_{X_2}(y-x)\, dx.$$

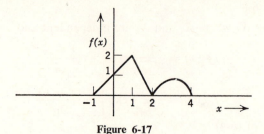

Figure 6-17

Let $y - x = x_2$; then $x = y - x_2$,

$$[f_{X_1} * f_{X_2}](y) = \int_{-\infty}^{\infty} f_{X_1}(y - x_2)f_{X_2}(x_2)\, dx_2$$

$$= \int_{-\infty}^{\infty} f_{X_2}(x_2)f_{X_1}(y - x_2)\, dx_2 = [f_{X_2} * f_{X_1}](y).$$

Since y is arbitrary the proof is complete.

Before we actually evaluate some integrals, the effect of different arguments in a function is shown. Let $f(x)$ be as shown in Figure 6-17. Then $f(-x)$ is as shown in Figure 6-18. Note that the effect of changing the sign of the argument causes the function to be reflected or flipped about zero. Observe that $f(1) = 2$ in both cases, except in Figure 6-18 when the argument is 1, x is -1. The student should check for other values of the argument. Now the effect of changing the argument from $-x$ to $y - x$ will be observed. $f(1 - x)$ is shown in Figure 6-19. Again note that $f(1) = 2$, and this occurs when $x = 0$. $f(2 - x)$ is shown in Figure 6-20. To generalize, $f(y - x)$ is $f(-x)$ shifted y units to the right.

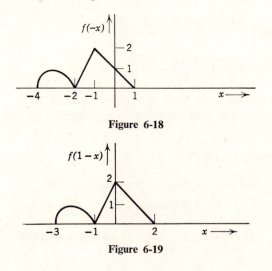

Figure 6-18

Figure 6-19

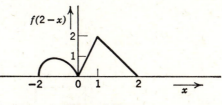

Figure 6-20

EXAMPLE 6-9

Now the convolution of two functions can be evaluated. Consider

$$Y = X_1 + X_2$$

where
$$f_{X_1}(x_1) = 2, \qquad 1 \leq x_1 \leq \tfrac{3}{2},$$
$$= 0, \qquad \text{elsewhere.}$$
$$f_{X_2}(x_2) = \tfrac{1}{2}, \qquad 2 \leq x_2 \leq 4,$$
$$= 0, \qquad \text{elsewhere.}$$

To evaluate the convolution, first $f_Y(0)$ will be found from Figure 6-21. Since

$$f_Y(0) = \int_{-\infty}^{\infty} f_{X_1}(x) f_{X_2}(-x)\, dx; \qquad f_Y(0) = 0.$$

Similarly $f_Y(y) = 0$ for $y \leq 3$ because the two functions $f_{X_1}(x)$ and $f_{X_2}(y - x)$ do not overlap. Next let $y = 3\tfrac{1}{4}$. Figure 6-22 shows this situation.

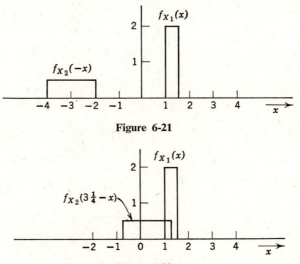

Figure 6-21

Figure 6-22

143

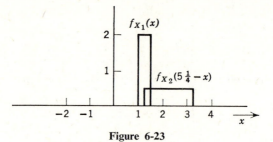

Figure 6-23

Note that

$$f_{X_2}(x_2) = \tfrac{1}{2} \quad \text{for} \quad 2 \le x_2 \le 4,$$

so
$$f_{X_2}(3\tfrac{1}{4} - x) = \tfrac{1}{2} \quad \text{for} \quad 2 \le 3\tfrac{1}{4} - x \le 4,$$

$$\text{or for} \quad -\tfrac{3}{4} \le x \le \tfrac{5}{4}.$$

$$f_Y(3\tfrac{1}{4}) = \int_1^{\frac{5}{4}} 2 \cdot \tfrac{1}{2} \, dx = \tfrac{1}{4}.$$

Generalizing for $3 \le y \le 3\tfrac{1}{2}$

$$f_Y(y) = \int_1^{y-2} 2 \cdot \tfrac{1}{2} \, dx = y - 3, \qquad 3 \le y \le \tfrac{7}{2}.$$

For $\tfrac{7}{2} \le y \le 5$

$$f_Y(y) = \int_1^3 2 \cdot \tfrac{1}{2} \, dx = \tfrac{1}{2}, \qquad \tfrac{7}{2} \le y \le 5.$$

Consider $y = 5\tfrac{1}{4}$; then Figure 6-23 applies.

$$f_Y(5\tfrac{1}{4}) = \int_{\frac{5}{4}}^{\frac{3}{2}} (2 \cdot \tfrac{1}{2}) \, dx = \tfrac{1}{4}.$$

Generalizing for $5 \le y \le \tfrac{11}{2}$

$$f_Y(y) = \int_{y-4}^{\frac{3}{2}} (2 \cdot \tfrac{1}{2}) \, dx = \tfrac{11}{2} - y, \qquad 5 \le y \le \tfrac{11}{2}.$$

The complete plot of f_Y is shown in Figure 6-24.

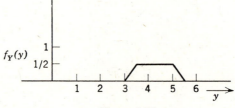

Figure 6-24

Note that f_Y is a density function (it is positive and has an area of 1) and that the smallest argument for which it is nonzero is 3, which checks with the sum of the smallest possible values of $X_1 + X_2$. Similarly the largest values $(1\frac{1}{2} + 4) = 5\frac{1}{2}$ checks.

EXAMPLE 6-10

Let $Y = X_1 + X_2$ where X_1 and X_2 are independent, and

$$f_{X_1}(x_1) = e^{-x_1}, \quad x_1 \geq 0, \qquad f_{X_2}(x_2) = 2e^{-2x_2}, \quad x_2 \geq 0,$$
$$= 0, \qquad x_1 < 0. \qquad = 0, \qquad x_2 < 0.$$

See Figure 6-25.

$$f_Y(y) = \int_0^y \exp(-x_1) 2 \exp[-2(y - x_1)] \, dx_1$$
$$= 2 \exp(-2y) \int_0^y \exp(x_1) \, dx_1 = 2 \exp(-2y)[\exp(y) - 1],$$

$$f_Y(y) = 2[\exp(-y) - \exp(-2y)], \qquad y \geq 0.$$

6-4 CHARACTERISTIC FUNCTION OF THE SUM OF RANDOM VARIABLES

If X_1 and X_2 are independent and $Y = X_1 + X_2$, then

$$\varphi_Y(j\omega) = \varphi_{X_1}(j\omega)\varphi_{X_2}(j\omega). \tag{6-6}$$

That is, if independent random variables are added, their characteristic functions are multiplied. Thus convolution of density functions corresponds to multiplication of characteristic functions. This is analogous to the fact that the convolution of the input and impulse response in the time domain

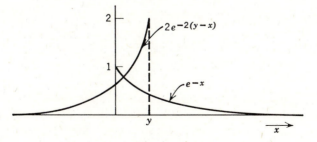

Figure 6-25

corresponds to multiplication of the Laplace or Fourier transforms. We now show that 6-6 is true.

$$\varphi_Y(j\omega) = E(e^{j\omega Y}) = \iint\limits_{-\infty}^{\infty} e^{j\omega(x_1+x_2)} f_{X_1}(x_1) f_{X_2}(x_2)\, dx_1\, dx_2$$

$$= \int_{-\infty}^{\infty} e^{j\omega x_1} f_{X_1}(x_1)\, dx_1 \int_{-\infty}^{\infty} e^{j\omega x_2} f_{X_2}(x_2)\, dx_2$$

$$= \varphi_{X_1}(\omega) \varphi_{X_2}(\omega).$$

EXAMPLE 6-11

Let X_1 and X_2 be independent random variables that are normally distributed with means μ_1 and μ_2 and variances of σ_1^2 and σ_2^2, respectively. Find the distribution function of Y where $Y = X_1 + X_2$.

From Chapter V

$$X_i(j\omega) = \exp\left[j\omega\mu_i - \tfrac{1}{2}\sigma_i^2\omega^2\right], \qquad i = 1, 2.$$

Thus
$$\varphi_Y = [\exp(j\omega\mu_1 - \tfrac{1}{2}\sigma_1^2\omega^2)][\exp(j\omega\mu_2 - \tfrac{1}{2}\sigma_2^2\omega^2)],$$

$$\varphi_Y = \exp\left[j\omega(\mu_1 + \mu_2) - \tfrac{1}{2}\omega^2(\sigma_1^2 + \sigma_2^2)\right].$$

Therefore from the uniqueness theorem for characteristic functions Y has a normal distribution with mean $(\mu_1 + \mu_2)$ and variance $\sigma_1^2 + \sigma_2^2$. That is

$$F_Y(y) = \int_{-\infty}^{y} \frac{1}{\sqrt{2\pi(\sigma_1^2 + \sigma_2^2)}} \exp\left[-\frac{(x - \mu_1 - \mu_2)^2}{2(\sigma_1^2 + \sigma_2^2)}\right] dx.$$

6-5 SUM OF n RANDOM VARIABLES

If n random variables X_i are independent and if

$$Y = \sum_{i=1}^{n} X_i,$$

then using induction and (6-4) and (6-6)

$$f_Y = f_{X_1} * f_{X_2} * \cdots * f_{X_n} \tag{6-7a}$$

and
$$\varphi_Y(j\omega) = \prod_{i=1}^{n} \varphi_{X_i}(j\omega). \tag{6-7b}$$

Let us now consider the resistance of a circuit consisting of resistors in series. All resistors are assumed to have a uniform density function between 9 and 11 ohms (10 ohms $\pm$ 10%) as shown in Figure 6-26a.

Figure 6-26b shows circuits and the corresponding density function for

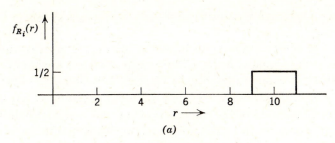

Figure 6-26a

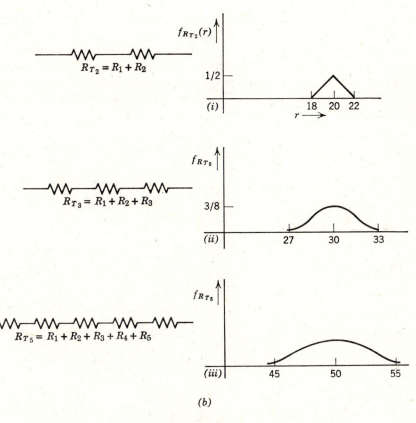

Figure 6-26b

147

total resistance when each resistor has a density function as shown in Figure 6-26a.

As more and more resistors are added in series, the density function begins to look more and more normal. Suppose five resistors are put in series, and each resistor is equally likely to be between 1.95 and 2.05 ohms. The mean value of each resistance is 2 and the standard deviation is $(20 \sqrt{3})^{-1}$. The exact density function of the resistance of the series circuit is plotted in Figure 6-27 along with the normal density function which has the same mean (10) and the same variance $(\frac{1}{240})$. Note the close correspondence.

The *central limit theorem* states that under certain general conditions if

$$Y = \sum_{i=1}^{n} X_i,$$

and if the X_i's are independent, then f_Y approaches (as $n \to \infty$) a normal density function. Furthermore the mean is equal to the sum of the individual means and variance is equal to the sum of the individual variances as is shown below. The proof of the central limit theorem is beyond the scope of this text. As illustrated by Figure 6-27 the normal density is a good approximation to f_Y for relatively small values of n if the individual densities are not too irregular.

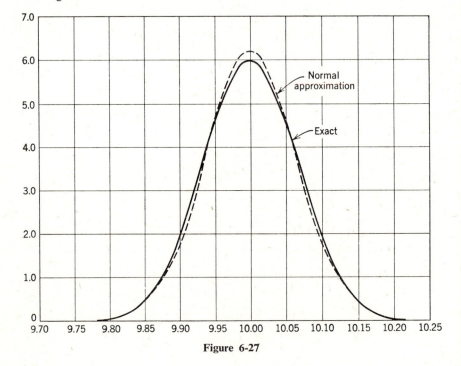

Figure 6-27

Since only the mean and variance are needed to define a normal distribution, let us find the mean and variance of a sum of random variables.

$$Y = \sum_{i=1}^{n} X_i,$$

$$\mu_Y = E[Y] = E\left[\sum_{i=1}^{n} X_i\right] = \sum_{i=1}^{n} E[X_i] = \sum_{i=1}^{n} \mu_i. \tag{6-8}$$

Thus the mean is the sum of the means and independence is not required for the above result.

$$E[(Y - \mu_Y)^2] = E\left[\left(\sum_{i=1}^{n} X_i - \mu_Y\right)^2\right] = E\left\{\left[\sum_{i=1}^{n} (X_i - \mu_i)\right]^2\right\},$$

$$= E\left[\sum_{i=1}^{n}\sum_{j=1}^{n} (X_i - \mu_i)(X_j - \mu_j)\right]$$

$$= \sum_{i=1}^{n}\sum_{j=1}^{n} E[(X_i - \mu_i)(X_j - \mu_j)]$$

$$= E[(X_1 - \mu_1)^2] + E[(X_1 - \mu_1)(X_2 - \mu_2)] + \cdots$$

$$+ E[(X_1 - \mu_1)(X_n - \mu_n)] + E[(X_2 - \mu_2)(X_1 - \mu_1)]$$

$$+ E[(X_2 - \mu_2)^2] + \cdots + E[(X_2 - \mu_2)(X_n - \mu_n)] + \cdots$$

$$\vdots$$

$$+ E[(X_n - \mu_n)(X_1 - \mu_1)] + \cdots + E[(X_n - \mu_n)^2].$$

If the random variables are independent

$$E[(X_i - \mu_i)(X_j - \mu_j)] = E[X_i - \mu_i]E[X_j - \mu_j] = 0, \qquad i \neq j.$$

Then $\qquad \sigma_Y^2 = E[(Y - \mu_Y)^2] = \sum_{i=1}^{n} E[(X_i - \mu_i)^2] = \sum_{i=1}^{n} \sigma_{X_i}^2. \tag{6-9}$

EXAMPLE 6-12

A circuit consists of 10 resistors in series. Each resistor has a mean of 10 and a variance of 1/10. Find the density of the total resistance R_T of the circuit.

R_T has approximately a normal density function with mean 100 and variance 1. That is

$$f_{R_T}(r) = \frac{1}{\sqrt{2\pi}} \exp\left[-\frac{(r - 100)^2}{2}\right].$$

Linear Combinations

From the work done to this point it is possible to find the distribution of

$$Y = \sum_{i=1}^{n} a_i X_i.$$

First define $Z_i = a_i X_i$. Then the density of Z_i can be found from the density of X_i by the methods of Section 6.1. Specifically

$$f_{Z_i}(z) = \left| \frac{1}{a_i} \right| f_{X_i}\!\left(\frac{z}{a_i}\right),$$

$$E[Z_i] = E[a_i X_i] = a_i E[X_i] = a_i \mu_{X_i},$$

$$\sigma_{Z_i}^{2} = E[(Z_i - \mu_{Z_i})^2] = E[(a_i X_i - a_i \mu_{X_i})^2]$$

$$= a_i^2 E[(X_i - \mu_{X_i})^2] = a_i^2 \sigma_{X_i}^2.$$

Then

$$Y = \sum_{i=1}^{n} Z_i$$

is just the problem solved in this section.

To summarize if $Y = \sum_{i=1}^{n} a_i X_i$

$$\mu_Y = E\left[\sum_{i=1}^{n} a_i X_i \right] = \sum_{i=1}^{n} a_i \mu_{X_i}. \tag{6-10}$$

If the X_i's are independent

$$\sigma_Y^2 = E[(Y - \mu_Y)^2] = \sum_{i=1}^{n} \sigma_{Z_i}^2$$

or

$$\sigma_Y^2 = \sum_{i=1}^{n} a_i^2 \sigma_{X_i}^2. \tag{6-11}$$

The distribution of the linear combination can be approximated by a normal distribution with the mean and variance as specified by (6-10) and (6-11). In the next section a general function of random variables is considered.

6-6 DISTRIBUTION OF $Y = g(X_1, X_2, \ldots, X_n)$

If

$$Y = g(X_1, X_2, \ldots, X_n),$$

then

$$P(Y \le y) = F_Y(y) = P\{g(X_1, X_2, \ldots, X_n) \le y\},$$

and if the densities exist

$$F_Y(y) = \int_{-\infty}^{y} f_Y(\lambda) \, d\lambda$$

$$= \underset{g(x_1, x_2, \ldots, x_n) \leq y}{\int \int \cdots \int} f_{X_1, X_2, \ldots, X_n}(x_1, x_2, \ldots, x_n) \, dx_1 \, dx_2 \cdots dx_n. \quad (6\text{-}12)$$

This general formulation is difficult to use. We now illustrate the general formulation when $n = 2$. Later a practical approximation is developed.

Consider

$$Y = g_1(X_1, X_2).$$

Define a function $\qquad Z = g_2(X_1, X_2).$

Now solve for $X_1 = h_1(Y, Z)$, $X_2 = h_2(Y, Z)$. Just as in the case of the transformation of a single random variable, there is no assurance that a unique solution exists. Assume for illustration there are three such solutions: h_{11}, h_{21}, h_{12}, h_{22}; and h_{13}, h_{23}. Then we have the result

$$f_{Y,Z}(y, z) = \sum_{i=1}^{3} f_{X_1, X_2}[h_{1i}(y, z), h_{2i}(y, z)] \, |J_i| \qquad (6\text{-}13)$$

where
$$J_i = \begin{vmatrix} \dfrac{\partial h_{1i}(y, z)}{\partial y} & \dfrac{\partial h_{1i}(y, z)}{\partial z} \\[2ex] \dfrac{\partial h_{2i}(y, z)}{\partial y} & \dfrac{\partial h_{2i}(y, z)}{\partial z} \end{vmatrix}.$$

This result can be derived from (6-12).

Having the joint density of Y and Z it is easy to find the marginal density of Y by

$$f_Y(y) = \int_{-\infty}^{\infty} f_{Y,Z}(y, z) \, dz. \qquad (6\text{-}14)$$

A special case of some interest illustrates the application of (6-13). It also illustrates the difficulty of this approach.

EXAMPLE 6-13

Let two resistors, having independent resistances, X_1 and X_2, uniformly distributed between 9 and 11 ohms, be placed in parallel. Find the probability density function of resistance Y of the parallel combination.

$Y = X_1 X_2 / (X_1 + X_2)$ and introducing another function

$$Z = X_2.$$

151

Solving for X_1 and X_2 results in

$$X_1 = \frac{YZ}{Z - Y}, \qquad X_2 = Z.$$

In this case there is only one solution; thus (6-13) reduces to

$$f_{Y,Z}(y, z) = f_{X_1,X_2}\left(\frac{yz}{z - y}, z\right) |J|$$

where

$$J = \begin{vmatrix} \dfrac{\partial\left(\dfrac{yz}{z - y}\right)}{\partial y} & \dfrac{\partial\left(\dfrac{yz}{z - y}\right)}{\partial z} \\[4mm] \dfrac{\partial z}{\partial y} & \dfrac{\partial z}{\partial z} \end{vmatrix} = \frac{z^2}{(z - y)^2}.$$

Now

$$f_{X_1 X_2}(x_1, x_2) = f_{X_1}(x_1) f_{X_2}(x_2) = \tfrac{1}{4}, \qquad 9 \leq x_1 \leq 11, \qquad 9 \leq x_2 \leq 11,$$
$$= 0, \qquad \text{elsewhere.}$$

Thus

$$f_{X_1 X_2}\left(\frac{yz}{z - y}, z\right) = \tfrac{1}{4}, \qquad y, z = ?,$$
$$= 0, \qquad \text{elsewhere}$$

but the limits of y, z where $f_{X_1,X_2} = 1/4$ are more difficult to determine. Figure 6-28 will help. Thus

$$f_{Y,Z}(y, z) = \frac{1}{4} \frac{z^2}{(z - y)^2} \qquad \text{in the region shown in Figure 6-28}b,$$
$$= 0, \qquad\qquad\qquad\qquad\qquad\qquad \text{elsewhere.}$$

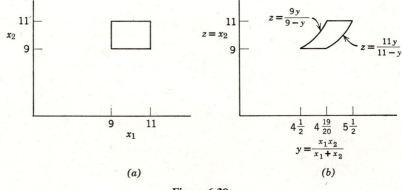

(a)

(b)

Figure 6-28

With the use of (6-14)

$$f_Y(y) = \int_9^{9y/(9-y)} \frac{1}{4} \frac{z^2}{(z-y)^2}\, dz, \qquad 4\tfrac{1}{2} \le y \le 4\tfrac{19}{20},$$

$$= \int_{11y/(11-y)}^{11} \frac{1}{4} \frac{z^2}{(z-y)^2}\, dz, \qquad 4\tfrac{19}{20} \le y \le 5\tfrac{1}{2},$$

$$= 0, \qquad\qquad\qquad\qquad \text{elsewhere.}$$

Carrying out the integration results in

$$f_Y(y) = \frac{y-9}{2} + \frac{y^2}{2(9-y)} + y \ln \frac{y}{9-y}, \qquad 4\tfrac{1}{2} \le y \le 4\tfrac{19}{20},$$

$$= \frac{11-y}{2} - \frac{y^2}{2(11-y)} + y \ln \frac{11-y}{y}, \qquad 4\tfrac{19}{20} \le y \le 5\tfrac{1}{2},$$

$$= 0. \qquad\qquad\qquad\qquad \text{elsewhere.}$$

An alternate method of working this problem is outlined in Problem 6-12. Unfortunately this method is not much simpler.

Since this simple problem becomes so complicated, there is not much hope for nonlinear functions of more than two variables. For this reason approximations and monte carlo techniques are usually substituted. These are discussed in the following sections.

6-7 APPROXIMATING THE DISTRIBUTION OF $Y = g(X_1, X_2, \ldots, X_n)$

A practical approximation based on a first-order Taylor series is discussed. Consider

$$Y = g(X_1, X_2, \ldots, X_n).$$

If Y is represented by its first-order Taylor series expansion about the point $\mu_1, \mu_2, \ldots, \mu_n$

$$Y \simeq g(\mu_1, \mu_2, \ldots, \mu_n) + \sum_{i=1}^{n}\left[\frac{\partial g}{\partial X_i}(\mu_1, \mu_2, \ldots, \mu_n)\right][X_i - \mu_i]. \quad (6\text{-}15)$$

Equation 6-15 is exactly of the form of $Y = a_0 + \sum_{i=1}^{n} a_i(X_i - \mu_i)$ where

$$a_0 = g(\mu_1, \ldots, \mu_n)$$

and

$$a_i = \frac{\partial g}{\partial X_i}(\mu_1, \ldots, \mu_n).$$

153

Thus

$$\mu_Y = E[Y] \simeq a_0 = g(\mu_1, \ldots, \mu_n). \qquad (6\text{-}16)$$

And if the X_i's are independent

$$\sigma_Y^2 \simeq \sum_{i=1}^{n} a_i^2 \sigma_i^2,$$

$$\sigma_Y^2 \simeq \sum_{i=1}^{n} \left[\frac{\partial g}{\partial X_i} (\mu_1, \ldots, \mu_n) \right]^2 \sigma_i^2. \qquad (6\text{-}17)$$

Equations 6-16 and 6-17 are good approximations as long as the stated assumption of independence is valid, and the first-order Taylor series is a good approximation. The Taylor series will be a good approximation if g is not too far from linear within the region that is within one standard deviation of the mean.

Only the mean and variance of Y are given by (6-16) and (6-17). These moments do not in general describe the distribution of Y. However if Y is assumed to be normally distributed, the mean and variance completely describe the distribution. The assumption of the output being normal is often reasonable in practice, as illustrated by the previous discussion of the central limit theorem.

EXAMPLE 6-14

Consider a four-stage amplifier. For simplicity assume that the gain (G) of one stage is represented by the equation

$$G = g_m R_L$$

where g_m is the tube transconductance and R_L is load resistance. Then the equation for gain (T) of the four-stage amplifier is

$$T = \prod_{i=1}^{4} G_i = \prod_{i=1}^{4} g_{m_i} R_{L_i}.$$

Now suppose that the tubes are such that g_m has a normal distribution with mean value of 900 μ mhos and a standard deviation of 100 μ mhos. Assume the load resistors have a mean of 20 K ohms and a standard deviation of 1 K ohms.

Let us find the distribution of the total db gain D.

$$D = 20 \log_{10} T$$

$$= 20 \log_{10} (g_{m_1} R_{L_1} g_{m_2} R_{L_2} g_{m_3} R_{L_3} g_{m_4} R_{L_4})$$

$$= 20 \left(\sum_{i=1}^{4} \log_{10} g_{m_i} + \sum_{i=1}^{4} \log_{10} R_{L_i} \right).$$

154

From (6-16)

$$\mu_D \simeq 20 \left[\sum_{i=1}^{4} \log_{10}(9 \times 10^{-4}) + \sum_{i=1}^{4} \log_{10}(2 \times 10^4) \right],$$

$$\simeq 80(-4 + .954 + 4 + .301),$$

$$\simeq 100 \; db.$$

From (6-17)

$$\sigma_D{}^2 \simeq \sum_{i=1}^{4} \left[\frac{\partial D}{\partial g_{m_i}} (\mu_{g_{m_1}}, \ldots, \mu_{R_{L_4}}) \right]^2 \sigma^2_{g_{m_i}}$$

$$+ \sum_{i=1}^{4} \left[\frac{\partial D}{\partial R_{L_i}} (\mu_{g_{m_1}}, \ldots, \mu_{R_{L_4}}) \right]^2 \sigma^2_{R_{L_i}}$$

$$= \sum_{i=1}^{4} \left[20(\log_{10} e) \frac{1}{\mu_{g_{m_i}}} \right]^2 \sigma^2_{g_{m_i}} + \sum_{i=1}^{4} \left[20(\log_{10} e) \frac{1}{\mu_{R_{L_i}}} \right]^2 \sigma^2_{R_{L_i}}$$

$$= 4(400)(\log_{10} e)^2 \frac{10^{-8}}{81 \times 10^{-8}} + (4)(400)(\log_{10} e)^2 \frac{10^6}{4 \times 10^8} \; .$$

Thus

$$\sigma_D \simeq 40(\log_{10} e)\sqrt{(\tfrac{1}{9})^2 + (\tfrac{1}{20})^2} \simeq 2.12 \; db.$$

The total *db* gain will be nearly a normal distribution, because the sum of eight independent random variables approaches a normal distribution. Thus using a table of the normal distribution function, 99.7% of the amplifiers will have a gain within $100 \pm 3\sigma = 100 \pm 6.36 \; db$ or within 93.64 to 106.36 *db*.

Dependent Random Variables

Independent random variables have been stressed. If the random variables are not independent, and the $E[X_i X_j]$ are known, then a similar equation can be developed as follows. As before, let

$$a_0 = g(\mu_1, \ldots, \mu_n),$$

$$a_i = \frac{\partial g}{\partial X_i} (\mu_1, \ldots, \mu_n),$$

$$g(X_1, \ldots, X_n) \simeq a_0 + \sum_{i=1}^{n} a_i(X_i - \mu_i).$$

Then, duplicating (6-16)

$$E[Y] = E[g(X_1, \ldots, X_n)] \simeq a_0.$$

155

Thus independence is not required for the first result. Now

$$E[(Y - \mu_Y)^2] \simeq E\left\{ \left[\sum_{i=1}^{n} a_i(X_i - \mu_i) \right]^2 \right\}$$

$$\simeq \sum_{j=1}^{n} \sum_{i=1}^{n} a_i a_j E[(X_i - \mu_i)(X_j - \mu_j)].$$

$$\sigma_Y^2 \simeq \sum_{i=1}^{n} a_i^2 \sigma_{X_i}^2 + \sum_{\substack{j=1 \\ i \neq j}}^{n} \sum_{i=1}^{n} a_i a_j \sigma_{ij} \qquad (6\text{-}18)$$

where
$$\sigma_{ij} = E[(X_i - \mu_i)(X_j - \mu_j)] = E[X_i X_j] - \mu_i \mu_j.$$

6-8 SYNTHETIC SAMPLING (MONTE CARLO TECHNIQUE)

The methods discussed so far to find the distribution of

$$Y = g(X_1, X_2, \ldots, X_n)$$

have been approximations or too involved to be practical for large problems. In this section a very simple and intuitively satisfying method is presented. Its only drawbacks are that it requires a digital computer and general parametric results are not obtained, thus limiting the applicability to synthesis.

It is assumed that $Y = g(X_1, \ldots, X_n)$ is known and that the joint density $f_{X_1, X_2, \ldots, X_n}$ is known. Now if a sample value of each random variable were known (say $X_1 = x_{11}$, $X_2 = x_{12}, \ldots, X_n = x_{1n}$), then a sample value of Y could be computed [say $y_1 = g(x_{11}, x_{12}, \ldots, x_{1n})$]. Then if another set of sample values were chosen for the random variables (say $X_1 = x_{21}, \ldots, X_n = x_{2n}$), then $y_2 = g(x_{21}, x_{22}, \ldots, x_{2n})$ could be computed.

If one had the time one could compute many such sample values of Y. The computer actually supplies the speed that makes many such calculations possible. There is just one problem. How does the computer select the different values of $X_1, X_2, \ldots, X_n$?

If each of the random variables had a uniform distribution between 0 and 1, numbers for each random variable could be chosen from a table of random numbers. Actually, computer routines generate pseudorandom numbers which may be used.

Consider the following case. Let the random variables $X_1, X_2, \ldots, X_{20}$ be independent and each uniformly distributed between zero and one, and let $Y = g(X_1, X_2, \ldots, X_{20})$ be a known function. Then the computer program to compute an approximation to the distribution of Y consists of the following basic steps.

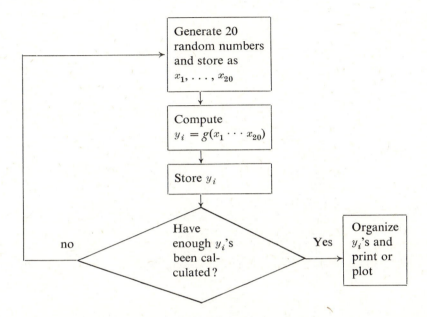

If a plot is desired it may be convenient to plot a usual bar chart or histogram. This method simply calls for breaking the range of Y into say 30 mutually exclusive cells of the same size and plotting vertically the number of samples that fell into that cell. (See Chapter VIII for more details of histograms.)

The case of uniformly distributed variables was considered. Now let X_i have a distribution function F_{X_i}. To obtain a random sample of X_i the following procedure may be used. Select a random sample of U which is uniformly distributed between 0 and 1. Call this random sample u_1. Then $F_X^{-1}(u_1)$ is the random sample of X (see Example 6-7).

For example, suppose that X is uniformly distributed between 10 and 20. Then

$$F_{X_i}(x) = 0, \qquad\qquad x < 10,$$
$$= (x - 10)/10, \qquad 10 \leq x < 20,$$
$$= 1, \qquad\qquad x \geq 20.$$

This is shown in Figure 6-29.

Notice $F_X^{-1}(u) = 10u + 10$. Thus if the value .250 were the random sample of U, then the corresponding random sample of X would be 12.5.

157

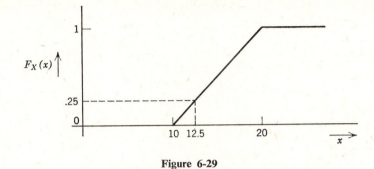

Figure 6-29

As another example suppose that X is normally distributed with mean 0 and variance 1. Then a random value of .250 for U would correspond to $-.67$ for a value of X. This result follows from a table of the normal distribution function. Practically, most computers automatically generate standard normal random variables. If this is the case then random sample of a normal random variable Y with mean μ and a variance σ^2 may be generated from a standard normal random variable X by recalling that

$$X = \frac{Y - \mu}{\sigma}$$

or
$$Y = X\sigma + \mu. \tag{6-19}$$

Equation 6-19 can be easily checked by finding the characteristic function of Y.

The only difference when the random variables are dependent is that the dependence must be taken into account when the random samples are generated. We assume that the dependence is expressed in terms of conditional distributions. If this is not the case the joint distributions can always be reduced to the required conditional distributions.

For illustration consider three dependent random variables: X_1, X_2, and X_3. We first generate a random sample of X_1 by the same methods discussed above. Call this sample x_{11}. We next generate a random sample of X_2 using $F_{X_2 | X_1 = x_{11}}$, by the same method used before. Call this sample x_{12}. Then we use $F_{X_3 | X_1 = x_{11}, X_2 = x_{12}}$ to generate x_{13}. Thus nothing changes except that the conditional distribution functions are used in generating the random samples.

With a fast digital computer thousands of simulations can be run in reasonable times. Monte Carlo solutions often involve 10,000 or more simulations. An example is given in the next chapter.

158

6-9 SUMMARY

The purpose of this chapter was to consider the important engineering problem of finding the distribution of $Y = g(X_1, X_2, \ldots, X_n)$ where the distribution of the X_i's is known.

First the problem of $Y = g(X)$ was considered and this problem was solved by (6-2). Examples were given to illustrate its application.

Next $Y = \sum_{i=1}^{n} X_i$ was considered, and it was shown that in the case of independent random variables the solution involved convolution of the density functions or multiplication of the characteristic functions. The case where Y is a linear combination of the X_i's was shown to be only a slight generalization of this problem.

The general problem was then considered and although a general method of solution was outlined, the difficulty of solution was illustrated by an example, and approximations were suggested.

Two approximations for the general solution were described. First a Taylor series approximation, moments, and the central limit theorem were used. Then a Monte Carlo method was suggested.

References B1, D3, D4, and P2 provide additional reading.

6-10 PROBLEMS

1. X has a normal density function with mean 1 and variance 2.

$$Y = \tfrac{1}{2} X - 1.$$

Find the density of Y.

2. The power P dissipated in a resistor is $P = I^2 R$. Assume $R = 2$ and I is a random variable with a normal density.

$$f_I(i) = \frac{1}{\sqrt{2\pi}} e \frac{-i^2}{2}.$$

Find the density function of P.

3. The output of a full wave rectifier is $Y = |X|$. Find the density function of Y when X has a uniform density from -1 to $+1$.

4. The output of a square law detector is

$$Y = aX^2, \qquad a > 0.$$

Find the density function of Y in terms of the density function f_X of X.

159

5. Find the density function of $Y = X_1 + X_2$, where X_1 and X_2 are independent.

(a) if $f_{X_1}(x) = \dfrac{1}{\sqrt{2\pi}} e^{\dfrac{-(x-2)^2}{2}}$

$$f_{X_2}(x) = \dfrac{1}{\sqrt{2\pi 2}} e^{\dfrac{-(x+3)^2}{8}}$$

(b) if $f_{X_1}(x) = 1, \qquad 0 \le x \le 1,$
$\qquad\qquad = 0, \qquad \text{otherwise.}$
$\qquad f_{X_2}(x) = 2x, \qquad 0 \le x \le 1,$
$\qquad\qquad = 0, \qquad \text{otherwise.}$

6. Consider the circuit shown in Figure P6-1. Let the switch be closed for 3 sec. Using convolution, find the voltage across the 1 ohm resistor. Hint:

$$h(t) = L^{-1}\left\{\frac{1}{1+s}\right\} = e^{-t}, \qquad t \ge 0,$$
$$= 0, \qquad\qquad t < 0.$$
$$v(t) = 4, \qquad 0 \le t \le 3,$$
$$= 0, \qquad\qquad \text{otherwise.}$$
$$V_r = h * v.$$

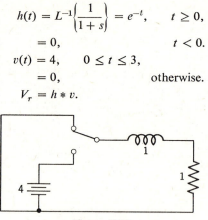

Figure P6-1

7. Consider the circuit shown in Figure P6-2. Assume $R_1 = R_2 = 10$, and V is a random variable with density function

$$f_V(v) = \frac{1}{\sqrt{2\pi}} e^{-v^2/2}.$$

Find the density function of I.

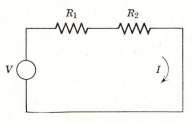

Figure P6-2

8. Find the density function of the series combination of R_1 and R_2 if R_1 and R_2 are independent random variables with

$$f_{R_1}(r) = 1, \quad 9 \leq r \leq 10,$$
$$= 0, \quad \text{otherwise.}$$
$$f_{R_2}(r) = \tfrac{1}{2}, \quad 6 \leq r \leq 8,$$
$$= 0, \quad \text{otherwise.}$$

9. In the circuit of Figure P6-2, find the approximate mean and variance of I if

$$\mu_{R_1} = 100, \quad \mu_{R_2} = 100, \quad \mu_V = 10,$$
$$\sigma_{R_1}^2 = 10, \quad \sigma_{R_2}^2 = 10, \quad \sigma_V^2 = 9.$$

10. Ten resistors are in series. Each resistance is an independent random variable and equally likely to be 1 K $\pm$ 10%. Find the approximate probability that the series combination will be 10 K $\pm$ 5%.

11. To evaluate the approximation $\mu_Y \simeq g(\mu_1)$ consider $Y = X^2$. Let

$$f_X(x) = \frac{1}{\sqrt{2\pi}\sigma} e^{-(x-10)^2/2\sigma^2}.$$

(a) Find

$$E[Y] = \int_{-\infty}^{\infty} x^2 f_X(x)\, dx.$$

(b) Approximate $E[Y]$ by the approximation suggested in the text.
(c) Expand Y in a complete Taylor series and compute $E[Y]$.
(d) Evaluate the approximation in terms of the ratio σ/μ. This ratio is often called the coefficient of variation and is often much less than one in engineering problems.

12. Consider two independent resistors R_1 and R_2 in parallel. Each resistor is equally likely to be between 9 and 11 ohms (10 Ω $\pm$ 10%).

(a) Find the density function of conductance $C_i = 1/R_i$, $i = 1, 2$.
(b) Find the total conductance C, where $C = C_1 + C_2$.
(c) Find the resistance R of the parallel combination and compare with Example 6.13 in the text.

13. Find the approximate density function of the resistance of the parallel circuit of Problem 12 by using moments and assuming the output is normal. Sketch the approximate and exact densities to compare the approximation.

14. The expected value of $Y = g(X)$ is by definition

$$E[Y] = \int_{-\infty}^{\infty} g(x) f_X(x)\, dx.$$

Also

$$E[Y] = \int_{-\infty}^{\infty} y f_Y(y)\, dy.$$

161

Assuming $x = g^{-1}(y)$ is uniquely defined for all y, using the substitution $y = g(x)$ in the first integral, show

$$f_Y(y) = f_X[g^{-1}(y)] \left| \frac{dg^{-1}(y)}{dy} \right| .$$

15. Let X be a standard normal random variable. Let $Y = X\sigma + \mu$. Using characteristic functions, show Y is normal with mean μ and variance σ^2.

....

CHAPTER VII

APPLICATIONS OF RANDOM VARIABLES TO SYSTEM PROBLEMS

The purpose of this chapter is to illustrate further the application of the theory presented in the last three chapters. The applications are chosen to illustrate and thus reinforce the theory, and at the same time to introduce some practical engineering studies.

7-1 INTRODUCTION TO SENSITIVITY STUDIES

In almost every system study there is some obligation on the part of the person doing the study to evaluate how sensitive the answer is to small changes in the parameters. The obvious reason is that most parameters are not known exactly; thus if the answer changes significantly when the parameters change within their predicted limits, then little confidence can be placed in the answer. On the other hand, if the answer changes very little while the parameters vary within their limits, then the answer seems dependable. In this section deterministic measures of sensitivity are reviewed and compared with the approximate variance as computed by the method of the previous chapter.

In all cases we assume that the output Y is some function of n parameters $X_1, \ldots, X_n$. In the case of a deterministic model, $X_1, \ldots, X_n$ will be called parameters. In the case of a probabilistic model, $X_1, \ldots, X_n$ will be called random variables. We now consider the deterministic model.

Deterministic Measure

The basic idea of a sensitivity (of the output to a change in parameter X_i) measure is

$$\frac{\Delta Y}{\Delta X_i}\bigg|_{X_j=\text{nominal value}, \, j \neq i \,\, j=1,\ldots,n}$$

Assume Y is a linear function of X_i, $i = 1, 2, \ldots, n$, that is

$$Y = \sum_{i=1}^{n} W_i X_i.$$

Now let X_{ij} be the ith sample value of X_j; then the ith sample value Y_i is

$$Y_i = \sum_{j=1}^{n} W_j X_{ij}.$$

Thus

$$\Delta Y = Y_1 - Y_2 = \sum_{j=1}^{n} W_j X_{1j} - \sum_{j=1}^{n} W_j X_{2j}$$

$$= \sum_{j=1}^{n} W_j \Delta X_j.$$

In this linear case, $\Delta Y/\Delta X_j = W_j$ if all other parameters are constant. Thus if Y is a linear function of the parameters, then the change in output is simply the sum of the change in each X_j multiplied by W_j, the sensitivity coefficient. Moreover, in this case $\Delta Y/\Delta X_j = \partial Y/\partial X_j$ and the partial derivative can be used as a sensitivity measure. Thus

$$\Delta Y = \sum_{j=1}^{n} \frac{\partial Y}{\partial X_j} \Delta X_j. \tag{7-1}$$

With the assumption that within small changes of each parameter a linear function is a good approximation to the true function, (7-1) is often used as an approximation for nonlinear equations.

The measure of sensitivity expressed by the partial derivative is the basis of sensitivity measures. However it is often modified to account for two factors. First it is argued that a change of say 1 V in Y is not too important if Y has a nominal value of 1000 V, while the same change is much more important if the nominal value of Y is 2 V. For this reason a sensitivity measure S_i^Y is often defined as

$$S_i^Y = \frac{\partial Y}{\partial X_i} \frac{1}{Y_n}, \tag{7-2}$$

where Y_n is the nominal value of Y.

164

One additional modification is often made. Since we often consider parameters whose tolerance ΔX_i is proportional to the nominal value of X_i (e.g., 10% resistors) to take into account a factor proportional to ΔX, we have a final definition of sensitivity

$$S_{X_i}^Y = \frac{\partial Y}{\partial X_i} \frac{1}{Y_n} X_i. \tag{7-3}$$

With this final definition and assuming a linear function and $C \Delta X_i = X_i$, the sum of the sensitivities as defined by (7-3) results in a number proportional to the percent change in Y. Indeed

$$\sum_{i=1}^{n} S_{X_i}^Y = \frac{C}{Y_n} \sum_{i=1}^{n} \frac{\partial Y}{\partial X_i} \Delta X_i = C \frac{\Delta Y}{Y_n}. \tag{7-4}$$

Comments on Deterministic Measures of Sensitivity

Although a deterministic approach to sensitivity studies has much to recommend it, there is an important drawback. We find by probability theory and by experience that the probability of everything being at its worst case simultaneously is small, and the more independent parameters within a system, the smaller the probability. Thus a deterministic approach is very conservative. Although conservative designs have some advantages, unrealistic evaluations do not. That is, if one is forced to meet worst case conditions for all parameters within a big system, this requires tighter tolerances (and thus more expensive components), and sometimes requires a design change (e.g., a feedback loop or an extra amplifier) which is unnecessary.

This is not meant to discount worst case sensitivity studies. They are useful. However we do suggest that the alternative measure of expected mean square error has the advantage of considering the probability of being at various positions including the worst case.

Expected Squared Error

The measure of variation of output because of variation of parameters suggested here is the expected deviation from the mean or the expected squared error.

The expected squared error (or mean square error) treating the parameters as random variables, is [see (6-18)]

$$\sigma_Y^2 \simeq \sum_{i=1}^{n} \left[\frac{\partial Y}{\partial X_i} (\mu_1, \ldots, \mu_n) \right]^2 \sigma_{X_i}^2$$

$$+ \sum_{i=1}^{n} \sum_{\substack{j=1 \\ i \neq j}}^{n} \frac{\partial Y}{\partial X_i} (\mu_1, \ldots, \mu_n) \frac{\partial Y}{\partial X_j} (\mu_1, \ldots, \mu_n) \rho_{X_i X_j} \sigma_{X_i} \sigma_{X_j}, \tag{7-5}$$

165

where
$$\mu_i = E[X_i],$$

$$\sigma_{X_i}^2 = E[(X_i - \mu_i)^2],$$

$$\rho_{X_i X_j} = \frac{E[(X_i - \mu_i)(X_j - \mu_j)]}{\sigma_{X_i}\sigma_{X_j}}.$$

If the random variables $X_1, \ldots, X_n$ are uncorrelated ($\rho_{X_i X_j} = 0$), then the double sum in (7-5) is zero. In this case (7-5) is similar to (7-4). That is, the mean square error is a sum (in the square sense) of terms close to sensitivities.

EXAMPLE 7-1. Series and Parallel Resistances

In this example the sensitivity of resistance in series and parallel circuits is compared. It is assumed that all resistors are $\pm b$ percent resistors and that the distributions are such that the mean is equal to the nominal value, with the standard deviation K percent of the mean. The standard deviations of resistance of three circuits which have the same nominal resistance will be used as a measure of sensitivity. See Figure 7.1. For these circuits to have the same mean resistance, the mean values of the individual resistances are related as follows:

$$\mu_{R_1} = 2\mu_{R_2} = 2\mu_{R_3} = \frac{\mu_{R_4}}{2} = \frac{\mu_{R_5}}{2}.$$

The standard deviation of circuit A is simply

$$\sigma_A = K\mu_{R_1}.$$

$R_B = R_2 + R_3$. Thus, using (6-17) and assuming independence,

$$\frac{\partial R_B}{\partial R_2} = \frac{\partial R_B}{\partial R_3} = 1.$$

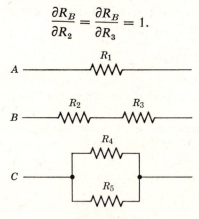

Figure 7-1

Thus

$$\sigma_B = \sqrt{\sigma_B{}^2} = \sqrt{\sigma_{R_2}{}^2 + \sigma_{R_3}{}^2} = \sqrt{(K\mu_{R_2})^2 + (K\mu_{R_3})^2}$$

$$= K\left[\left(\frac{\mu_{R_1}}{2}\right)^2 + \left(\frac{\mu_{R_1}}{2}\right)^2\right]^{1/2} = K\frac{\mu_{R_1}}{\sqrt{2}}$$

$$= \frac{\sigma_A}{\sqrt{2}}.$$

For circuit C

$$R_C = \frac{R_4 R_5}{R_4 + R_5}.$$

$$\frac{\partial R_C}{\partial R_4} = \frac{R_4 R_5 + R_5{}^2 - R_4 R_5}{(R_4 + R_5)^2} = \left(\frac{R_5}{R_4 + R_5}\right)^2.$$

$$\frac{\partial R_C}{\partial R_5} = \left(\frac{R_4}{R_4 + R_5}\right)^2.$$

$$\sigma_C \simeq \left[\left(\frac{\mu_{R_5}}{\mu_{R_4} + \mu_{R_5}}\right)^4 \sigma_{R_4}{}^2 + \left(\frac{\mu_{R_4}}{\mu_{R_4} + \mu_{R_5}}\right)^4 \sigma_{R_5}{}^2\right]^{1/2}$$

$$\simeq \left[\left(\frac{2\mu_{R_1}}{4\mu_{R_1}}\right)^4 (K\mu_{R_4})^2 + \left(\frac{2\mu_{R_1}}{4\mu_{R_1}}\right)^4 (K\mu_{R_4})^2\right]^{1/2}.$$

$$\sigma_C \simeq K[(\tfrac{1}{2})^4 (2\mu_{R_1})^2 + (\tfrac{1}{2})^4 (2\mu_{R_1})^2]^{1/2}$$

$$\simeq K\mu_{R_1}[(\tfrac{1}{2})^2 + (\tfrac{1}{2})^2]^{1/2}$$

$$\simeq K\frac{\mu_{R_1}}{\sqrt{2}} = \frac{\sigma_A}{\sqrt{2}}.$$

Thus, by combining two resistors in either series or parallel, the standard deviation is smaller by a factor of $\sqrt{2}$ than the standard deviation of a single resistor.

Other sensitivity calculations are given in the problems. We now mention a practical method for determining the partial derivatives needed for a measure of sensitivity.

Experimental Determination of Partial Derivatives

In many practical problems an analytical model either is unknown or the model is only a gross approximation of the actual hardware, that is, $Y = g(X_1, X_2, \ldots, X_n)$ is not known. In these instances experimental methods may be used to find the partial derivatives.

167

The mean output can be measured with all variables at their mean values. The partial derivatives may be approximated by holding all but the Kth variable at its mean value and then varying the selected variable X_K about one standard deviation and measuring the variation of the output. That is

$$\frac{\partial Y}{\partial X_K} \simeq \frac{\Delta Y}{\Delta X_K}.$$

Summary

In summary the partial derivative is the basic idea in sensitivity studies. These partial derivatives, which may be found experimentally, are often multiplied by a measure of the variation in X and, in the deterministic case, divided by the nominal value of Y. The mean square error approximation provides a method of combining sensitivity measures with either independent or nonindependent random variables. The effect of series and parallel resistances is illustrated in an example.

One of the primary uses of measures of sensitivity is in setting tolerances. This is discussed in the following section.

7-2 INTRODUCTION TO TOLERANCE STUDIES

Experience shows that parameters of a physical device vary. There are two distinct types of variation. The first is variation that takes place in one device, and the second is the variation that takes place within a population of devices manufactured from the same specifications. Here we discuss the second type of variation. Perhaps the most elementary example of this variation is parts produced to conform to mechanical specifications. All specifications of this type include an allowable variation or tolerance.

In this section we discuss how tolerances are related to distributions, and then apply the results of the previous chapters to analyze the effect of tolerances on design. Such analysis must be a factor in the design of any equipment that is going to be mass produced; however in practice, often the analysis of the effects of tolerances is only intuitive.

Deterministic Tolerance Analysis

The idea is to determine the tolerance on the output Y as a function of the tolerances on the parameters $X_1, X_2, \ldots, X_n$. The concept is quite simple. We find the value within the tolerance range of each X_i that will make Y the lowest. (Usually it is either the high or the low value in the range of X_i.)

168

The resulting values of $X_1, \ldots, X_n$ result in an output Y_L. A similar procedure results in the highest value of output Y_H. The two values Y_L and Y_H describe the output tolerance caused by the tolerance in the parameters $X_1, \ldots, X_n$.

Such an approach can be carried out analytically if the function relating the output to the parameters is known explicitly. Such studies are almost always performed when setting mechanical tolerances to ensure that all parts will go together.

If the output function is not known analytically, then such studies often can be done experimentally by determining the values of parameters that lead to high (low) output, putting these values into the system and measuring the output.

Probabilistic Tolerance Studies

The basic idea of probabilistic tolerance studies is to find the distribution of the output from the distribution of the "parameters" and compute the probability of being outside certain limits. Since an exact solution is usually not possible, the two approximations introduced in the last chapter are used.

The approximation based on a series expansion was discussed in the last section on sensitivity studies. To apply sensitivity studies to tolerances we must infer standard deviations from specifications. Then the variance of the output must be related to tolerance of the output.

Relation Between Tolerance and Standard Deviation

When specifications are written they almost always include an upper and a lower limit on important parameters. In addition they usually include some desired (also called target or bogey) value. These two or three values do not uniquely define a distribution. In fact, they do not even define the first two moments. Our task is to assume some reasonable distributions and with these assumptions, determine the standard deviation in terms of the tolerances.

First, assuming that the parameter of interest is uniformly distributed between the lower limit L and the upper limit U, we get the result (see Problem 7-5)

$$\sigma = \frac{U - L}{\sqrt{12}} \simeq \frac{U - L}{3.5}. \tag{7-6}$$

Other common assumptions involve a truncated normal distribution. The relation between the tolerance limits and the standard deviation in this case depends on the limits at which the screening takes place relative to the spread of the distribution. For instance, consider that 100 ohm resistors as produced

169

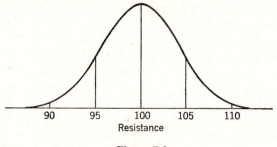

Resistance

Figure 7-2

have a density function as shown in Figure 7-2. If $\pm 10\%$ resistors are purchased from this population, then approximately

$$3\sigma = \frac{U - L}{2}$$

or
$$\sigma = \frac{U - L}{6}.$$
$$(7\text{-}7)$$

If $\pm 1\%$ resistors were purchased from the population shown in Figure 7-2 then the screened population would be approximately uniformly distributed, so (7-6) would be applicable.

In the two cases discussed the standard deviation can vary between 1/3.5 and 1/6 times the difference in the tolerance limits. These are not absolute limits but do cover usual cases. One unusual case can be described by the case where 5% resistors are screened from a lot of resistors and the remainder of the population is sold as 10% resistors. The resulting density function is shown in Figure 7-3. In this case the standard deviation is larger than 1/3.5 times the difference in tolerance limits. Obviously the standard deviation can also be much smaller than 1/6 of the tolerance if the acceptance limits are set well beyond the spread of the distribution, but this does not usually occur in practice.

As a conclusion, samples of the product can determine the standard deviation, but for design planning the rather conservative relation

$$\sigma = \frac{U - L}{3.5}$$

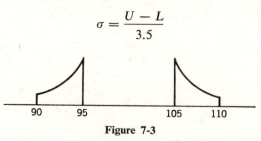

Figure 7-3

is usually a good assumption. When the standard deviation of the parts are known, then the mean and variance of the output can be computed using the approximation developed in the last chapter.

Again something must be assumed to describe the output distribution and the probability of being out of tolerance. It is suggested that if enough variables are involved and the function is approximately a linear combination, then a normal density may be assumed. Then the probability of being outside tolerance limits can be computed.

We now show an example of a tolerance problem which is solved using a Monte Carlo approach.

EXAMPLE 7-2

This example, a simplification of a problem that actually occurred in practice, was worked by the method shown, and the results actually obtained in manufacture corresponded with the theoretical results.

The simplified version is shown in Figure 7-4. The bar will fit within the bracket if $Y_a < Y_b$ and there will be interference (or no fit) if $Y_b < Y_a$. In the actual case Y_a and Y_b involved 41 dimensions and the configuration was more complicated than simply a sum of lengths. Actually the equations for Y_a and Y_b involved arcs and angles, thus various trigonometric functions were involved.

The problem was solved by finding the probability distribution of $Z = Y_b - Y_a$ by Monte Carlo sampling. Note that if $Z > 0$ there is no problem, while if $Z < 0$ there will be no fit and the parts cannot be assembled.

The problem arose because just as production and assembly were about to start it was discovered that interference was possible. (As customary, the intent was to tighten the tolerances on each part until no interference is possible at the worst case, but the designer made a mistake in his worst case calculations.) Then the question was, what is the probability of interference? If it is low enough then it would be better to have a few that would not fit, rather than wait and spend the extra money to redesign some of the parts.

To find the probability density of Z via a Monte Carlo technique, one must have $Z = g(X_1, \ldots, X_{41})$ and must know the joint distribution of

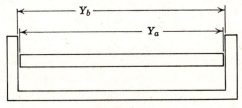

Figure 7-4

171

$X_1, \ldots, X_{41}$. The function g was found from the drawings using trigonometry. A part of the equation is shown in Figure 7-5 simply to illustrate the

$$
Y = (R + W)\sin\left\{\theta - \left[\sin^{-1}\left(\frac{H_{21}}{R}\right) + 2\sin^{-1}\right.\right.
$$

$$
\times \left(\frac{\sqrt{(\sqrt{r^2 - H_{11}^2} - \sqrt{r^2 - (S - V)^2})^2 + [(S - V) - H_{11}]^2} - \frac{H_{22} - H_{12}}{2}}{2(R + W)}\right)\right]\right\}
$$

$$
+ \sqrt{(S - V)^2 + (Z + r - \sqrt{r^2 - (S - V)^2})^2}
$$

$$
\cdot \sin^{-1}\left\{\tan^{-1}\left(\frac{Z + r - \sqrt{r^2 - (S - V)^2}}{(S - V)}\right) + \tan^{-1}\left(\frac{\sqrt{r^2 - (S - V)^2} - \sqrt{r^2 - V^2}}{S}\right)\right.
$$

$$
- \left[180° - \left[\left[\left(180° - 2\sin^{-1}\frac{\sqrt{S^2 + (\sqrt{r^2 - (S - V)^2} - \sqrt{r^2 - V^2})^2}}{2(R + W)}\right)\right]\right.\right.
$$

$$
+ \left[90° - \left[\theta - \left(\sin^{-1}\left(\frac{H_{21}}{R + W}\right) + 2\sin^{-1}\right.\right.\right.
$$

$$
\times \left(\frac{\sqrt{(\sqrt{r^2 - H_{11}^2} - \sqrt{r^2 - (S - V)^2})^2 + [(S - V) - H_{11}]^2} - \frac{H_{22} - H_{12}}{2}}{2(R + W)}\right)\right)\right]\right]\right]\right\}
$$

Figure 7-5

form. The various dimensions were assumed to be independent and equally likely between their upper and lower tolerance limits. The result of 8000 simulations is shown in Figure 7-6.

Note that the results appear nearly normal and that interference occurred 71 times in 8000 simulations. Based on these results it was decided to produce units without a design change and to rebuild those few on which interference did occur. The results of the actual assembly operation corresponded very well with the prediction that 71/8000 would not fit.

Summary

Tolerances of parameters can be combined by Monte Carlo methods or by Taylor series approximation. Both produce the probability of being outside certain limits and are less conservative than deterministic tolerance studies. Note that the analysis described above is the basis for deciding which tolerances to assign to what parts.

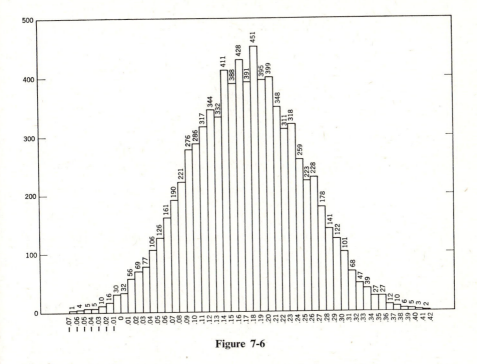

Figure 7-6

7-3 CLOSURE TIME OF SWITCHING CIRCUITS

Consider that the distribution of closure times of switches (e.g., relays, fluid logic, solid state devices) is known. We wish to know the closure time of networks of these switches.

We first model the circuit with a mathematical function and then solve the basic mathematical problems in more general terms (so we can use it later).

Consider two switches in series as shown in Figure 7-7. What is the closure time T of the series circuit in terms of T_1, closure time of the first switch and T_2, closure time of the second switch?

There will be a closed circuit from a to b at the time when both switches are closed or when the latter of the two switches closes. That is

$$T = \max(T_1, T_2).$$

a ———○　○——— b
$T = \max(T_1, T_2)$

Figure 7-7

173

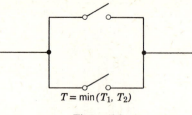

$$T = \min(T_1, T_2)$$

Figure 7-8

Similarly if the two switches are placed in parallel (see Figure 7-8), then

$$T = \min (T_1, T_2).$$

Thus we investigate $Z = \max (X, Y)$ and $W = \min (X, Y)$. Following this, an example demonstrates how series parallel networks are handled, and problems at the end of this chapter explore other extensions.

If $Z = \max (X, Y)$, then

$$P(Z \le z_0) = P\{(X \le z_0) \cap (Y \le z_0)\}$$
$$F_Z(z_0) = F_X(z_0)F_{Y|X \le z_0}(z_0) = F_Y(z_0)F_{X|Y \le z_0}(z_0).$$

If X and Y are independent

$$F_Z(z_0) = F_X(z_0)F_Y(z_0). \tag{7-8}$$

The density function of Z is

$$f_Z(z_0) = \frac{dF_Z(z_0)}{dz_0} = F_X(z_0)f_Y(z_0) + f_X(z_0)F_Y(z_0). \tag{7-9}$$

If $W = \min (X, Y)$, then

$$P(W \le z_0) = P\{(X \le z_0) \cup (Y \le z_0)\},$$
$$F_W(z_0) = F_X(z_0) + F_Y(z_0) - F_{X,Y}(z_0, z_0).$$

If X and Y are independent

$$F_W(z_0) = F_X(z_0) + F_Y(z_0) - F_X(z_0)F_Y(z_0). \tag{7-10}$$

The density function of W is

$$f_W(z_0) = \frac{dF_W(z_0)}{dz_0} = f_X(z_0) + f_Y(z_0) - F_X(z_0)f_Y(z_0) - f_X(z_0)F_Y(z_0)$$
$$= f_X(z_0)[1 - F_Y(z_0)] + f_Y(z_0)[1 - F_X(z_0)]. \tag{7-11}$$

174

EXAMPLE 7-3

Consider the circuit shown in Figure 7-9. Let the time at which the ith switch closes be denoted by X_i. Suppose X_1, X_2, X_3, X_4 are independent, identically distributed random variables each with distribution function F. As time increases from zero switches will close until there is an electrical path from A to C. Let

$U =$ time when circuit is first completed from A to B;
$V =$ time when circuit is first completed from B to C;
$W =$ time when circuit is first completed from A to C.

Find the following:

1. The distribution function of U.
2. The distribution function of W.
3. If $F(x) = x$, $0 \le x \le 1$ (i.e., uniform), what are the mean and variance of X_i, U, and W?

1. When either of two switches in parallel closes, a complete circuit results. Therefore the minimum time of closure of the two switches is the time of circuit closure.
$$U = \min (X_1, X_3), \qquad V = \min (X_2, X_4),$$
$$F_U(\lambda) = F_{X_1}(\lambda) + F_{X_3}(\lambda) - F_{X_1}(\lambda)F_{X_3}(\lambda) = 2F(\lambda) - [F(\lambda)]^2.$$

2. $W = \max (U, V)$,
$$F_W(\lambda) = F_U(\lambda)F_V(\lambda) = \{2F(\lambda) - [F(\lambda)]^2\}^2.$$

3. $E[X_i] = \displaystyle\int_0^1 x\,dx = \tfrac{1}{2}.$

$E[X_i^2] = \displaystyle\int_0^1 x^2\,dx = \tfrac{1}{3}.$

$\sigma_{X_i}^2 = \tfrac{1}{3} - (\tfrac{1}{2})^2 = \tfrac{1}{12}.$

$f_U(\lambda) = \dfrac{dF_U(\lambda)}{d\lambda} = \dfrac{d(2\lambda - \lambda^2)}{d\lambda} = 2 - 2\lambda, \qquad 0 \le \lambda \le 1.$

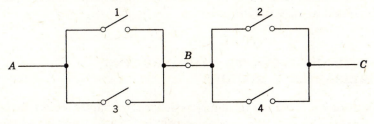

Figure 7-9

175

This may also be found from

$$f_U(\lambda) = f_{X_1}(\lambda) + f_{X_3}(\lambda) - F_{X_1}(\lambda)f_{X_3}(\lambda) - F_{X_3}(\lambda)f_{X_1}(\lambda) = 2 - 2\lambda,$$
$$0 \leq \lambda \leq 1.$$

$$E(U) = \int_0^1 \lambda(2 - 2\lambda) \, d\lambda = \lambda^2 - \tfrac{2}{3}\lambda^3\big|_0^1 = \tfrac{1}{3}.$$

$$E(U^2) = \int_0^1 \lambda^2(2 - 2\lambda) \, d\lambda = \frac{2\lambda^3}{3} - \frac{1}{2}\lambda^4\bigg|_0^1 = \frac{1}{6}.$$

$$\sigma_U{}^2 = \tfrac{1}{6} - (\tfrac{1}{3})^2 = \tfrac{1}{18}.$$

$$F_W(x) = (2x - x^2)^2 = x^4 - 4x^3 + 4x^2, \qquad 0 \leq x \leq 1.$$

$$f_W(x) = \frac{d(x^4 - 4x^3 + 4x^2)}{dx} = 4x^3 - 12x^2 + 8x, \qquad 0 \leq x \leq 1.$$

$$E[W] = \int_0^1 x(4x^3 - 12x^2 + 8x) \, dx = \frac{4}{5}x^5 - 3x^4 + \frac{8}{3}x^3\bigg|_0^1$$

$$= \frac{12 - 45 + 40}{15} = \frac{7}{15}.$$

$$E[W^2] = \int_0^1 x^2(4x^3 - 12x^2 + 8x) \, dx = \frac{2}{3}x^6 - \frac{12}{5}x^5 + 2x^4\bigg|_0^1$$

$$= \frac{10 - 36 + 30}{15} = \frac{4}{15}.$$

$$\sigma_W{}^2 = \frac{4}{15} - \left(\frac{7}{15}\right)^2 = \frac{60 - 49}{225} = \frac{11}{225}.$$

Note that the circuit mean time of closure 7/15 is slightly less than the initial mean time 1/2, while the variance is reduced from $1/12 \simeq .083$ to $11/225 \simeq .049$.

In general series and parallel combinations of switches change the mean value (the mean is decreased by paralleled switches and increased by series switches) and reduces the variance.

7-4 INTRODUCTION TO TIME VARYING RELIABILITY STUDIES

In this section we consider reliability as defined in Chapter III as a function of time. That is, there is a group of equipment in the field, and we are interested in the probability that a piece of equipment performs satisfactorily for a prescribed time t.

In reliability studies of this type, data are usually gathered by life testing N pieces of equipment. That is, N pieces of equipment are put into operation that simulates their actual use as closely as possible and the time of failure of each piece of equipment is recorded. A plot of such data might look like that shown in Figure 7-10.

Let $n(t)$ be the number of pieces of equipment that are still operating at time t. Then the number failing from t to $t + \Delta t$, $n(t) - n(t + \Delta t)$ is the quantity plotted in Figure 7-10. Thus the natural definition of failure rate $r(t)$ is the negative of the derivative of $n(t)$

$$r(t) = -\frac{dn(t)}{dt}. \tag{7-12}$$

It should be pointed out that the failure rate is not a good measure of the rate at which a given piece of equipment tends to fail because the number in the sample is continually decreasing. To measure the average rate at which a single piece of equipment is going to fail, the hazard rate $h(t)$ is defined by

$$h(t) = \frac{r(t)}{n(t)} = \frac{-dn(t)/dt}{n(t)}. \tag{7-13}$$

The hazard rate $h(t)$ is often called the force of mortality. One can also show (see Problem 7-10) that the conditional probability density function of T given $\{T \geq t\}$ is $h(t)$.

We now relate reliability R, time of failure T, and hazard rate. The results in this section are for $t \geq 0$. By definition

$$R(t) = \frac{n(t)}{N}. \tag{7-14}$$

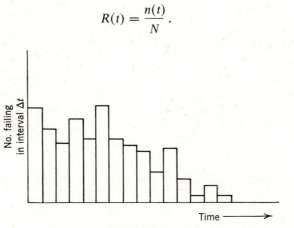

Figure 7-10

Also the distribution function F_T of time of failure is

$$F_T(t) = 1 - R(t) = \frac{N - n(t)}{N}. \tag{7-15}$$

Using the definitions

$$n(t) = N[1 - F_T(t)]. \tag{7-16}$$

Use of (7-16) in (7-13) results in

$$h(t) = \frac{N f_T(t)}{N[1 - F_T(t)]} = \frac{f_T(t)}{1 - F_T(t)}. \tag{7-17}$$

Equation 7-17 may be solved for $f_T(t)$ as follows:

$$\int_0^t h(x)\,dx = \int_0^t \frac{F'_T(x)\,dx}{1 - F_T(x)}$$

$$= -\ln\,[1 - F_T(x)]\big|_0^t$$

$$= -\ln\,[1 - F_T(t)]$$

because $F_T(0) = 0$ (none had failed at time 0). Thus

$$F_T(t) = 1 - \exp\left[-\int_0^t h(x)\,dx\right] \tag{7-18}$$

and

$$f_T(t) = h(t)\exp\left[-\int_0^t h(x)\,dx\right]. \tag{7-19}$$

Note that the reliability $R(t)$ is

$$R(t) = P(T > t) = 1 - F_T(t) = \exp\left[-\int_0^t h(x)\,dx\right]. \tag{7-20}$$

A typical plot of $h(t)$ is shown in Figure 7-11.

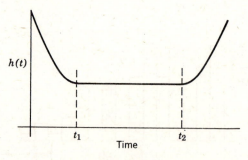

Figure 7-11

Such a curve is called (for obvious reasons) a bathtub curve. The failures occurring before t_1 are called burn-in or infantile failure and are attributed to equipment that is poorly made such that very little use will disclose this weakness or poor workmanship. The failures that occur after t_2 are called wear-out failures. The region between t_1 and t_2 is known as the chance failure region.

Note that if one does assume that $h(t) = \lambda$, a constant, such as is the case between t_1 and t_2 in Figure 7-11, then

$$F_T(t) = 1 - e^{-\lambda t}; \tag{7-21}$$

$$f_T(t) = \lambda e^{-\lambda t}; \tag{7-22}$$

$$R(t) = e^{-\lambda t}. \tag{7-23}$$

Note that $E[T] = \int_0^\infty t\lambda e^{-\lambda t} = 1/\lambda$. Thus $\tau = 1/\lambda$ is called the mean time to failure.

The exponential distribution of T is often assumed in practice. Sometimes it is "derived" from a "Poisson process" as follows.

Assume the number of failures N in a certain piece of equipment in a time interval t has a Poisson distribution, that is

$$P(N = K) = e^{-\lambda t}\frac{(\lambda t)^K}{K!} \quad \text{for} \quad K = 0, 1, \ldots$$

where the expected number of failures (i.e., the parameter of the Poisson distribution) is λt.

Thus $R(t) = P$ (there are no failures up to t) $= P(N = 0) = e^{-\lambda t}$. This is the result obtained above when λ was the constant hazard rate. In the Poisson model λ is interpreted as the average rate of occurrences of failures within a piece of equipment.

CAUTION. Throughout the rest of this introductory section the exponential distribution is used as an example. However we caution that assuming an exponential distribution does not mean that the actual equipment is required to have a constant hazard rate. Such assumptions must be checked by experiments!

System Reliability Studies

We consider a system composed of n components. The ith component, $i = 1, \ldots, n$, has a reliability $R_i(t)$. We consider series, parallel, and standby connections of components.

The logic and the mathematics in the case of a series system are just the same as those discussed in Chapter III. However we should clarify what is meant by a series system. A series system is one consisting of a number of components, each of which must function properly in order for the system to function. Thus the system reliability $R_S(t)$ is

$$R_S(t) = \prod_{i=1}^{n} R_i(t) \tag{7-24}$$

if the events are independent. Note this is exactly the same as the result from Chapter III. Using the expression (from 7-20)

$$R_i(t) = \exp\left[-\int_0^t h_i(x)\,dx\right], \qquad t \geq 0.$$

Then

$$R_S(t) = \exp\left\{-\int_0^t \left[\sum_{i=1}^n h_i(x)\right]dx\right\}, \qquad t \geq 0. \tag{7-25}$$

If $h_i(t) = \lambda_i$ then

$$R_S(t) = \exp\left[-\left(\sum_{i=1}^n \lambda_i\right)t\right], \qquad t \geq 0. \tag{7-26}$$

Thus in the case that the components have exponential times to failure, the system also has an exponential distribution with hazard rate equal to the sum of the hazard rates of the individual components. A problem asks for this same result by considering times to failure.

Now consider parallel systems, where a system of 2 components is a parallel system if either functioning properly constitutes a successful system. Such a combination is also called active redundancy. The word active indicates that both components are used simultaneously.

The definition can be generalized to more than two components as was done in Chapter III but such redundancy is not often used in continuously operating equipment. However, if such a system were to be considered, the results of Chapter III could be used.

The system reliability in the case of a parallel system is

$$R_S(t) = R_1(t) + R_2(t) - R_1(t)R_2(t), \tag{7-27}$$

assuming independence and using the previous definitions. Once again, this result follows directly from Chapter III. Thus

$$R_S(t) = \exp\left[-\int_0^t h_1(x)\,dx\right] + \exp\left[-\int_0^t h_2(x)\,dx\right]$$
$$- \exp\left\{-\int_0^t [h_1(x) + h_2(x)]\,dx\right\}. \tag{7-28}$$

If $h_i(t) = \lambda_i$,

$$R_S(t) = e^{-\lambda_1 t} + e^{-\lambda_2 t} - e^{-(\lambda_1 + \lambda_2)t}. \tag{7-29}$$

Note that if the components have an exponential time to failure, an active redundant circuit does not have an exponential time to failure.

The last type of system problem considered in this section is standby redundancy. In this case there are two components and either one functioning properly results in system success; however the second is not turned on until the first ceases to function. The mechanism (or human) which senses that the first has failed and turns on the second is assumed to be perfect in this introductory analysis.

In this case we consider the system time to failure as was done only in problems for the first two cases considered. The system time to failure is the time to failure of the first plus the time to failure of the second:

$$T_S = T_1 + T_2. \tag{7-30}$$

Thus assuming independence

$$f_{T_S} = f_{T_1} * f_{T_2}, \tag{7-31}$$

or using characteristic functions

$$\varphi_{T_S} = \varphi_{T_1}\varphi_{T_2}. \tag{7-32}$$

If

$$\begin{aligned} f_{T_i}(t) &= \lambda_i e^{-\lambda_i t}, & t \geq 0, \\ &= 0, & t < 0. \end{aligned} \tag{7-33}$$

Then using either (7-31) or (7-32), if $\lambda_1 = \lambda_2 = \lambda$

$$\begin{aligned} f_{T_S}(t) &= \lambda^2 t e^{-\lambda t}, & t \geq 0, \\ &= 0, & t < 0; \end{aligned} \tag{7-34}$$

and if $\lambda_2 \neq \lambda_1$ then

$$\begin{aligned} f_{T_S}(t) &= \frac{\lambda_1 \lambda_2}{\lambda_2 - \lambda_1} [e^{-\lambda_1 t} - e^{-\lambda_2 t}], & t \geq 0, \\ &= 0, & t < 0. \end{aligned} \tag{7-35}$$

In the case $\lambda_1 = \lambda_2$ (which would be the case if identical equipment is used in standby), then

$$F_{T_S}(t) = \int_0^t \lambda^2 x e^{-\lambda x} \, dx = 1 - e^{-\lambda t}(\lambda t + 1), \qquad t \geq 0. \tag{7-36}$$

Thus

$$R_S(t) = 1 - F_{T_S}(t) = e^{-\lambda t}(\lambda t + 1), \qquad t \geq 0. \tag{7-37}$$

Summary

This concludes a brief introduction to reliability of continuously operating equipment. Many interesting questions (e.g., various kinds of failure, partial failure or drifts, repair, maintenance) are untouched, but the introduction supplied here provides the basic tools for analyzing the reliability of time varying systems. Reference S3 contains a thorough treatment.

7-5 PROBLEMS

1. $Y = 2X_1 + 10X_2 - X_3X_4 + (X_5/X_6) + \sin X_7$.

   (a) Compute $\partial Y/\partial X_i$, $i = 1, \ldots, 7$.
   (b) Assume that each X_i has a nominal value of 10. Compute the nominal value of Y.
   (c) Is the nominal value computed in (b) the exact mean? Hint: Recall the approximation developed for the mean in the preceding chapter.
   (d) Assume

   $$\sigma_{X_1} = \sigma_{X_2} = \sigma_{X_3} = .1,$$
   $$\sigma_{X_4} = \sigma_{X_5} = \sigma_{X_6} = \sigma_{X_7} = 1,$$

   and that $\rho_{X_iX_j} = 0$, for all $i \neq j$. Compute the approximate mean square error and compare contributions of each X_i.
   (e) Assume the same parameters as in part (d) except assume that

   $$\rho_{X_2X_3} = 1, \qquad \rho_{X_5X_6} = 1, \qquad \rho_{X_1X_7} = 1,$$

   and all other correlation coefficients are zero. What is the approximate mean square error?

2. Two temperature-sensitive resistors R_1 and R_2 are to be used in series in a certain kind of equipment that is to be used in varying climates. Each resistor is chosen independently from a collection that has a normal distribution with mean $1000 + T$ and variance 100, where T is the value of temperature. We assume that the resistors are conditionally independent given that the temperature is t, and that they both experience the same temperature.

   (a) Find the joint distribution of R_1 and R_2 given $T = t$, and from this identify the conditional mean of R_1 and the conditional mean and variance of $R = R_1 + R_2$.
   (b) Assuming that T is uniformly distributed between 50 and 150, find $E[R]$ and σ_R^2.
   (c) Find the correlation coefficient of R_1 and R_2.

3. The output Y of a certain system depends on eight parameters. The partial derivatives were measured experimentally and are

$$\frac{\partial Y}{\partial X_1} = .5, \qquad \frac{\partial Y}{\partial X_2} = .6, \qquad \frac{\partial Y}{\partial X_3} = .4, \qquad \frac{\partial Y}{\partial X_4} = .5,$$

$$\frac{\partial Y}{\partial X_5} = -.3, \qquad \frac{\partial Y}{\partial X_6} = -.4, \qquad \frac{\partial Y}{\partial X_7} = -.2, \qquad \frac{\partial Y}{\partial X_8} = .9.$$

The total tolerance on the variables is given as

$$\text{for } X_1 \text{ through } X_6, \quad \pm 1;$$
$$\text{for } X_7, \quad \pm 2;$$
$$\text{for } X_8, \quad \pm \tfrac{1}{2}.$$

(a) What is the deterministic tolerance of Y?

(b) What is the probabilistic tolerance on Y? Note that assumptions are necessary and the probability of Y being outside certain limits is what is required.

4. The allowable tolerance of gain of a 10-stage amplifier is ± 10 db.

(a) What is the allowable tolerance of each stage using a deterministic approach?

(b) If each stage is uniformly distributed within ± 1 db of its mean, what is the probability the 10-stage amplifier has a gain that is within ± 7 db of its mean?

5. Show that if X is uniformly distributed between L and U then

$$\mu = \frac{U + L}{2},$$

$$\sigma^2 = \frac{(U - L)^2}{12}.$$

6. A steel beam will support its load if the strength S exceeds the load L. If S is a normal random variable with mean 2000 psi and standard deviation 400 psi and L is a normal random variable with mean 1000 psi and standard deviation 300 psi, what is the probability that the beam will support the load?

7. In many deterministic approaches to tolerance studies, the tolerance is considered proportional to the mean value. Such an assumption is consistent with the idea of talking about $\pm 1\%$, $\pm 10\%$, or in general $\pm b\%$ components. In the terms of parameters of distributions, the tolerance, hence the standard deviation being proportional to the mean value, implies a constant coefficient of variation where the coefficient of variation C is

$$C = \sigma/\mu.$$

Give examples where the assumption that the coefficient of variation is constant would be reasonable, and give examples where this assumption would not be reasonable.

8. X_1, X_2, and X_3 are independent random variables with identical distribution F. Find the distribution function of

$$Y = \min (X_1, X_2, X_3) \quad \text{and} \quad Z = \max (X_1, X_2, X_3).$$

9. Using the same definitions as in Example 7-3, find the mean and variance of W, the time of closure from A to C, with the circuit modified to that shown below in Figure P7-1.

183

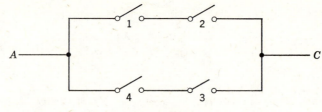

Figure P7-1

10. Given $f_T(t)$ which is zero for $t < 0$, and $F_T(t) = \int_0^t f_T(\lambda)\, d\lambda$, find
    (a) $P[(T \le x) \cap (T \ge t)]$,
    (b) $F_{T|\{T \ge t\}}(x)$,
    (c) $f_{T|\{T \ge t\}}(x)$.

11. Is the hazard rate $h(t)$ a density function? Hint: Check $\int_0^\infty h(t)\, dt$ and note that

$$F_T(\infty) = 1 - \exp\left[-\int_0^\infty h(x)\, dx\right].$$

12. If a system is composed of two components, each of which must function properly, the time of failure of the system is the minimum of the times of failure of the two components, that is, $T_S = \min(T_1, T_2)$. Find f_{T_S}, F_{T_S}, and $R_{T_S} = 1 - F_{T_S}$ in terms of f_{T_1} and f_{T_2}. Generalize to n components in series.

    Let $f_{T_i}(t) = \lambda_i e^{-\lambda_i t}$, $t \ge 0$,
               $= 0$, $t < 0$.

13. Consider a parallel system of two components. Then $T_P = \max(T_1, T_2)$. Find f_{T_P}, F_{T_P}, and R_{T_P} in terms of f_{T_1} and f_{T_2}. Let

$$f_{T_i} = \lambda_i e^{-\lambda_i t}, \quad t \ge 0,$$
$$= 0, \quad t < 0.$$

14. Show that (7-34) and (7-35) follow from (7-31) or (7-32).

CHAPTER VIII

··

DISTRIBUTIONS FROM DATA

The next three chapters consider the question of how data are used to estimate unknowns or to make decisions. Examples of estimation problems are, estimate the reliability of a component; estimate the gain of a control system; estimate the mean and variance of the period of a certain type of oscillator. Typical decision problems are, decide if a signal is present or not; decide which of three proposed designs should be built; decide if a certain lot of components should be purchased or not.

Thus these three chapters form an introduction to statistics. Previous chapters have dealt with probabilistic models. We now consider using data in applying these models.

This chapter shows how a distribution of an unknown random variable can be generated from data. In the next two chapters these distributions are used to estimate unknown quantities and to make decisions, respectively.

Two methods of generating distributions from data are discussed. In either case data in the form of n independent samples $X_1, X_2, \ldots, X_n$ of a random variable X are assumed to be available. By n independent samples we mean that the n random variables $X_1, X_2, \ldots, X_n$ have the same distribution and are independent given the parameters of that distribution. First, Bayes' rule is used to "learn" unknown parameters of a model. The unknown parameters are considered to be random variables and a distribution of the unknowns which depends on the data is generated.

Note that the distribution found is not the distribution of X, the random variable being sampled. Rather the distribution of X depends on the unknown parameters which are considered to be random variables; the distribution of these random variables is what is found. Second the samples of X are used to create empirically a distribution function of X.

8-1 BAYES' RULE

In Chapter II Bayes' rule was stated in the (2-46) form:

$$P(A_j \mid B) = \frac{P(B \mid A_j)P(A_j)}{\sum\limits_{i=1}^{n} P(B \mid A_i)P(A_i)} \qquad (8\text{-}1)$$

where $\qquad \bigcup\limits_{i=1}^{n} A_i = S.$ and $\quad A_i \cap A_j = \varnothing, \quad i \neq j.$

Equation 8-1 is the basis of generating the distribution of an unknown from data. B will represent some data, while A_j will represent some event that influences the data. $P(A_j)$ is called the prior probability and $P(A_j \mid B)$ is called the posterior probability. We now consider an example taken from engineering that illustrates the use of Bayes' theorem.

EXAMPLE 8-1. THE UNRELIABLE TRANSISTOR TESTER

A lot of transistors is 99% good and 1% defective. In order to be "sure" to use a good transistor in the circuit, we test a selected transistor from the lot in a transistor tester. The tester is not completely accurate, however it is reasonably good as represented by the following conditional probabilities:

$$P(\text{says good} \mid \text{actually good}) = .95,$$
$$P(\text{says bad} \mid \text{actually bad}) = .95.$$

The question is, given that a transistor is tested and the tester says it is bad, what is the probability that it is actually bad? In the following the answer is worked out using Bayes' theorem, but the reader should first guess the answer. Is your guess $P(\text{bad} \mid \text{says bad}) = .95$?

Using Bayes' theorem:

$$P(\text{bad} \mid \text{says bad}) = \frac{P(\text{says bad} \mid \text{bad})P(\text{bad})}{P(\text{says bad} \mid \text{bad})P(\text{bad}) + P(\text{says bad} \mid \text{good})P(\text{good})}$$

$$= \frac{(.95)(.01)}{(.95)(.01) + (.05)(.99)} \simeq \frac{1}{6}.$$

Probabilities computed via Bayes' theorem are often called personal probabilities. However it is well to justify the above answer, which may be surprising, on the basis of relative frequency.

If there were a lot of 100 transistors we would expect about one bad one and the tester would probably call it bad. In addition out of the 99 good ones we would expect to call 5% or about 5 of the 99 transistors bad. Thus of the 6 that the transistor tester called bad, on the average, one would actually be a bad transistor. This corresponds to our answer of 1/6 found by Bayes' rule.

We now consider a slightly more general form of (8-1). Using (2-37) from Chapter II

$$P(ABC) = P(A \mid BC)P(B \mid C)P(C) = P(B \mid AC)P(A \mid C)P(C).$$

Thus
$$P(A \mid BC)P(B \mid C) = P(B \mid AC)P(A \mid C)$$

or
$$P(A \mid BC) = \frac{P(B \mid AC)P(A \mid C)}{P(B \mid C)}. \tag{8-2}$$

This is of the same form as Bayes' rule and, simply stated, says that Bayes' rule is applicable if all of the probabilities are conditional on another event C. We expect this form since, as stated earlier, conditional probabilities behave just like probabilities.

Continuous Form of Bayes' Rule

If X and Y are both continuous random variables, then using (4-26) from Chapter IV

$$f_{XY} = f_{X|Y} f_Y = f_{Y|X} f_X$$

or
$$f_{Y|X} = \frac{f_{X|Y} f_Y}{f_X}. \tag{8-3}$$

Furthermore, using (4-33) from Chapter IV

$$f_{Y|X}(y \mid x) = \frac{f_{X|Y}(x \mid y) f_Y(y)}{\displaystyle\int_{-\infty}^{\infty} f_{X|Y}(x \mid \lambda) f_Y(\lambda) \, d\lambda}. \tag{8-4}$$

Equation 8-4 is the continuous form of Bayes' rule. Slightly more general forms can be written using (4-38) and (4-39):

$$f_{Y|Z_1, Z_2, \ldots, Z_n} = \frac{f_{Z_1, \ldots, Z_n|Y} f_Y}{f_{Z_1, \ldots, Z_n}} \tag{8-5}$$

where

$$f_{Z_1, \ldots, Z_n}(z_1, \ldots, z_n) = \int_{-\infty}^{\infty} f_{Z_1, \ldots, Z_n|Y}(z_1, \ldots, z_n \mid \lambda) f_Y(\lambda) \, d\lambda. \tag{8-6}$$

Also
$$f_{Y|Z_1, \ldots, Z_n} = \frac{f_{Z_n|Y, Z_1, \ldots, Z_{n-1}} f_{Y|Z_1, \ldots, Z_{n-1}}}{f_{Z_n|Z_1, \ldots, Z_{n-1}}} \tag{8-7}$$

where

$$f_{Z_n|Z_1, \ldots, Z_{n-1}}(z_n \mid z_1, \ldots, z_{n-1}) = \int_{-\infty}^{\infty} f_{Z_n|Y, Z_1, \ldots, Z_{n-1}}(z_n \mid \lambda, z_1, \ldots, z_{n-1})$$
$$\times f_{Y|Z_1, \ldots, Z_{n-1}}(\lambda \mid z_1, \ldots, z_{n-1}) \, d\lambda. \tag{8-8}$$

187

Mixed Form of Bayes' Rule

We now consider the case where Y is a continuous random variable and there is some observed event B, which has positive probability.

$$P\{B \cap (y < Y \le y + \Delta y)\} = P(B \mid y < Y \le y + \Delta y)P(y < Y \le y + \Delta y)$$
$$= P(y < Y \le y + \Delta y \mid B)P(B).$$

Taking limits as Δy approaches zero produces (assuming the limits exist)

$$P(B \mid Y = y)f_Y(y) = f_{Y|B}(y)P(B), \qquad (8\text{-}9)$$

and dividing both sides by $P(B)$

$$f_{Y|B}(y) = \frac{P(B \mid Y = y)f_Y(y)}{P(B)} \qquad (8\text{-}10)$$

where

$$P(B) = \int_{-\infty}^{\infty} P(B \mid Y = \lambda)f_Y(\lambda)\, d\lambda. \qquad (8\text{-}11)$$

The last integral is justified as follows:

$$P(B) = P(B)\int_{-\infty}^{\infty} f_{Y|B}(\lambda)\, d\lambda$$

because the integral of a conditional density function is 1. Using (8-9)

$$P(B) = \int_{-\infty}^{\infty} P(B)f_{Y|B}(\lambda)\, d\lambda = \int_{-\infty}^{\infty} P(B \mid Y = y)f_Y(y)\, dy.$$

Bayes' Rule to Form a Distribution of Unknown Parameters

In the following three sections it is assumed that there is some true value of an unknown parameter. We model the unknown parameter as a random variable which has some subjective distribution function. Then Bayes' rule is used to find how the data modify the assumed distribution. Thus (8-1), (8-4), and (8-10) are the basis of using data to formulate a distribution of an unknown parameter. These subjective distributions are used to estimate the value of the parameter and to make decisions that depend on the unknown parameter.

8-2 DISTRIBUTION OF PROBABILITY

Consider the problem of generating the distribution of the probability of some event A. We assume that the probability remains constant from one trial to the next and that the results of the different trials are independent. Thus we are assuming Bernoulli trials. This was chosen as the first case because it is important in applications, and the mathematics are not very involved.

From the results of these trials we want to "learn" the true value p_0 of $P(A) = P$. We assume that we have an original assumption of the probability of the event A gained from engineering experience. We want to modify this original assumption based upon the data received such that we learn the true value.

The first assumption discussed is that P is either p_1 or P is p_2. A physical situation that corresponds to this assumption is that we have a lot of components from one manufacturer which has reliability p_1 and another lot from another manufacturer which has reliability p_2. We have confused the labels and we don't know which group of components we actually have in the box. We know it is not a mixture; it is either one or the other.

We run an experiment and observe if event A occurs. The question then is what are the probabilities.*

$$Pr(P = p_1 \mid A); \qquad Pr(P = p_2 \mid A).$$
$$Pr(P = p_1 \mid \bar{A}); \qquad Pr(P = p_2 \mid \bar{A}).$$

Using Bayes' theorem in the form of (8-1)

$$Pr(P = p_1 \mid A) = \frac{Pr(A \mid P = p_1)Pr(P = p_1)}{Pr(A \mid P = p_1)Pr(P = p_1) + Pr(A \mid P = p_2)Pr(P = p_2)}.$$

Since by assumption

$$Pr(A \mid P = p_1) = p_1,$$

then

$$Pr(P = p_1 \mid A) = \frac{p_1 b}{p_1 b + p_2(1 - b)}$$

where

$$b = Pr(P = p_1),$$
$$1 - b = Pr(P = p_2).$$

Similarly

$$Pr(P = p_2 \mid A) = \frac{p_2(1 - b)}{p_1 b + p_2(1 - b)} = 1 - Pr(P = p_1 \mid A).$$

Also

$$Pr(P = p_1 \mid \bar{A}) = \frac{Pr(\bar{A} \mid P = p_1)Pr(P = p_1)}{Pr(\bar{A} \mid P = p_1)Pr(P = p_1) + Pr(\bar{A} \mid P = p_2)Pr(P = p_2)}$$

$$= \frac{(1 - p_1)b}{(1 - p_1)b + (1 - p_2)(1 - b)},$$

$$Pr(P = p_2 \mid \bar{A}) = \frac{(1 - p_2)(1 - b)}{(1 - p_1)b + (1 - p_2)(1 - b)}.$$

* Pr is used to represent probability to avoid confusion with P, which is the probability of event A.

189

Problems at the end of the chapter illustrate what happens when numerical values are used for p_1 and p_2, and the reader should check these results against what he believes should happen.

We now consider the same problem when there is more than one repetition of the experiment. We begin by performing two trials (repetitions of the experiment).

Assume that on the first trial A occurs and on the second trial $\bar{A}$ occurs. We then want

$$Pr(P = p_1 \mid A, \bar{A})$$

where $(A, \bar{A})$ represents that A occurs on the first trial and $\bar{A}$ occurs on the second trial. There are two ways of looking at the problem which are mathematically equivalent. We first consider $Pr(P = p_1 \mid A)$ as the starting point or prior probability and compute as follows:

$$Pr(P = p_1 \mid A, \bar{A})$$
$$= \frac{Pr(\bar{A} \mid P = p_1)Pr(P = p_1 \mid A)}{Pr(\bar{A} \mid P = p_1)Pr(P = p_1 \mid A) + Pr(\bar{A} \mid P = p_2)Pr(P = p_2 \mid A)}. \quad (8\text{-}12)$$

Note that this is a special case of (8-2) where, because of the Bernoulli trial assumption, $Pr(\bar{A} \mid P = p_1, A) = Pr(\bar{A} \mid P = p_1)$.

We now use previous results in (8-12).

$$Pr(P = p_1 \mid A, \bar{A}) = \frac{(1 - p_1)\dfrac{p_1 b}{p_1 b + p_2(1 - b)}}{(1 - p_1)\dfrac{p_1 b}{p_1 b + p_2(1 - b)} + (1 - p_2)\dfrac{p_2(1 - b)}{p_1 b + p_2(1 - b)}}$$

$$= \frac{(1 - p_1)p_1 b}{(1 - p_1)p_1 b + (1 - p_2)p_2(1 - b)}. \quad (8\text{-}13)$$

We now take the other point of view. Rather than updating the previous result we update the distribution of P using the results of both experiments. Again using (8-1) with $B = A\bar{A}$

$$Pr(P = p_1 \mid A, \bar{A})$$
$$= \frac{Pr(A, \bar{A} \mid P = p_1)Pr(P = p_1)}{Pr(A, \bar{A} \mid P = p_1)Pr(P = p_1) + Pr(A, \bar{A} \mid P = p_2)Pr(P = p_2)}$$

$$= \frac{p_1(1 - p_1)b}{p_1(1 - p_1)b + p_2(1 - p_2)(1 - b)}.$$

This is in exact agreement with (8-13) and shows, at least in this special case, that both points of view are equivalent. Moreover it can be shown that as long as the trials are independent the two points of view are mathematically equivalent.

We also note that

$$Pr(P = p_2 \mid A, A) = \frac{p_2{}^2(1 - b)}{bp_1{}^2 + (1 - b)p_2{}^2}.$$

Next the general case of n trials with the event A occurring k times and the event $\bar{A}$ occurring $n - k$ times is considered. We note without proof that we don't need to know on which trials A occurs and on which trials $\bar{A}$ occurs, but only how many times each occurs.

$$Pr(P = p_1 \mid k \text{ in } n) = \frac{Pr(k \text{ in } n \mid P = p_1)Pr(P = p_1)}{\begin{array}{l} Pr(k \text{ in } n \mid P = p_1)Pr(P = p_1) \\ \quad + Pr(k \text{ in } n \mid P = p_2)Pr(P = p_2) \end{array}}$$

$$= \frac{\binom{n}{k}p_1{}^k(1 - p_1)^{n-k}b}{\binom{n}{k}p_1{}^k(1 - p_1)^{n-k}b + \binom{n}{k}p_2{}^k(1 - p_2)^{n-k}(1 - b)}$$

$$= \frac{p_1{}^k(1 - p_1)^{n-k}b}{p_1{}^k(1 - p_1)^{n-k}b + p_2{}^k(1 - p_2)^{n-k}(1 - b)}. \tag{8-14}$$

A problem at the end of the chapter considers the case where there are three possible values of P. This is easily generalized to any finite number of possible values of P.

Continuous P

We now assume that P can be any value between zero and one (rather than just two isolated values) and use Bayes' theorem to "learn" the true value p_0 of P from the data.

Then with the use of (8-10) and (8-11) and the assumption that B represents the event that n experiments were run and the event A was observed k times

$$f_{P|k;n}(\rho) = \frac{Pr(k \text{ in } n \mid P = \rho)f_P(\rho)}{\displaystyle\int_{-\infty}^{\infty} Pr(k \text{ in } n \mid P = \lambda)f_P(\lambda)\, d\lambda}.$$

191

The original assumption is that the prior density of P is uniformly distributed between zero and one, that is

$$f_P(\rho) = 1, \qquad 0 \leq \rho \leq 1,$$
$$= 0, \qquad \text{elsewhere.}$$

Thus

$$f_{P|k;\,n}(\rho) = \frac{\binom{n}{k}\rho^k(1-\rho)^{n-k}1}{\int_0^1 \binom{n}{k}\lambda^k(1-\lambda)^{n-k}\,d\lambda}$$

$$= \frac{\rho^k(1-\rho)^{n-k}}{\int_0^1 \lambda^k(1-\lambda)^{n-k}\,d\lambda}, \qquad 0 \leq \rho \leq 1, \qquad (8\text{-}15)$$

$$= 0, \qquad\qquad \text{elsewhere.}$$

The integral in the denominator of (8-15) is called the beta function, and for $\eta \geq \gamma \geq 0$

$$\int_0^1 x^\gamma(1-x)^{\eta-\gamma}\,dx = \frac{\Gamma(\gamma+1)\Gamma(\eta-\gamma+1)}{\Gamma(\eta+2)} \qquad (8\text{-}16)$$

where $\Gamma(x)$ is the gamma function, that is

$$\Gamma(x) = \int_0^\infty t^{x-1}e^{-t}\,dt, \qquad x \geq 1. \qquad (8\text{-}17)$$

It is easily shown by integration by parts that

$$\Gamma(x+1) = x\Gamma(x). \qquad (8\text{-}18)$$

Direct integration of (8-17) shows

$$\Gamma(1) = 1.$$

Using (8-18) it is easily seen that

$$\Gamma(2) = 1,$$
$$\Gamma(3) = 2!$$
$$\Gamma(n) = (n-1)!.$$

Thus, if k and $n-k$ are non-negative integers,

$$\int_0^1 \lambda^k(1-\lambda)^{n-k}\,d\lambda = \frac{k!\,(n-k)!}{(n+1)!}. \qquad (8\text{-}19)$$

Using (8-19) in (8-15)

$$f_{P|k;n}(\rho) = \frac{(n+1)!}{k!\,(n-k)!}\rho^k(1-\rho)^{n-k}, \qquad 0 \leq \rho \leq 1, \qquad (8\text{-}20)$$

$$= 0, \qquad\qquad \text{elsewhere.}$$

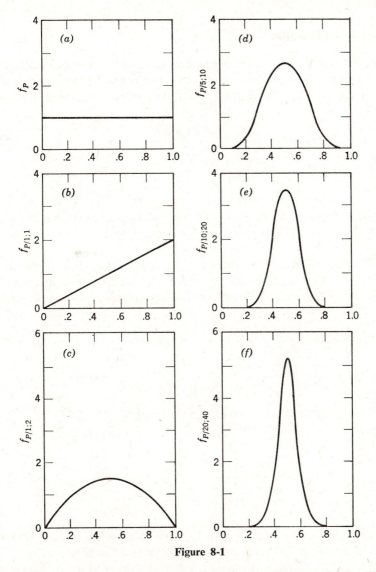

Figure 8-1

Plots of $f_{P|k;n}$ are shown in Figure 8-1. A study of Figure 8-1b will reveal that after one experiment in which the event A occurred the posterior probability density function of P is zero at zero and increases to its maximum value at one. This is certainly a reasonable result. Figure 8-1c–f illustrates, with the ratio of $k/n = 1/2$, that the posterior density function of P has a peak at 1/2, and, moreover, that the peak becomes more pronounced as more data are obtained.

Other Prior Densities

The work to this point can be generalized to consider other prior densities. The only restriction on a prior density of probability is the logical one that it should be zero outside the interval [0, 1]. Any other density may be used, resulting in

$$f_{P|k;n}(\rho) = \frac{\rho^k(1 - \rho)^{n-k}f_P(\rho)}{\displaystyle\int_0^1 \lambda^k(1 - \lambda)^{n-k}f_P(\lambda)\, d\lambda}, \qquad 0 \le \rho \le 1, \qquad (8\text{-}21)$$

$$= 0, \qquad\qquad\qquad \text{elsewhere.}$$

For many forms of f_P, (8-21) may be difficult to evaluate by hand computation, but with digital computers an approximate evaluation may be easily obtained.

It is now shown that if the prior density of P is beta, then the posterior density is beta. That is, the posterior density is of the same form as the prior, only the parameters change. A beta probability density function is described by

$$f_P(\rho) = \frac{\Gamma(\eta + 2)}{\Gamma(\gamma + 1)\Gamma(\eta - \gamma + 1)}\, \rho^\gamma(1 - \rho)^{\eta-\gamma}, \qquad 0 \le \rho \le 1, \quad (8\text{-}22)$$

$$= 0, \qquad\qquad\qquad\qquad \text{elsewhere,}$$

for $\eta \ge \gamma \ge 0$.

If (8-22) is used in (8-21), we obtain the result

$$f_{P|k;n}(\rho)$$

$$= \frac{\Gamma(\eta + n + 2)}{\Gamma(k + \gamma + 1)\Gamma(\eta + n - k - \gamma + 1)}\, \rho^{\gamma+k}(1 - \rho)^{\eta+n-\gamma-k}, \qquad 0 \ge \rho \ge 1,$$

$$= 0 \qquad\qquad\qquad\qquad \text{elsewhere.} \qquad (8\text{-}23)$$

The only difficulty in obtaining (8-23) from (8-21) is the evaluation of the integrals which involves simple properties of gamma functions.

Note from (8-23) that if the prior is a beta density with parameters η and γ, the posterior is a beta density with parameters $n + \eta$ and $\gamma + k$. That is, the beta density is a Bayesian reproducing density if the data have a binomial distribution. For this reason the binomial and beta are called Bayesian conjugate distributions.

At this point, with the binomial assumption, the reader should know how to change his original assumptions on the basis of data. This is the central idea of statistics, that is, how to use data. The posterior density functions of probability are used to generate an estimate of P in the next chapter and are used to make decisions in the following chapter. Two other well-known situations are discussed now.

8-3 EXPONENTIAL SITUATION

In this case we assume that the random variable T has an exponential probability density function:

$$f_T(t) = Ae^{-At}, \qquad t \geq 0, \tag{8-24}$$
$$= 0, \qquad t < 0.$$

The problem is to find a distribution of A, which is considered to be a random variable, after observations of values of T. In this context the exponential distribution should be written as

$$f_{T|A}(t \mid a) = ae^{-at}, \qquad t \geq 0, \tag{8-25}$$
$$= 0, \qquad t < 0.$$

First we consider the simple case where the prior distribution on A is uniform between 2 and 4 and there is one observation T_1 of T. That is, we feel that A is somewhere between 2 and 4, and as far as we know any value is equally likely.

Then using Bayes' rule in the form of (8-4)

$$f_{A|T_1}(a \mid t_1) = \frac{f_{T_1|A}(t_1 \mid a) f_A(a)}{\displaystyle\int_{-\infty}^{\infty} f_{T_1|A}(t_1 \mid \lambda) f_A(\lambda) \, d\lambda}$$

$$= \frac{ae^{-t_1 a}\frac{1}{2}}{\displaystyle\int_{2}^{4} \lambda e^{-t_1 \lambda}\frac{1}{2} \, d\lambda}, \qquad 2 \leq a \leq 4,$$

$$= 0, \qquad \text{elsewhere.}$$

Evaluating the integral in the denominator

$$f_{A|T_1}(a \mid t_1) = \frac{t_1^2 ae^{-t_1 a}}{(2t_1 + 1)e^{-2t_1} - (4t_1 + 1)e^{-4t_1}}, \qquad 2 \leq a \leq 4, \tag{8-26}$$

$$= 0, \qquad \text{elsewhere.}$$

Consider now a second observation T_2. Then using (8-7) and (8-8)

$$f_{A|T_1, T_2}(a \mid t_1, t_2) = \frac{ae^{-t_2 a}f_{A|T_1}(a \mid t_1)}{\int_2^4 \lambda e^{-t_2 \lambda}f_{A|T_1}(\lambda \mid t_1)\, d\lambda}, \qquad 2 \le a \le 4, \quad (8\text{-}27)$$

$$= 0, \qquad\qquad\qquad \text{elsewhere.}$$

This procedure may be continued for as many observations as are available to get the final posterior distribution of A.

If the prior density of A is of the form

$$f_A(a) = \mu e^{-\mu a} \qquad a \ge 0,$$
$$= 0, \qquad\qquad a < 0,$$

which is a special case of the gamma probability density function, then the posterior density is also gamma. Indeed

$$f_{A|T_1}(a \mid t_1) = \frac{ae^{-t_1 a}\mu e^{-\mu a}}{\int_0^\infty \lambda e^{-t_1 \lambda}\mu e^{-\mu \lambda}\, d\lambda}, \qquad a \ge 0,$$

$$= \frac{ae^{-(\mu+t_1)a}}{\int_0^\infty \lambda e^{-(\mu+t_1)\lambda}\, d\lambda}$$

$$= a(\mu + t_1)^2 e^{-(\mu+t_1)a}, \qquad a \ge 0, \qquad (8\text{-}28)$$

$$= 0, \qquad\qquad\qquad a < 0.$$

Similarly

$$f_{A|T_1, \ldots, T_n}(a \mid t_1, \ldots, t_n)$$

$$= \frac{a^n}{n!}\left(\mu + \sum_{i=1}^n t_i\right)^{n+1} \exp\left[-\left(\mu + \sum_{i=1}^n t_i\right)a\right], \qquad a \ge 0, \quad (8\text{-}29)$$

$$= 0, \qquad\qquad\qquad\qquad\qquad\qquad\qquad a < 0.$$

Thus with this reproducing density the posterior density function is simple to write and can easily be expressed in recursive form. A slightly more general form is given in problem 16.

8-4 NORMAL SITUATION

Assume that X is a normal random variable with mean Y and variance N^2. Observations $X_1, X_2, \ldots, X_n$, which are assumed to be conditionally (upon Y) independent observations of X, are to be used to learn the distribution of the mean Y. The variance N^2 is assumed to be known. The prior distribution of Y will be assumed to be normal with mean μ and variance σ^2. The assumption that Y is normal will result in a reproducing distribution.

We now show that $f_{Y|X}(y \mid x)$ is normal with

$$E[Y \mid X] = \frac{X\sigma^2 + \mu N^2}{N^2 + \sigma^2} \tag{8-30}$$

and variance

$$\text{Var } [Y \mid X] = \frac{N^2\sigma^2}{N^2 + \sigma^2}. \tag{8-31}$$

By Bayes' rule (8-3)

$$f_{Y|X}(y \mid x_1) = \frac{f_{X|Y}(x_1 \mid y)f_Y(y)}{\displaystyle\int_{-\infty}^{\infty} f_{X|Y}(x_1 \mid \lambda)f_Y(\lambda)\, d\lambda},$$

$f_{Y|X}(y \mid x)$

$$= \frac{\dfrac{1}{\sqrt{2\pi}\,N}\exp\left\{\dfrac{-(x-y)^2}{2N^2}\right\}\dfrac{1}{\sqrt{2\pi}\,\sigma}\exp\left\{\dfrac{-(y-\mu)^2}{2\sigma^2}\right\}}{\displaystyle\int_{-\infty}^{\infty}\dfrac{1}{2\pi N\sigma}\exp\left\{\dfrac{-(x-\lambda)^2}{2N^2}\right\}\exp\left\{\dfrac{-(\lambda-\mu)^2}{2\sigma^2}\right\} d\lambda}$$

$$= \frac{\exp\left\{-\dfrac{1}{2N^2\sigma^2}[y^2(N^2+\sigma^2) - 2y(x\sigma^2 + \mu N^2) + x^2\sigma^2 + \mu^2 N^2]\right\}}{\displaystyle\int_{-\infty}^{\infty}\exp\left\{-\dfrac{1}{2N^2\sigma^2}[\lambda^2(N^2+\sigma^2) - 2\lambda(x\sigma^2 + \mu N^2) + x^2\sigma^2 + \mu^2 N^2]\right\} d\lambda}.$$

If we multiply above and below by

$$\exp\left[\frac{x^2\sigma^2 + \mu^2 N^2}{2N^2\sigma^2} - \frac{a^2}{2N^2\sigma^2/(N^2+\sigma^2)}\right],$$

$$f_{Y|X}(y \mid x) = \frac{\exp\left\{-\left(\dfrac{2N^2\sigma^2}{N^2+\sigma^2}\right)^{-1}\left[y^2 - 2y\left(\dfrac{x\sigma^2 + \mu N^2}{N^2+\sigma^2}\right) + a^2\right]\right\}}{\displaystyle\int_{-\infty}^{\infty}\exp\left[-\left(\dfrac{2N^2\sigma^2}{N^2+\sigma^2}\right)^{-1}\left(\lambda^2 - 2\lambda\,\dfrac{x\sigma^2 + \mu N^2}{N^2+\sigma^2} + a^2\right)\right] d\lambda}.$$

Choosing a^2 to complete the square and recognizing (area under a normal density $= 1$)

$$\int_{-\infty}^{\infty}\exp\left\{-\frac{1}{2v^2}(\lambda - b)^2\right\} d\lambda = \sqrt{2\pi}\,v,$$

then

$$f_{Y|X}(y \mid x) = \left(2\pi\,\frac{N^2\sigma^2}{N^2+\sigma^2}\right)^{-1/2}\exp\left[-\frac{N^2+\sigma^2}{2N^2\sigma^2}\left(y - \frac{x\sigma^2 + \mu N^2}{N^2+\sigma^2}\right)^2\right].$$

197

This completes the proof that $f_{Y|X}$ is normal with mean and variance as stated in (8-30) and (8-31). We now state without the formal steps of an induction proof that given m observations $x_1, x_2, \ldots, x_m$, the posterior density of the mean Y is normal with

$$\mu_{Y_m} = \frac{x_m \sigma_{Y_{m-1}}^2 + \mu_{Y_{m-1}} N^2}{N^2 + \sigma_{Y_{m-1}}^2} \tag{8-32}$$

and

$$\sigma_{Y_m}^2 = \frac{\sigma_{Y_{m-1}}^2 N^2}{\sigma_{Y_{m-1}}^2 + N^2}. \tag{8-33}$$

These recursive relationships can also be expressed in nonrecursive form:

$$\mu_{Y_m} = \frac{\sigma^2 \sum\limits_{i=1}^{m} x_i + \mu N^2}{m\sigma^2 + N^2}, \tag{8-34}$$

$$\frac{1}{\sigma_{Y_m}^2} = \frac{m}{N^2} + \frac{1}{\sigma^2} = \frac{m\sigma^2 + N^2}{\sigma^2 N^2}. \tag{8-35}$$

EXAMPLE 8-2

The mean Y of a normal random variable X has a prior mean of 3 and a variance of 4. The variance of X is 1. Given the observations of X which are (2, 1, 3, 0, 2, 4, 3, 1). Find μ_{Y_m}, σ_{Y_m} for $m = 1, 2, 3, \ldots, 8$. From (8-30)

$$\mu_{Y_1} = \frac{2 \cdot 4 + 3 \cdot 1}{1 + 4} = \frac{11}{5}.$$

Using (8-31)

$$\sigma_{Y_1}^2 = \frac{1 \cdot 4}{4 + 1} = \frac{4}{5}.$$

Similarly using (8-32) and (8-33)

$$\mu_{Y_2} = \frac{4/5 + 11/5}{1 + 4/5} = \frac{15}{9},$$

$$\sigma_{Y_2}^2 = \frac{1 \cdot 4/5}{9/5} = \frac{4}{9}.$$

The rest of this example is covered in a problem. The answers are

$$\mu_{Y_3} = \tfrac{27}{13}, \qquad \sigma_{Y_3}^2 = \tfrac{4}{13}, \qquad \mu_{Y_7} = \tfrac{63}{29},$$
$$\mu_{Y_4} = \tfrac{27}{17}, \qquad \sigma_{Y_4}^2 = \tfrac{4}{17}, \qquad \mu_{Y_8} = \tfrac{67}{33},$$
$$\mu_{Y_5} = \tfrac{35}{21}, \qquad \sigma_{Y_5}^2 = \tfrac{4}{21}, \qquad \sigma_{Y_7}^2 = \tfrac{4}{29},$$
$$\mu_{Y_6} = \tfrac{51}{25}, \qquad \sigma_{Y_6}^2 = \tfrac{4}{25}, \qquad \sigma_{Y_8}^2 = \tfrac{4}{33}.$$

We can also use Bayes' rule to learn the variance. This is requested in problems 8-17, 8-18, and 8-19. Use of Bayes' rule to learn both the mean and the variance is also possible, but not considered in this introductory book.

8-5 GENERAL COMMENTS ON BAYESIAN STATISTICS

In the three previous sections we used Bayes' rule to modify an original distribution of an unknown based upon the data. We now discuss this procedure in general, emphasizing the important concepts applicable to any problem. We discuss the components of Bayes' rule, give an interpretation of Bayes' rule, and describe the mathematical behavior of the posterior density.

Bayes' rule will be discussed assuming that the data and the unknown are both continuous. Only minor modifications are necessary if either is discrete. In this case Bayes' rule is

$$f_{X|D}(x \mid d) = \frac{f_{D|X}(d \mid x)f_X(x)}{\displaystyle\int_{-\infty}^{\infty} f_{D|X}(d \mid \lambda)f_X(\lambda)\, d\lambda}, \tag{8-36}$$

where X is a random variable that represents an unknown about which we wish to learn on the basis of data D.

The Components of Bayes' Rule

The application of Bayes' rule requires that the two factors of the numerator on the right side of (8-36) be known. That is, we must know or assume the prior distribution of the unknown and the conditional distribution of the data given the unknown.

The critical component of Bayes' rule is $f_{D|X}$ and this is called the likelihood function. If $f_{D|X}$ is known, then Bayes' rule can be applied, and furthermore $f_{D|X}$ eventually determines the form of the posterior distribution of X.

In learning the probability of an event where the event can be observed in a series of independent trials, it is easy to argue that the likelihood function is the binomial probability mass function. However in all other cases of interest, the form of $f_{D|X}$ is not known. We cannot know that a random variable has a Poisson or a normal or a uniform distribution. We must be willing to assume that this is the case from our experience. To say it in a way more familiar to engineers, we must be willing to model the real problem as if $f_{D|X}$ were known.

The remaining necessary factor is the prior distribution of the unknown. The choice of a prior distribution is subjective, and the primary guide is that experience must guide this choice. However there are two mathematical

199

considerations in the choice of a prior density. First if the prior is zero at any place, the posterior will also be zero at that place. Thus if the unknown actually is x_0 and if $f_X(x_0) = 0$, then the posterior would also be 0 at x_0. This suggests that the prior density should not be zero at any feasible value of the unknown. The other consideration is strictly one of mathematical convenience; if possible the prior should be such that a reproducing distribution results. For the three likelihood functions studied, reproducing distributions were given.

Classical statistical procedures have been based on the assumption that the conditional distribution or likelihood function is known but the prior is unknown. Although this situation is not stressed in this book, we mention that the likelihood function by itself is often used in classical estimation procedures, and an example is given in the next chapter.

In the discussion in this section we have not mentioned the denominator of the right side of Bayes' rule. Close observation of Bayes' rule, (8-36), reveals that given the data, the denominator is simply a constant which ensures that the posterior density does integrate to 1. Thus the denominator is not important in analyzing the posterior distribution. Except for this normalizing constant, the posterior density is simply a product of the prior and the likelihood function.

Interpretation of Bayes' Rule

The probability space interpretation of the application of Bayes' rule is relatively simple but is sometimes confused. It is assumed that there is a sample space on which X and D are defined, and there is a distribution of X and a conditional distribution of D given any value x of X. A point in the sample space is chosen which fixes X at the values x_0. To apply Bayes' rule repeatedly in the way described in this book, it must be assumed that the same value x_0 of X prevails while all the data are taken. We repeat the above in the context of an engineering problem.

Assume we want to learn about the noise on a certain communication channel. We have observed the noise many times before, but we don't know what it will be like this time. Our job is to determine what the noise is now. We are willing to assume that the noise D is normal with a mean of X and a variance of 1. In other words the variance seems to be constant at about 1 but the mean changes. The prior of the mean X of the noise, from experience, is, say, normal with a mean of 0 and a variance of 1/4. We can now observe the noise and learn via Bayes' rule the present mean, providing we are willing to assume it does not change during the measurements. We learn about the mean of the noise at this time; we do not learn what the noise will be next year or next week.

Another example illustrates the engineering application of Bayes' rule. A company produces many computers. Each computer has a reliability X, which we assume is different from one computer to another. We purchase one computer. This fixes X at some unknown value x_0. To apply Bayes' rule we need to assume the distribution of the data D given X. We also need to assume the distribution of X over all computers. Thus we can learn, using Bayes' rule, about the reliability of the purchased computer, but not about the computers that we did not buy. We do not have the other computers or we may not be sure how many other computers there actually are; we only need to have a model that includes the other computers.

The Posterior Distribution

In this elementary book we do not have the tools to investigate mathematically the limiting form of (8-36). However this question has been investigated, and it has been shown that the posterior density approaches a narrow spike centered about the true value of the unknown after a large number of data has been taken. A little more precisely, if the prior and $f_{D|X}$ satisfy certain conditions, and if n is the number of observations, then (with probability 1)

$$\lim_{n \to \infty} F_{X|D}(x) = 0, \qquad x < x_0,$$
$$= 1, \qquad x > x_0,$$

where x_0 is the true value.

In order for the posterior distribution to converge to the true value as indicated above, it is necessary that x_0 is not excluded by the prior density and that the data contain information about the value of X. That is, if $f_{D|X} = f_D$ (i.e., the data are independent of X), then $f_{X|D} = f_X$, and the data do not cause a change in the distribution of X.

In the examples given earlier these conditions were satisfied, and thus it is possible to learn the probability of an event, the parameter of an exponential distribution, and the mean of a normal distribution.

8-6 EMPIRICAL DISTRIBUTION

In this section we consider the case where there is no model that describes the form of the distribution function. Only data are available. These data are used to form an empirical distribution function.

The empirical distribution function can be formed with no assumptions, and thus can be used in cases where Bayes' rule is not applicable. Some of the relative advantages and disadvantages of an empirical distribution and a Bayes' distribution are discussed at the close of this chapter.

Definition of the Empirical Distribution Function

Assume that the random variable X has a distribution function F_X and that a series of independent samples, $X_1, X_2, X_3, \ldots, X_n$, of X is available. From these samples we construct an empirical distribution function, $\hat{F}_{X|X_1,\ldots,X_n}$, which should be close to the common unknown distribution function F_X.

Before a definition is given the student is asked to make his own definition via the following example. Suppose the resistance of 20 resistors from a lot of resistors had been measured to be (the readings have been listed in ascending order) 10.3, 10.4, 10.5, 10.5, 10.6, 10.6, 10.6, 10.7, 10.8, 10.8, 10.9, 10.9, 10.9, 10.9, 11.0, 11.0, 11.0, 11.1, 11.1, 11.2. What is the probability that a resistor selected from the lot has a resistance less than or equal to 10.75?

If your answer is 8/20 because eight out of the 20 samples were less than 10.75, then you have essentially defined the empirical distribution function. More precisely

$$\hat{F}_{X|X_1,\ldots,X_n}(x \mid x_1, \ldots, x_n)$$

$$= \frac{\text{number of samples } x_1, \ldots, x_n \text{ no greater than } x}{n}. \quad (8\text{-}37)$$

Note that $\hat{F}_{X|X_1,\ldots,X_n}$ is a distribution function; that is,

$$\hat{F}_{X|X_1,\ldots,X_n}(-\infty \mid \cdots) = 0,$$
$$\hat{F}_{X|X_1,\ldots,X_n}(\infty \mid \cdots) = 1,$$
$$\hat{F}_{X|X_1,\ldots,X_n}(x \mid x_1, \ldots, x_n) \geq \hat{F}_{X|X_1,\ldots,X_n}(y \mid x_1, \ldots, x_n), \qquad \text{if } x > y$$

and

$$\lim_{\substack{\Delta x > 0 \\ \Delta x \to 0}} \hat{F}_{X|X_1,\ldots,X_n}(x + \Delta x \mid x_1, \ldots, x_n) = \hat{F}_{X|X_1,\ldots,X_n}(x \mid x_1, \ldots, x_n).$$

Problems at the end of this chapter call for construction of empirical distribution functions based on definition (8-37).

Note that a probability mass function can be easily derived from the empirical distribution function or equivalently directly from the data.

The empirical distribution function and the empirical probability mass function for the resistance data are shown in Figure 8-2.

Approximations

When many data are available, in order to simplify both data handling and visual interpretation, the data are often grouped into cells. That is, the range of data is divided into a number of cells of equal size and the number of

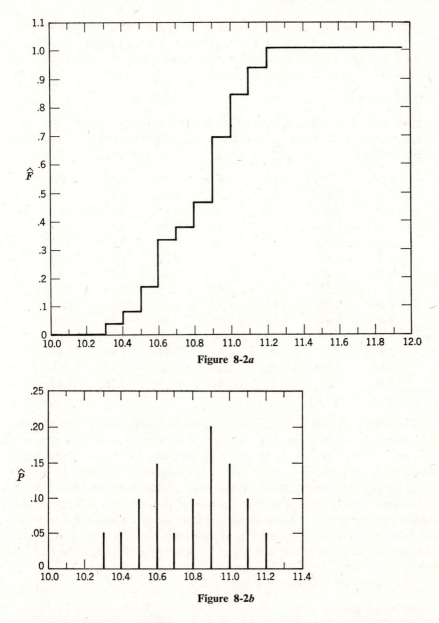

Figure 8-2a

Figure 8-2b

data points within each cell is tabulated. This approximation or grouping of the data results in some loss of information. However this loss is usually more than compensated for by the ease in data handling and interpretation if the goal is visual display.

203

When the grouped data are plotted as an approximate distribution function, the plot is usually called a cumulative frequency polygon. A graph of the grouped data plotted as an approximate probability mass function in the form of a bar graph is called a histogram.

EXAMPLE 8-3

The periods (in microseconds) of a certain type of solid state, free-running multivibrator are measured and recorded below. Plot a histogram of the periods.

3.42,	3.51,	3.61,	3.47,	3.36,	3.39,	3.56,
3.48,	3.40,	3.52,	3.59,	3.46,	3.57,	3.45,
3.54,	3.42,	3.50,	3.54,	3.48,	3.47,	3.65,
3.51,	3.46,	3.59,	3.47,	3.40,	3.38,	3.57,
3.48,	3.50,	3.42,	3.54,	3.55,	3.45,	3.43,
3.57,	3.49,	3.50,	3.62,	3.52,	3.61,	3.50,
3.59,	3.48,	3.54,	3.50,	3.41,	3.52,	3.51,
3.63,	3.53,	3.38,	3.49,	3.50,	3.50,	3.58,
3.50,	3.51,	3.47,	3.52,	3.43,	3.49,	3.42,
3.45,	3.44,	3.48,	3.57,	3.49,	3.53,	3.49,
3.51,	3.59,	3.35,	3.60,	3.48,	3.59,	3.61,
3.55,	3.57,	3.40,	3.51,	3.61,	3.49,	3.40,
3.59,	3.55,	3.56,	3.45,	3.56,	3.47,	3.58,
3.50,	3.46,	3.49,	3.41,	3.52,	3.50,	3.47,
3.61,	3.52.					

The idea is to split the data into groups. The range of the data is 3.35 to 3.65. Since there are about 100 readings, if each interval were chosen to be .030 units wide, there would be about 10 divisions and the resulting picture would be reasonable. The center cell of the histogram is chosen to be 3.485 to 3.515, and this, with the size chosen above, determines all cells. For ease of tabulation it is common practice to choose the ends of a cell to one more significant figure than is recorded in the data so that there is no ambiguity concerning into which cell a reading should be placed. As with most pictorial representations some of the choices as to how to display the data (e.g., cell size) are arbitrary.

The histogram is shown in Figure 8-3.

EXAMPLE 8-4

A random variable has a normal distribution with mean 2 and variance 9. A quantized empirical distribution function is shown along with the true distribution function in Figures 8-4 and 8-5 for samples of size 200 and 2000, respectively.

204

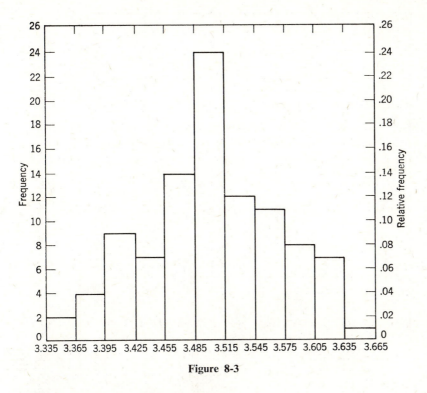

Figure 8-3

Joint Empirical Distribution Functions

We consider two random variables X and Y with n samples X_1, Y_1, X_2, Y_2, ... , X_n, Y_n. That is, for every outcome we observe, the values of both random variables are recorded. Then the joint empirical distribution function is

$$\hat{F}_{X, Y | X_1, Y_1, \ldots, X_n, Y_n}(x, y \mid x_1, y_1, \ldots, x_n, y_n)$$

$$= \frac{\text{number of samples where both } x_i \leq x \text{ and } y_i \leq y}{n}.$$

Higher dimensional empirical distribution functions can be defined in a similar fashion.

8-7 SUMMARY

We have studied two methods of developing distribution functions from data. The important difference is that the Bayesian approach requires that the

205

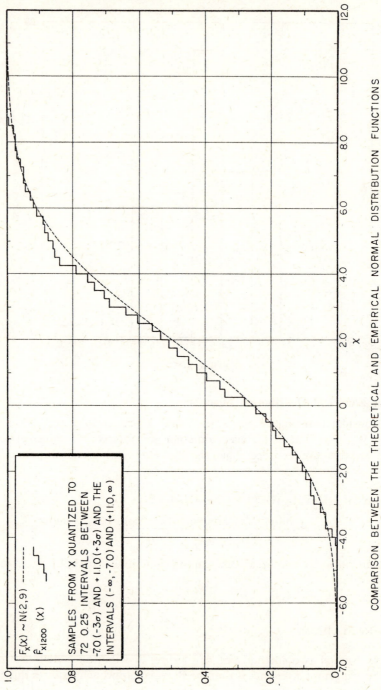

COMPARISON BETWEEN THE THEORETICAL AND EMPIRICAL NORMAL DISTRIBUTION FUNCTIONS

Figure 8-4

Legend within figure:

$F_X(x) \sim N(2,9)$ ------

$\hat{F}_{X1200}(x)$ ————

SAMPLES FROM X QUANTIZED TO 72 0.25 INTERVALS BETWEEN $-7.0\,(-3\sigma)$ AND $+11.0\,(+3\sigma)$ AND THE INTERVALS $(-\infty, -7.0)$ AND $(+11.0, \infty)$

206

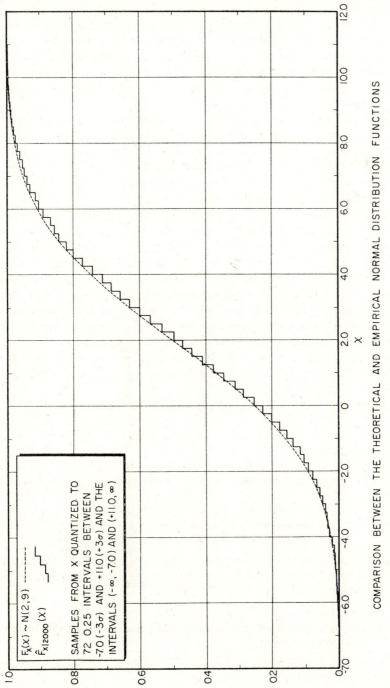

COMPARISON BETWEEN THE THEORETICAL AND EMPIRICAL NORMAL DISTRIBUTION FUNCTIONS

Figure 8-5

form of the likelihood function be known and described by one or more*
parameters, and a prior distribution on the unknown is needed. However the
prior distribution is not important if there are a large number of samples and
if the prior is not zero at the true value of the parameter to be learned. An
empirical distribution can be generated with no assumptions about the
distribution.

If the assumed form of the distribution is correct and if the prior distribution is not drastically wrong, then a Bayesian approach will settle down to
the true distribution faster than an empirical distribution function. On the
other hand, if the wrong distribution function is assumed or if the prior on
the parameter is zero at the true value of the parameter, then the Bayesian
approach will never result in the correct answer.

The practical answer as to which approach to use lies with the person
modeling the problem. If he feels he knows the form of the distribution, then
a Bayesian approach is recommended, and if he doesn't know or is unwilling
to assume the form of the distribution, the empirical technique is available.

Note that with a Bayesian approach subjective distributions of unknown
parameters are found. The empirical distribution is derived directly from
samples of the random variable. Thus the empirical distribution function
approximates the unknown distribution function while the Bayesian approach
approximates parameters of the unknown distribution function.

There are intermediate grounds between an empirical distribution function and a Bayesian approach. However none of these is covered in this
introductory text.

References D3, F2, G2, H1, and J1 provide additional reading.

8-8 PROBLEMS

1. Three shipments of transistors are received. The following probabilities are
   given:

   Shipment A — P(good transistor | A) = .96;
   Shipment B — P(good transistor | B) = .80;
   Shipment C — P(good transistor | C) = .76.

   Assume shipment A contains 2000 transistors, shipment B contains 1000
   transistors, and shipment C contains 1000 transistors. All transistors are mixed
   in the same bin, and one is selected at random.

   (a) What is the probability that it is good?
   (b) Given the selected transistor is good, what is the posterior probability that
   it came from shipment A?

* Only learning one parameter was discussed, but learning more than one parameter is
possible by the same method.

2. Refer to Example 8-1. What is the probability that a transistor is good given the tester says it is good?

3. Refer to Example 8-1. Assume that each transistor is tested twice and that the results of the test are conditionally independent given the true state of the transistor, that is,

$$P(\text{both tests say good} \mid \text{good}) = (.95)^2;$$

$$P(\text{both tests say bad} \mid \text{bad}) = (.95)^2$$

$$P(\text{one test says good and one says bad} \mid \text{good}) = 2(.95)(.05).$$

What is

(a) $P(\text{bad} \mid \text{both tests say bad})$;
(b) $P(\text{bad} \mid \text{one says bad and one says good})$;
(c) $P(\text{bad} \mid \text{both tests say good})$?

Note that the assumption that the test results are conditionally independent may not be reasonable. Discuss this assumption.

4. Show that

$$f_X(x) = \int_{-\infty}^{\infty} f_{X|Y}(x \mid y) f_Y(y) \, dy.$$

5. The probability that a circuit is good given the input voltage V is shown in Figure P8-1. If V is equally likely to be anywhere between 10 and 30 V, find $P(\text{Good})$.

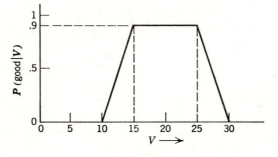

Figure P8-1

6. A shipment of 50,000 screws is received from a certain supplier. If the supplier's process is in control the fraction defective is 0.1 %. However about 1 shipment in 100 comes from a lot when the manufacturing process is out of control. When the process is out of control, 10 % of the screws are defective! A shipment is received and 20 screws are checked with the result that 0 are defective. What is the posterior probability that the received shipment came from a lot made when the process was in control?

209

7. Find, under the Bernoulli trial assumption

$$\Pr(P = p_i \mid k \text{ in } n) \quad \text{for} \quad i = 1, 2, 3$$

if $\qquad \Pr(P = p_1) = a_1, \qquad \Pr(P = p_2) = a_2, \qquad \Pr(P = p_3) = a_3$

where $\qquad\qquad\qquad a_1 + a_2 + a_3 = 1.$

Show $\qquad\qquad \sum_{i=1}^{3} \Pr(P = p_i \mid k \text{ in } n) = 1.$

8. Let m and n be positive integers.

$$f(x) = x^m, \qquad x \geq 0, \qquad g(x) = x^n, \qquad x \geq 0,$$
$$= 0, \qquad x < 0. \qquad\qquad = 0, \qquad x < 0.$$

$$h = f * g \qquad \text{and} \qquad h(1) = \int_0^1 x^m (1 - x)^n \, dx.$$

Find $h(1)$ by finding the Laplace transforms of f and g, multiplying the transforms and taking the inverse transform. Compare the answer with (8-19).

9. The probability that a thumbtack when tossed comes to rest with the point up is P. Assuming that a thumbtack is selected from a population of thumbtacks that are equally likely to have P any value between zero and one, find the probability density of P for the selected thumbtack which has been tossed 100 times with point up 90 times.

10. Show $f_{P|k,n}$ as given by (8-20) is a probability density function.

11. The reliability of a missile has prior density

$$f_R(r) = 11r^{10}, \qquad 0 \leq r \leq 1,$$
$$= 0, \qquad \text{elsewhere.}$$

Ten such missiles are tested and one fails. What is the posterior density of reliability?

12. Plot (8-29) when

$$t_i = 10, \qquad i = 1, \ldots, n, \qquad \mu = 5$$

for $\qquad\qquad n = 1, 2, 3, 4, 5, \text{ and } 10.$

13. Complete Example 8-2.

14. The number of defects in 100 yards of metal is Poisson distributed; that is,

$$P(X = k \mid \Lambda = \lambda) = e^{-\lambda} \frac{\lambda^k}{k!}.$$

Given one observation was 10 defects, find

$$f_{\Lambda|X}(\lambda \mid 10)$$

if f_Λ is equally likely between 0 and 15.

210

15. Show that if X is Poisson distributed given Λ, and if f_Λ has a density of the form

$$f_\Lambda(\lambda) = \frac{a^{b+1}\lambda^b}{\Gamma(b+1)} e^{-a\lambda} \qquad \lambda \geq 0,$$

$$= 0, \qquad \lambda < 0,$$

then

$$f_{\Lambda|X_1,\ldots,X_n}(\lambda \mid x_1, \ldots, x_n) = \frac{(a+n)^{b'+1}\lambda^{b'} e^{-(a+n)\lambda}}{\Gamma(b'+1)}$$

where

$$b' = b + \sum_{i=1}^n x_i.$$

16. If

$$f_{X|A}(x \mid a) = ae^{-ax}, \qquad x \geq 0,$$

$$= 0, \qquad x < 0.$$

$$f_A(a) = \frac{a^k e^{-ab} b^{k+1}}{\Gamma(k+1)} \qquad a \geq 0,$$

$$= 0, \qquad a < 0.$$

Show that if $X_1, X_2, \ldots, X_n$ are independent samples of X

$$f_{A|X_1,X_2,\ldots,X_n}(a \mid x_1, x_2, \ldots, x_n)$$

$$= \frac{a^{k+n}\exp\{-a(b + \Sigma\, x_i)\}(b + \Sigma\, x_i)^{k+n+1}}{\Gamma(k+n+1)}, \qquad a \geq 0,$$

$$= 0, \qquad a < 0.$$

17. The noise in a communication system is normal with a mean of zero. The variance is 1 or 1/2 or 2, and the relative frequency of each value is about the same. Assume there is a sample of the present noise and that sample is 1/4. Given this sample find the posterior probability that the present noise has a variance of 1.

18. Assume that X is normally distributed with mean 0 and unknown variance σ^2. Assume that σ^2 has a uniform density between 1 and 2. Find the posterior density on σ^2 after one observation x_1 of X. It is not necessary to evaluate the integral in the denominator. Express the answer as

$$f_{\sigma^2|X}(\lambda \mid x) = Kg(x, \lambda), \qquad 1 \leq \lambda \leq 2,$$

$$= 0, \qquad \text{elsewhere.}$$

19. Assume that N is normal with mean zero and unknown variance σ^2. If the prior density of σ^2 is

$$f_{\sigma^2}(\lambda) = \frac{1}{\lambda^2} e^{-1/\lambda}, \qquad \lambda > 0,$$

$$= 0, \qquad \lambda \leq 0.$$

211

Given the independent observations, $n_1, \ldots, n_m$ of N, show

$$f_{\sigma^2|N_1,\ldots,N_m}(\lambda \mid n_1, \ldots, n_m) = \frac{\left(\dfrac{\sum\limits_{i=1}^{m} n_i^2 + 2}{2}\right)^{\frac{m}{2}+1}}{(\lambda)^{\frac{m}{2}+2} \; \Gamma\left(\dfrac{m}{2} + 1\right)} \exp -\left(\dfrac{\sum\limits_{i=1}^{m} n_i^2 + 2}{2\lambda}\right), \qquad \lambda > 0,$$

$$= 0, \qquad\qquad\qquad\qquad\qquad \lambda \le 0.$$

20. Show that for large m (8-34) becomes

$$\mu_{Y_m} = \frac{\sum\limits_{i=1}^{m} X_i}{m}$$

and (8-35) results in

$$\sigma_{Y_m}^2 = N^2/m.$$

21. X is uniformly distributed between 0 and b but b is unknown. Assume that the prior distribution on b is uniform between 2 and 4. Find the posterior distribution of b after one observation, $X = 2.5$.

22. The diameters in inches of 50 rivet heads are given in the table below. Construct an empirical distribution function and a histogram.

.338	.342
.341	.350
.337	.346
.353	.354
.351	.348
.330	.349
.340	.335
.340	.336
.328	.360
.343	.335
.346	.344
.354	.334
.355	.333
.329	.325
.324	.328
.334	.349
.355	.333
.366	.326
·343	.326
.341	.355
.338	.354
.334	.337
.357	.331
.326	.337
.333	.333

CHAPTER IX

··

ESTIMATION

In this chapter we first consider how to estimate the parameters needed in a probabilistic model on the basis of samples of the random variables. For instance we consider estimating the probability of an event, the mean and variance of a random variable, and the parameter of an exponential distribution. We also consider estimating the value of a random variable having observed the value of another random variable that is probabilistically related to it. For example we estimate the height of a man, having observed his weight, or we estimate the signal at one time, having observed the signal at an earlier time. Finally, the first two kinds of estimation are combined to consider curve fitting or regression analysis. That is, given a set of points how does one draw the "best" line through the points?

9-1 PARAMETER ESTIMATION

In the last chapter we studied methods of developing the distribution of an unknown quantity which was considered to be a random variable. Although such a distribution function is needed in decision theory, often in practical problems one good guess of the unknown is what is called for. For instance one might like the "best" guess of the probability that a missile will work, or the "best" guess of the average resistance, or the "best" guess of a state variable in a control system. This best guess will be generated from data and perhaps from engineering experience (prior information). The "best" guess of the unknown parameter may then be used in either deterministic or probabilistic models.

With a deterministic model we usually assume that the parameters of the model are fixed. That is, if one talks about the gain of a certain transistor, then that gain is usually assumed to be some fixed quantity. If we can measure the parameter with very little error, we often assume the measurement to be the true value. However if there is significant random error in the measurement, that is, it is not a repeatable measurement, then we usually assume that the measurement M is a random variable determined by an unknown parameter and measurement error. For example,

$$M = m_0 + \text{error}.$$

Our purpose in this chapter is to determine estimates of the parameters of the distribution of M which can then be related to the unknown parameter of the deterministic model. For instance, if $M = m_0 + \text{error}$ and if $E[\text{error}] = 0$, then a natural estimate of m_0 is the estimate of the mean of M. This particular model is so common that we generally confuse estimating the mean of the measurements with estimating the unknown parameter.

With a probabilistic model we either want to estimate probabilities or we want to estimate other parameters of a distribution (e.g., μ and σ of a normal distribution). Thus with probabilistic models we also seek estimates of parameters of a distribution. For example, if we are considering the resistance R of a resistor to be used in an amplifier and there are many amplifiers to be built, we estimate the mean and variance of R rather than estimate R. If we are interested in the resistance of a certain resistor and have measurements with errors of that resistance, then we estimate the mean of the measurements which we may take as the estimate of the resistance.

Thus we are interested in both cases in estimating some unknown parameter from data. In the last chapter it was assumed under a Bayesian philosophy that the unknown parameter was a random variable, and data were used to obtain a conditional distribution of the random variable. In this chapter we treat the unknown as a parameter. There is no basic inconsistency because, as will be seen, the mean or most likely value of the distribution of an unknown (random variable) will be used as an estimate of the unknown (parameter).

It is assumed that we have access to independent samples (measurements) $X_1, X_2, \ldots, X_n$ of a random variable X. We want to derive "good" estimates* of parameters of the distribution of X. In order to do this we discuss criteria of "good" estimators,* and then derive estimators from the empirical distribution of X and from the posterior distribution of the parameter we are estimating.

---

* An estimator is a random variable that is a function of the observations. When the point in the sample space is fixed and there is a number for each observation and thus the estimator becomes a number, we call that number the estimate. Roughly, an estimator is a formula and an estimate is a number.

214

9-2 GOOD ESTIMATORS

There is an unknown parameter θ and there are some observations X_1, $X_2, \ldots, X_n$. We then form a function of the observations, $g(X_1, X_2, \ldots, X_n)$, which we hope is close to θ. We will call

$$\hat{\theta} = g(X_1, X_2, \ldots, X_n)$$

an estimator of θ. What properties do we wish $\hat{\theta}$ to possess? It seems natural that we would wish that

$$\hat{\theta} = \theta,$$

but because $\hat{\theta}$ is a random variable (since it is a function of random variables) we recognize that we will have to have a probabilistic description of a good estimator of θ.

The above considerations form the basis of criteria for best estimators. We now investigate some criteria.

Unbiased

An estimator $\hat{\theta}$ of θ is called unbiased if

$$E[\hat{\theta}] = \theta. \tag{9-1}$$

This says that the mean of $g(X_1, X_2, \ldots, X_n)$ is the same as the unknown quantity.

For example if $X_1, \ldots, X_n$ were independent samples from the same distribution, then

$$\hat{\theta} = g(X_1, \ldots, X_n) = \frac{X_1 + X_2 + \cdots + X_n}{n} = \bar{X}$$

is an unbiased estimator of the mean because

$$E[\hat{\theta}] = \frac{E[X_1 + X_2 \cdots + X_n]}{n} = \frac{n\mu}{n} = \mu$$

where μ is the mean. Note that the estimator given above is a natural one; however another unbiased estimator is simply the first reading. Indeed

$$E[X_1] = \mu.$$

Thus some other criteria of estimators are needed to separate good estimators from those not so good. We now define one such criterion.

215

Mean Square Error

If the estimator $\hat{\theta}$ has a mean of θ, then we also desire $\hat{\theta}$ to have a small spread. The spread can be measured in various ways. For instance measures of the spread are $E[|\hat{\theta} - \theta|]$, [maximum $\hat{\theta}$ − minimum $\hat{\theta}$], or $E[(\hat{\theta} - \theta)^2]$. Although any of these or other measures might be used, the most common measure is the mean square error, or

$$E[(\hat{\theta} - \theta)^2].$$

If $\hat{\theta}$ is unbiased then the mean square error is simply the variance of $\hat{\theta}$. If $E[\hat{\theta}] = \mu$, then

$$E[(\hat{\theta} - \theta)^2] = (\theta - \mu)^2 + \sigma_{\hat{\theta}}^2. \tag{9-2}$$

This result may be shown as follows:

$$\begin{aligned}
E[(\hat{\theta} - \theta)^2] &= E[(\hat{\theta} - \mu + \mu - \theta)^2] \\
&= E[(\hat{\theta} - \mu)^2] + 2(\mu - \theta)E(\hat{\theta} - \mu) + E[(\mu - \theta)^2] \\
&= \sigma_{\hat{\theta}}^2 + 2(\mu - \theta)(\mu - \mu) + (\mu - \theta)^2 \\
&= \sigma_{\hat{\theta}}^2 + (\mu - \theta)^2.
\end{aligned}$$

Note that one does not know θ or there would be no need of sampling and estimating. Thus often only the variance $\sigma_{\hat{\theta}}^2$ is used as the measure of the mean square error of an estimator. This results in the criterion that a minimum variance estimator is a good estimator.

Return to the example of estimating the mean of a random variable X when independent samples, $X_1, X_2, \ldots, X_n$ are available. Although

$$E\left[\sum_{i=1}^{n} \frac{X_i}{n}\right] = E[X_1] = \mu;$$

the variances of the estimators $\bar{X}$ and X_1 are different. Indeed

$$\text{Var}\,[X_1] = \sigma^2,$$

$$\text{Var}\,[\bar{X}] = \text{Var}\left[\sum_{i=1}^{n} \frac{X_i}{n}\right] = \frac{\sigma^2}{n}.$$

The average has a lower variance and by the criteria just introduced is a better estimator than a single reading.

Other Criteria of Estimators

The criteria given above are the ones to be used in this book. Other criteria for estimators can be defined and we mention that if estimation were viewed as the final decision, then a definition of a best estimator for each particular decision could be derived using the techniques discussed in Chapter X.

9-3 METHODS OF DETERMINING ESTIMATORS

We discuss three methods of determining estimators of parameters of the distribution of a random variable:

1. Moments of the empirical distribution.

2. Estimates from posterior Bayes distribution.

3. Maximum likelihood estimators.

The first two methods are illustrated in the following sections. The third method, maximum likelihood estimation, is defined as follows: The maximum likelihood estimator $\hat{\theta}_L(X_1 \cdots X_n)$ is the value of θ that maximizes the likelihood function $f_{X_1 \ldots X_n|\theta}$.

From the Bayesian point of view, a maximum likelihood estimator is the maximum value of the posterior density, assuming that the prior density is a constant over all possible values of the parameter.

The maximum likelihood estimator is usually found by differentiating the likelihood function (or the logarithm of the likelihood function) with respect to θ and setting the derivative equal to zero.

We now consider examples of the three methods.

9-4 MOMENTS OF THE EMPIRICAL DISTRIBUTION FUNCTION

If one assumes that the empirical distribution function is the true distribution, then moments can be calculated by the methods already discussed for discrete random variables. These moments, called sample moments, will be good estimators of the true moments if the empirical distribution function is a good approximation to the true distribution. This will be true if enough independent samples are taken and the samples are representative of the population.

Given the samples $X_1, X_2, \ldots, X_n$, the empirical distribution function of X is discrete and has a mean

$$E[X \mid X_1, X_2, \ldots, X_n] = \sum_{i=1}^{n} X_i \frac{1}{n}. \tag{9-3}$$

This mean of the samples is called $\bar{X}$.

217

Higher moments of X may be estimated by the moments computed from the empirical distribution function.

The Kth moment calculated from the empirical distribution function is

$$E[X^K \mid X_1, \ldots, X_n] = \frac{\sum_{i=1}^{n} X_i^K}{n}. \tag{9-4}$$

Similarly the Kth central moment is for $K = 2, 3, \ldots,$

$$E[(X - \bar{X})^K \mid X_1, \ldots, X_n] = \frac{\sum_{i=1}^{n} (X_i - \bar{X})^K}{n}. \tag{9-5}$$

In particular, when $K = 2$, the second central moment $\hat{\mu}_2$ is an estimate of σ^2 derived from the empirical distribution function.

$$\hat{\mu}_2 = \sum_{i=1}^{n} \frac{(X_i - \bar{X})^2}{n}$$

$$= \sum_{i=1}^{n} \frac{\left(X_i - \sum_{j=1}^{n} \frac{X_j}{n}\right)^2}{n}. \tag{9-6}$$

Properties of $\bar{X}$ and $\hat{\mu}_2$

$\bar{X}$ is an unbiased estimator of the true mean. Indeed,

$$E[\bar{X}] = E\left[\sum \frac{X_i}{n}\right] = \sum_{i=1}^{n} \frac{E(X_i)}{n} = \frac{n\mu}{n} = \mu$$

where $\qquad \mu = E[X_i] = E[X].$

We find the mean square error or variance of $\bar{X}$ and show that the variance approaches zero as the number of samples gets large. That is,

$$\sigma_{\bar{X}}^2 = E[(\bar{X} - \mu)^2]$$

$$= E\left[\left(\sum_{i=1}^{n} \frac{X_i}{n} - \mu\right)^2\right] = E\left[\left(\frac{\sum_{i=1}^{n} (X_i - \mu)}{n}\right)^2\right]$$

$$= \frac{1}{n^2} \sum_{i=1}^{n} E(X_i - \mu)^2 + \sum_{\substack{i=1 \\ i \neq j}}^{n} \sum_{j=1}^{n} E[(X_i - \mu)(X_j - \mu)].$$

The last double sum is zero if the samples are independent. Thus

$$\sigma_{\bar{X}}^2 = \frac{1}{n^2} \sum_{i=1}^{n} \sigma^2 = \frac{\sigma^2}{n} \tag{9-7}$$

where $\sigma^2 = E[(X - \mu)^2]$. As $n \to \infty$, $\sigma_{\bar{X}}^2 \to 0$, or

$$\lim_{n \to \infty} E[(\bar{X} - \mu)^2] = 0. \tag{9-8}$$

We now compute the expected value of $\hat{\mu}_2$ to see if it is an unbiased estimator. First we note (assuming independence)

$$E\left[(X_i - \mu)\left(\sum_{j=1}^{n} X_j - \mu\right)\right]$$
$$= E[(X_i - \mu)^2] + \sum_{\substack{j=1 \\ j \neq i}}^{n} E(X_i - \mu)E(X_j - \mu) = \sigma^2. \tag{9-9}$$

Next,

$$E[\hat{\mu}_2] = E\left[\frac{1}{n}\sum_{i}(X_i - \bar{X})^2\right] = \frac{1}{n}\sum_{i} E(X_i - \bar{X})^2,$$

$$E[(X_i - \bar{X})^2] = E\left[\left(X_i - \frac{1}{n}\sum_{j=1}^{n} X_j\right)^2\right]$$

$$= E\left[\left(X_i - \mu + \mu - \frac{1}{n}\sum_{j} X_j\right)^2\right]$$

$$= E\left\{\left[X_i - \mu - \frac{1}{n}\sum_{j}(X_j - \mu)\right]^2\right\}$$

$$= E[(X_i - \mu)^2] - \frac{2}{n} E\left[(X_i - \mu)\sum_{j=1}^{n}(X_j - \mu)\right] + \frac{1}{n^2}\sum_{j=1}^{n} \sigma^2.$$

Using (9-9) in the center term

$$E[(X_i - \bar{X})^2] = \sigma^2 - \frac{2}{n}\sigma^2 + \frac{1}{n}\sigma^2.$$

But

$$E[\hat{\mu}_2] = \frac{1}{n}\sum_{i} E(X_i - \bar{X})^2 = E[(X_i - \bar{X})^2]$$

$$E[\hat{\mu}_2] = \sigma^2 - \frac{2}{n}\sigma^2 + \frac{1}{n}\sigma^2 = \sigma^2\left(1 - \frac{1}{n}\right) = \sigma^2\frac{(n-1)}{n}. \tag{9-10}$$

Thus $\hat{\mu}_2$ is a biased estimator of σ^2. However it is easy to see that

$$s^2 = \frac{1}{n-1} \sum_i (X_i - \bar{X})^2$$

is an unbiased estimator of σ^2.

Notice that for large n, $\hat{\mu}_2$ and s^2 are nearly the same. For small n the intuitive idea is that one degree of freedom is used in determining $\bar{X}$. That is, knowing $\bar{X}, X_1, \ldots, X_{n-1}$, then X_n can be determined. Thus the sum should be divided by $n - 1$ rather than by n.

We do not propose to judge whether $\hat{\mu}_2$ or s^2 should be used. $\hat{\mu}_2$ is the estimator derived from the empirical distribution function, while s^2 is an unbiased estimator.

In general, moments computed from the empirical distribution function will be good estimators in the sense that as n gets large the estimates will approach the moments they are estimating. However the central moments will be biased.

EXAMPLE 9-1

The measurements of the gain of transistors of a certain type are 9.4, 9.3, 9.3, 9.5, 9.6, 9.0, 9.5, 8.9, 9.2, 9.3. What is the estimate of the average gain? What is the estimate of the variance?

We use $\bar{X}$ and s^2 to estimate the mean and variance. Note that we assume that the readings are independent and that the measurements are samples of the true gain, that is, the average measurement error is zero.

$$\bar{X} = \frac{9.4 + 9.3 + 9.3 + 9.5 + 9.6 + 9.0 + 9.5 + 8.9 + 9.2 + 9.3}{10} = 9.3;$$

$$s^2 = \frac{(.1)^2 + 0^2 + 0^2 + (.2)^2 + (.3)^2 + (.3)^2 + (.2)^2 + (.4)^2 + (.1)^2 + 0^2}{9} \simeq .049.$$

EXAMPLE 9-2

A resistor is selected from a lot of resistors that are nominally 1 K ohm resistors. The resistor is measured on an ohmmeter that has a measurement error. However the error of measurement is not systematically on one side or the other of the true value of resistance. Moreover the errors of a series of readings can be considered independent. Given three readings, 1002, 1003, 1004, what is your estimate of the resistance? What is the mean square error of the estimate of the resistance in terms of the mean square error N of the ohmmeter?

We assume a reading R_i is

$$R_i = r_0 + E_i$$

where r_0 is the true resistance and E_i is a random variable with mean 0 and variance N.

Now $\bar{R} = (R_1 + R_2 + R_3)/3$ is a good estimator of the mean of R_i, hence of r_0.

Thus $(1002 + 1003 + 1004)/3 = 1003$ is the estimate of r_0. The variance of $\bar{R}$ is the same as the mean square error since $E[\bar{R}] = r_0$; thus from (9-7)

$$\sigma_{\bar{R}}^2 = \frac{N}{3}.$$

9-5 ESTIMATES FROM BAYESIAN DISTRIBUTIONS

In the last chapter an unknown was modeled as a random variable and a density function of the unknown was found. Moreover the density function tended to peak at the true value of the unknown with less spread as more data are available.

If one number is to be taken as a best guess of the unknown, it seems that the mean or the peak value of the posterior density would be a natural choice. We investigate both of these estimators derived from the posterior distribution. The mean $\hat{\theta}_n$ of the posterior distribution is

$$\hat{\theta}_n = E[\theta \mid X_1 = x_1, X_2 = x_2, \ldots, X_n = x_n]$$

$$= \int_{-\infty}^{\infty} y f_{\theta \mid X_1, \ldots, X_n}(y \mid x_1, \ldots, x_n) \, dy. \qquad (9\text{-}11)$$

With respect to the posterior distribution $\hat{\theta}_n$ has the property that

$$E[(\theta - \hat{\theta}_n)^2 \mid X_1 = x_1, \ldots, X_n = x_n]$$

$$\leq E[(\theta - a)^2 \mid X_1 = x_1, \ldots, X_n = x_n]. \qquad (9\text{-}12)$$

That is, $\hat{\theta}_n$ minimizes the expected squared error, with the expectation taken with respect to the posterior distribution of θ. We now prove (9-12). Note that the proof is simply a restatement that the expected squared error about the mean is smaller than the expected squared deviation about any other number.

$$E[(\theta - a)^2 \mid X_1 = x_1, \ldots, X_n = x_n]$$

$$= \int_{-\infty}^{\infty} (y - a)^2 f_{\theta \mid X_1, \ldots, X_n}(y \mid x_1, \ldots, x_n) \, dy$$

$$= \int_{-\infty}^{\infty} (y - \hat{\theta}_n + \hat{\theta}_n - a)^2 f_{\theta \mid X_1, \ldots, X_n}(y \mid x_1, \ldots, x_n) \, dy$$

$$= \int_{-\infty}^{\infty} (y - \hat{\theta}_n)^2 f_{\theta \mid X_1, \ldots, X_n}(y \mid x_1, \ldots, x_n) \, dy$$

$$+ \int_{-\infty}^{\infty} (\hat{\theta}_n - a)^2 f_{\theta \mid X_1, \ldots, X_n}(y \mid x_1, \ldots, x_n) \, dy$$

$$+ 2(\hat{\theta}_n - a) \int_{-\infty}^{\infty} (y - \hat{\theta}_n) f_{\theta \mid X_1, \ldots, X_n}(y \mid x_1, \ldots, x_n) \, dy$$

$$= E[(\theta - \hat{\theta}_n)^2 \mid X_1 = x_1, \ldots, X_n = x_n] + (\hat{\theta}_n - a)^2 + 2(\hat{\theta}_n - a)(\hat{\theta}_n - \hat{\theta}_n)$$

$$= E[(\theta - \hat{\theta}_n)^2 \mid X_1 = x_1, \ldots, X_n = x_n]$$

$$+ (\hat{\theta}_n - a)^2 \geq E[(\theta - \hat{\theta}_n)^2 \mid X_1 = x_1, \ldots, X_n = x_n].$$

In addition to the mean we consider one other point estimator derived from the posterior density, the maximum value of the posterior density which is called the mode. Calling the mode of the posterior density $\hat{\hat{\theta}}_n$

$$\hat{\hat{\theta}}_n = y \qquad \text{that maximizes} \qquad f_{\theta \mid X_1, \ldots, X_n}(y \mid x_1, \ldots, x_n). \qquad (9\text{-}13)$$

We now consider these two point estimators for two cases discussed in the last chapter.

Estimating Probability

With independent trials we found [(8-23)] that if the prior was beta, then the posterior was beta. We now find the mean $\hat{p}_n$.

$$\hat{p}_n = \int_0^1 \frac{\Gamma(\eta + n + 2)}{\Gamma(k + \gamma + 1)\Gamma(\eta + n - k - \gamma + 1)} \rho^{\gamma + k + 1}(1 - \rho)^{\eta + n - \gamma - k} \, d\rho$$

$$= \frac{\gamma + k + 1}{\eta + n + 2}. \qquad (9\text{-}14)$$

The mode $\hat{\hat{p}}_n$ is found by differentiating (8-23) with respect to ρ and setting the derivative equal to zero, that is,

$$0 = \frac{\partial f_{P|k;n}(\rho)}{\partial \rho}$$

$$= \frac{\Gamma(\eta + n + 2)}{\Gamma(k + \gamma + 1)\Gamma(\eta + n + k - \gamma - 1)}$$

$$\times [(\gamma + k)(1 - \rho) - (\eta + n - \gamma - k)\rho][\rho^{\gamma + k - 1}(1 - \rho)^{\eta + n - \gamma - k - 1}].$$

And $\hat{\hat{p}}_n$ is the value of ρ that satisfies the above equation, or

$$(\gamma + k)(1 - \hat{\hat{p}}_n) = (\eta + n - \gamma - k)\hat{\hat{p}}_n,$$

$$\hat{\hat{p}}_n = \frac{\gamma + k}{\eta + n}. \tag{9-15}$$

To evaluate these two point estimators, consider the case of an equally likely prior density, that is, $\eta = \gamma = 0$; then

$$\hat{p}_n = \frac{k + 1}{n + 2} \tag{9-16}$$

and

$$\hat{\hat{p}}_n = \frac{k}{n}. \tag{9-17}$$

Note that for large values of n, that is, many samples, $\hat{p}_n$ and $\hat{\hat{p}}_n$ are very nearly the same. Also note that $\hat{\hat{p}}_n$ is the usual intuitive estimator of probability; in fact the limit as n goes to infinity of (9-17) is sometimes taken as the definition of probability.

Note that if $\eta = \gamma = 0$ then $\hat{\hat{p}}_n$ is an unbiased estimator. The proof of this statement is given as a problem.

As $n \to \infty$, $\hat{p}_n \to \hat{\hat{p}}_n$, and this conclusion is independent of the prior assumption if η and γ are finite. The practical effect of prior distribution and the minor difference between $\hat{p}_n$ and $\hat{\hat{p}}_n$ occur for a small number of samples.

EXAMPLE 9-3

This example illustrates the intuitive justification for the mean $\hat{p}_n$ rather than the unbiased estimate $\hat{\hat{p}}_n$. Consider that a new type of motor is being built and 10 have been tested with 10 successes. Assume further that the prior on the reliability of the new motor was equally likely, that is, $\eta = \gamma = 0$. Then using (9-16) and (9-17)

$$\hat{\hat{p}}_n = \tfrac{10}{10} = 1,$$
$$\hat{p}_n = \tfrac{11}{12}.$$

There is often a strong feeling that nothing is perfect and thus $\frac{11}{12}$ is a better point estimate of the quality than 1. The reader must choose for himself. The differences are insignificant for a large number of samples, and a small number of samples always requires a large measure of judgement.

Estimating the Parameter of an Exponential Distribution

We consider now the exponential case as described by (8-29). In this case the mean $\hat{A}$ is

$$
\hat{A} = \int_0^\infty \frac{a^{n+1}}{n!} \left(\mu + \sum_{i=1}^n t_i \right)^{n+1} \exp \left[-\left(\mu + \sum_{i=1}^n t_i \right) a \right] da
$$

$$
= \frac{n+1}{\mu + \sum_{i=1}^n t_i} . \tag{9-18}
$$

$\hat{\hat{A}}$ is the value of a which satisfies

$$
0 = \frac{\partial f_{A|T_1,\ldots,T_n}(a \mid t_1, \ldots, t_n)}{\partial a} = \frac{\left(\mu + \sum_{i=1}^n t_i \right)^{n+1}}{n!}
$$

$$
\times \left\{ \left[n - a\left(\mu + \sum_{i=1}^n t_i \right) \right] a^{n-1} \exp \left[-\left(\mu + \sum_{i=1}^n t_i \right) a \right] \right\},
$$

or
$$
\hat{\hat{A}} = \frac{n}{\mu + \sum_{i=1}^n t_i} . \tag{9-19}
$$

$\hat{A}$ and $\hat{\hat{A}}$ are easily interpreted by regarding $1/A$ as the time constant or mean time to failure. Thus

$$
\frac{1}{\hat{A}} = \frac{\mu + \sum_{i=1}^n t_i}{n+1}
$$

is simply the average of the observed times and the prior guess of the average and $1/\hat{\hat{A}}$ is a similar average. Note that once again as the number of samples gets large the estimators becomes independent of μ, so long as μ is finite.

If the prior distribution is taken to be a more general form of a gamma distribution, more or less weight will be attached to the prior mean. See Problem 11.

224

EXAMPLE 9-4

The waiting time T before a time-shared computer can be used has an exponential distribution with unknown parameter "a". We model the unknown parameter as a random variable A, which has the prior density

$$f_A(a) = 10e^{-10a}, \qquad a \geq 0,$$
$$= 0, \qquad\qquad a < 0.$$

Assume samples of waiting times others have experienced today are 19, 15, 17, and 16; find the expected value of A given the data as an estimate of the parameter "a." Using (9-18)

$$\hat{A} = \frac{5}{10 + 19 + 15 + 17 + 16} = \frac{1}{15.4}.$$

Estimating the Mean of a Normal Random Variable

If the prior of the mean of a normal distribution is normal with a mean of μ and a variance of σ^2, then given the samples $X_1, \ldots, X_n$, the posterior of the mean is also normal with a mean of

$$\frac{\sigma^2 \sum_{i=1}^{m} X_i + \mu N^2}{m\sigma^2 + N^2}$$

where N^2 is the variance of the random variable from which samples are taken. This result is (8-34) from Chapter XIII.

Because the posterior distribution is normal, the mean and the mode agree. Thus

$$\hat{\mu} = \hat{\hat{\mu}} = \frac{\sigma^2 \sum_{i=1}^{m} X_i + \mu N^2}{m\sigma^2 + N^2}. \tag{9-20}$$

Thus the point estimators are weighted averages of the samples and the prior mean. Note that if $\sigma^2 \gg N^2$, that is, the error of the assumed mean is large with respect to the measurement error, then we simply average. That is,

$$\hat{\mu} = \hat{\hat{\mu}} \simeq \frac{\sum_{i=1}^{m} X_i}{m}.$$

Also note that in any case, so long as N^2 and σ^2 are finite

$$\lim_{m \to \infty} \hat{\mu}_m = \lim_{m \to \infty} \frac{\sum_{i=1}^{m} X_i}{m}.$$

225

EXAMPLE 9-5

Refer to Example 9-2. Assume the lot of resistors from which the measured resistor was selected is normally distributed with mean 1000 and variance 10. Assume the measurements are normally distributed with mean square measurement error N^2 of 20. Given the three readings, 1002, 1003, and 1004, estimate the value of resistance of the selected resistor.

We wish to estimate the mean of R where

$$R = r_0 + E.$$

The mean r_0 is a sample from a distribution which has a mean of 1000 and a variance of 10. Thus the mean of the prior is 1000 and the variance of the prior is 10. R has a variance of 20. Thus using (9-20)

$$\hat{r}_0 = \frac{10(1002 + 1003 + 1004) + 1000(20)}{3(10) + 20}$$

$$= \frac{10}{10 + (20/3)} 1003 + \frac{20}{10 + (20/3)} 1000(\tfrac{1}{3}) = 1001.8$$

9-6 MAXIMUM LIKELIHOOD ESTIMATES

A value $\hat{\theta}$ such that for all values ϕ

$$f_{X_1, X_2, \ldots, X_n | \theta}(x_1, x_2, \ldots, x_n \mid \hat{\theta}) \geq f_{X_1, X_2, \ldots, X_n | \theta}(x_1, x_2, \ldots, x_n \mid \phi),$$

is called a maximum likelihood estimate of θ. Such an estimate is justified on the basis that $\hat{\theta}$ best explains the samples that were obtained. From the Bayesian point of view, the maximum likelihood estimator is simply the mode of the posterior density assuming the prior density is a constant. In this sense maximum likelihood estimates are special cases of Bayesian estimation, so we conclude with an example.

EXAMPLE 9-6

Assume that there are n independent samples from a normal distribution with known variance σ^2. Find the maximum likelihood estimate of the mean.

$$f_{X_1, \ldots, X_n | \theta}(x_1, \ldots, x_n \mid \mu) = \prod_{i=1}^{n} \frac{1}{\sqrt{2\pi}\,\sigma} \exp\left[-\frac{(x_i - \mu)^2}{2\sigma^2}\right].$$

Finding the value that maximizes $\ln f$ is equivalent to finding the value that maximizes f. Thus

$$g = \ln f = n \ln \left(\frac{1}{\sqrt{2\pi}\,\sigma}\right) - \frac{1}{2\sigma^2} \sum_{i=1}^{n} (x_i - \mu)^2,$$

$$\frac{dg}{d\mu} = -\frac{1}{2\sigma^2} \sum_{i=1}^{n} -2(x_i - \mu) = 0,$$

or

$$n\mu = \sum_{i=1}^{n} x_i.$$

$$\mu = \frac{\sum_{i=1}^{n} x_i}{n}.$$

This is the intuitive answer. Moreover if we take the prior to have infinite variance, this is the Bayesian estimate [see (9-20)].

9-7 INDIRECT ESTIMATION

We consider a different type of estimation problem. We now assume that all of the needed parameters of a probabilistic model are known. We observe a sample of one random variable X and would like to estimate the corresponding value of another random variable Y which we cannot observe directly for one reason or another. For example we observe the velocity of a satellite on one side of the world and would like to know the velocity on the other side, or we observe the weight of a certain object and want to know its density, or we observe the noise now, and want to know what it will be later.

Consider that there is a probability space upon which X and Y are defined. Observing a sample value x of X determines an event $\{s:X = x\}$ in the probability space. The question now is, into what value or values does Y map the event, $\{s:X = x\}$?

The two extreme cases are easy. If Y and $\{s:X = x\}$ are such that only one value y of Y is possible for the given event, then the value of Y can be positively identified. If this is true for all values of X, that is, for all events, $\{s:X = x\}$, then there is a known relation between X and Y. At the other extreme, if X and Y are independent, then observation of a value of X does not help to estimate the value of Y.

The case of interest to us now falls between these two extremes. The observation of a value of X results in an event in the sample space that produces not an exact value of Y but a changed distribution of Y. This

227

changed distribution function is precisely what we have called $F_{Y|X}(y \mid x)$. We now seek the first and second moments of $F_{Y|X}$ in a manner called linear mean square estimation.

9-8 LINEAR MEAN SQUARE ESTIMATION

Having observed X we seek a linear estimator $\hat{Y} = a + bX$ of the value of Y such that

$$E[(Y - \hat{Y})^2]$$

is minimized. Note that our estimator $\hat{Y}$ is a linear function of X and that the criterion of fit is to minimize the mean square error; hence the name linear mean square estimation. For now we assume that all necessary moments of X and Y are known. In a later section this assumption is relaxed.

We now show that the best estimator is

$$b = \frac{\sigma_{XY}}{\sigma_X^2}, \tag{9-21}$$

$$a = \mu_Y - b\mu_X = \mu_Y - \frac{\sigma_{XY}}{\sigma_X^2}\mu_X, \tag{9-22}$$

where $\qquad \sigma_{XY} = E[(X - \mu_X)(Y - \mu_Y)].$

Taking derivatives and setting them equal to zero

$$\frac{\partial E[(Y - a - bX)^2]}{\partial a} = -2E[Y - a - bX] = 0; \tag{9-23}$$

$$\frac{\partial E[(Y - a - bX)^2]}{\partial b} = -2E[(Y - a - bX)X] = 0. \tag{9-24}$$

Equation 9-23 can be written

$$\mu_Y - a - b\mu_X = 0, \tag{9-25}$$

which is the same as (9-22); (9-24) becomes

$$E[XY] - a\mu_X - bE[X^2] = 0. \tag{9-26}$$

Recognizing that

$$E[X^2] = \sigma_X^2 + \mu_X^2$$

and $\qquad \sigma_{XY} = E[(X - \mu_X)(Y - \mu_Y)] = E[XY] - \mu_X\mu_Y,$

Equation 9-26 becomes

$$\sigma_{XY} + \mu_X\mu_Y = a\mu_X + b\sigma_X^2 + b\mu_X^2.$$

Using (9-25) in the above produces

$$\sigma_{XY} + \mu_X\mu_Y = \mu_X\mu_Y - b\mu_X{}^2 + b\sigma_X{}^2 + b\mu_X{}^2$$

or

$$b = \frac{\sigma_{XY}}{\sigma_X{}^2}.$$

Thus the solutions given above do provide at least a stationary point. We now show that it is indeed a minimum.

Assume $\hat{Y} = c + dX$; then

$$
\begin{aligned}
E[(Y - c - dX)^2] &= E\{[Y - a - bX + (a - c) + (b - d)X]^2\} \\
&= E\{(Y - a - bX)^2 + [(a - c) + (b - d)X]^2 \\
&\quad + 2[Y - a - bX][(a - c) + (b - d)X]\} \\
&= E[(Y - a - bX)^2] + E\{[(a - c) + (b - d)X]^2\} \\
&\quad + 2(a - c)E[Y - a - bX] + 2(b - d)E[(Y - a - bX)X].
\end{aligned}
$$

The last two terms are zero if a and b are the constants chosen as shown above [see (9-23) and (9-24)]. Thus

$$E[(Y - c - dX)^2] = E[(Y - a - bX)^2] + E\{[(a - c) + (b - d)X]^2\}.$$

Now $E\{[(a - c) + (b - d)X]^2\}$ is the expected value of a non-negative random variable (there are no real values of a, b, c, d, or X that make $\{[(a - c) + (b - d)X]^2\}$ negative) and thus

$$E[(Y - c - dX)^2] \geq E[(Y - a - bX)^2],$$

which proves that a and b as specified above are the best choice, and moreover, that the minimum mean square error with a and b so chosen is

$$E[(Y - a - bX)^2].$$

We now obtain a simpler expression for the minimum mean square error. Since

$$E[(Y - a - bX)a] = 0 = E[(Y - a - bX)bX],$$

$$
\begin{aligned}
E[(Y - a - bX)^2] &= E[(Y - a - bX)Y] \\
&= E[Y^2] - aE[Y] - bE[XY] \\
&= \sigma_Y{}^2 + \mu_Y{}^2 - \mu_Y{}^2 + \frac{\sigma_{XY}}{\sigma_X{}^2}\mu_X\mu_Y - \frac{\sigma_{XY}}{\sigma_X{}^2}(\sigma_{XY} + \mu_X\mu_Y) \\
&= \sigma_Y{}^2 - \frac{\sigma_{XY}{}^2}{\sigma_X{}^2},
\end{aligned}
$$

$$E[(Y - a - bX)^2] = \sigma_Y{}^2 - \frac{\rho^2\sigma_X{}^2\sigma_Y{}^2}{\sigma_X{}^2} = \sigma_Y{}^2(1 - \rho^2), \qquad (9\text{-}27)$$

where

$$\rho = \frac{\sigma_{XY}}{\sigma_X\sigma_Y}.$$

229

Thus the residual error (error after the best fit) is the original variance minus the original variance times the square of the correlation coefficient. If the correlation coefficient is near ± 1, then the squared error is nearly zero. Similarly if ρ is zero, then the variance of Y is not reduced; thus an observation of X used in a linear estimate is of no use in estimating Y.

EXAMPLE 9-7

The observed signal X is made up of the true signal Y and noise. It is known that $\mu_X = 10$, $\mu_Y = 11$, $\sigma_X^2 = 1$, $\sigma_Y^2 = 4$, $\rho = .9$. Find the best linear mean square estimator of Y in terms of X. What is the expected square error with this estimator? Find the estimate of the true signal if we observe the value 12.

Using (9-21) and (9-22)

$$b = \frac{(1)(2)(.9)}{1} = 1.8;$$

$$a = 11 - (1.8)(10) = -7.$$

Thus $\hat{Y} = -7 + 1.8\,X$.

Using (9-27)

$$E[(Y - \hat{Y})^2] = 4[1 - (.81)] = .76.$$

If $X = 12$ then

$$\hat{Y} = -7 + (1.8)(12) = 14.6.$$

Multivariable Linear Mean Square Estimation

We now assume that values of the n random variables $X_1, \ldots, X_n$ are observed and the value of Y is to be estimated with a linear estimator,

$$\hat{Y} = a_0 + \sum_{i=1}^{n} a_i X_i,$$

such that $E[(Y - \hat{Y})^2]$ is minimized.

Differentiating $E[(Y - \hat{Y})^2]$ with respect to each a_j, $j = 0, 1, \ldots, n$ results in

$$E\left[Y - a_0 - \sum_{i=1}^{n} a_i X_i\right] = 0, \tag{9-28}$$

$$E\left[\left(Y - a_0 - \sum_{i=1}^{n} a_i X_i\right) X_j\right] = 0, \qquad j = 1, \ldots, n. \tag{9-29}$$

Equation 9-28 becomes

$$a_0 = \mu_Y - \sum_{i=1}^{n} a_i \mu_{X_i}. \tag{9-30}$$

230

Use of (9-30) in the n equations represented by (9-29) produces

$$\sum_{i=1}^{n} a_i(E[X_i X_j] - \mu_{X_i}\mu_{X_j}) = E[Y X_j] - \mu_Y\mu_{X_j},$$

$$\sum_{i=1}^{n} a_i \sigma_{X_i X_j} = \sigma_{Y X_j}, \qquad j = 1, \ldots, n. \tag{9-31}$$

The set of n linear equations in n unknowns represented by (9-31) will usually have a unique solution.* In a practical problem computer solution is recommended.

Thus (9-30) and (9-31) provide an implicit solution to the multivariable linear mean square estimation problem. To prove that it is a solution the same technique used in the single variable case may be used. We now examine the mean square error when the a_i's are chosen in accordance with (9-28) and (9-29).

$$E\left[\left(Y - a_0 - \sum_{i=1}^{n} a_i X_i\right)\left(Y - a_0 - \sum_{j=1}^{n} a_j X_j\right)\right]$$

$$= E\left[\left(Y - a_0 - \sum_{i=1}^{n} a_i X_i\right)Y\right] - a_0 E\left[Y - a_0 - \sum_{i=1}^{n} a_i X_i\right]$$

$$- \sum_{j=1}^{n} a_j E\left[\left(Y - a_0 - \sum_{i=1}^{n} a_i X_i\right)X_j\right].$$

The last two terms are zero by (9-28) and (9-29). Thus

$$E[(Y - \hat{Y})^2] = E[Y^2] - a_0\mu_Y - \sum_{i=1}^{n} a_i E[X_i Y].$$

Using (9-30)

$$E[(Y - \hat{Y})^2] = \sigma_Y^2 + \mu_Y^2 - \mu_Y^2 - \sum_{i=1}^{n} a_i(E[X_i Y] - \mu_{X_i}\mu_Y),$$

$$E[(Y - \hat{Y})^2] = \sigma_Y^2 - \sum_{i=1}^{n} a_i \sigma_{X_i Y}. \tag{9-32}$$

EXAMPLE 9-8

The period Y of a certain monostable multivibrator is determined largely by the storage time X_1 of one transistor and the rise time X_2 of another

---

* The covariance matrix is a positive definite matrix which guarantees a unique solution, except in the case when some of the X_i's have a correlation coefficient of 1.

transistor. The needed moments are

$$\mu_Y = 1 \qquad \mu\text{sec}$$
$$\sigma_Y = .01 \qquad \mu\text{sec}$$
$$\mu_{X_1} = .02 \qquad \mu\text{sec}$$
$$\sigma_{X_1} = .005 \qquad \mu\text{sec}$$
$$\mu_{X_2} = .006 \qquad \mu\text{sec}$$
$$\sigma_{X_2} = .0009 \qquad \mu\text{sec}$$
$$\sigma_{X_1 X_2} = 0 \qquad \mu\text{sec}^2$$
$$\sigma_{X_1 Y} = +.00003 \qquad \mu\text{sec}^2$$
$$\sigma_{X_2 Y} = +.000004 \quad \mu\text{sec}^2.$$

Find the best linear estimator and the expected square error.
Using (9-31)

$$a_1(.005)^2 + a_2(0) = +.00003,$$
$$a_1(0) + a_2(.0009)^2 = +.000004$$

results in
$$a_1 = \frac{+.00003}{.000025} = +\frac{6}{5},$$

$$a_2 = \frac{+.000004}{.00000081} = +\frac{400}{81}.$$

Using (9-30)
$$a_0 = 1 - \tfrac{6}{5}(.02) - \tfrac{400}{81}(.006) \simeq .946.$$

Thus the best estimator of the period is

$$\hat{Y} = +\frac{6}{5} X_1 + \frac{400}{81} X_2 + .946$$

where X_1 is storage time of one transistor (the one that is on during the one-shot period) and X_2 is the rise time of the other transistor.

The mean square error with the constants chosen as above is, using (9-32),

$$E[(Y - \hat{Y})^2] = (.01)^2 - \left(+\frac{6}{5}\right)(+.00003) - \left(+\frac{400}{81}\right)(+.000004)$$

$$\simeq .0001 - .000036 - .000020 = .000044.$$

9-9 MEAN SQUARE ESTIMATION

Linear Estimator

Suppose we tried to find a linear estimator $\hat{Y} = a + bX$ when, unknown to us, Y is deterministically related to X by the equation

$$Y = X^2$$

and
$$f_X(x) = \tfrac{1}{2}, \qquad -1 \leq x \leq 1,$$
$$= 0, \qquad \text{elsewhere.}$$

In this case from the assumed density of X and the relation $Y = X^2$, the moments are

$$E[X] = 0,$$
$$E[X^2] = \sigma_X{}^2 = \int_{-1}^{1} \tfrac{1}{2}x^2 \, dx = \tfrac{1}{3},$$
$$E[Y] = E[X^2] = \tfrac{1}{3},$$
$$E[XY] = E[X^3] = \int_{-1}^{1} \tfrac{1}{2}x^3 \, dx = 0.$$

Thus using (9-21) and (9-22)
$$b = 0,$$
$$a = \tfrac{1}{3},$$

or
$$\hat{Y} = \tfrac{1}{3},$$

and using (9-27)
$$E[(Y - \hat{Y})^2] = \sigma_Y{}^2.$$

Thus the best fit is a constant, and the expected squared error is simply the variance. The best linear estimator is obviously a very poor estimator in this case. A little thought makes it only reasonable that if one chooses a poor model, then the results will not be good.

Suppose we tried the estimator

$$\hat{Y} = a_0 + a_1 X + a_2 X^2.$$

This estimator would result in a perfect fit for the example given above. (See problem 9-20.) However is this a linear estimator? Most would answer no, but suppose we define $X = X_1$ and $X^2 = X_2$; then

$$\hat{Y} = a_0 + a_1 X_1 + a_2 X_2$$

is a linear estimator of the type discussed before.

233

The answer lies in the fact that if one is willing to consider multivariate equations, then any polynomial in X can be used, and in addition models such as

$$\hat{Y} = a_0 + a_1 X + a_2 X^2 + a_3 \cos X + a_4[X^3 - 3X + 19 \tan^{-1}(X)]$$

can be used. This model is linear in the a_i's and thus can be used with the theory developed.

Nonlinear Minimum Mean Square Error Estimators

In the last section limitations of linear estimators were discussed. We now show that the minimum mean square error estimator of Y, given $X_1, \ldots, X_n$, is

$$Y^* = E[Y \mid X_1, \ldots, X_n]. \tag{9-33}$$

Indeed

$$E[(Y - \hat{Y})^2] = E\{E[(Y - \hat{Y})^2 \mid X_1, \ldots, X_n]\}$$

$$= \int_{-\infty}^{\infty} \cdots \int E[(Y - \hat{Y})^2 \mid X_1 = x_1, \ldots, X_n = x_n]$$

$$\times f_{X_1, \ldots, X_n}(x_1, \ldots, x_n) \, dx_1, \ldots, dx_n.$$

Thus in order to minimize the multiple integral, because $f_{X_1, \ldots, X_n} \geq 0$, we must minimize the conditional expected value of $(Y - \hat{Y})^2$. We show that Y^* as given by (9-33) does this.

$$E[(Y - \hat{Y})^2 \mid X_1 = x_1, \ldots, X_n = x_n]$$

$$= E[(Y - Y^* + Y^* - \hat{Y})^2 \mid X_1 = x_1, \ldots, X_n = x_n]$$

$$= E[(Y - Y^*)^2 \mid X_1 = x_1, \ldots, X_n = x_n]$$

$$+ E[(Y^* - \hat{Y})^2 \mid X_1 = x_1, \ldots, X_n = x_n]$$

$$+ 2E[(Y - Y^*)(Y^* - \hat{Y}) \mid X_1 = x_1, \ldots, X_n = x_n].$$

Note that given $X_1 = x_1, \ldots, X_n = x_n$, the second term is a non-negative constant and the third term is

$$2(Y^* - \hat{Y})\{E[Y \mid X_1 = x_1, \ldots, X_n = x_n] - Y^*\} = 0.$$

Thus

$$E[(Y - \hat{Y})^2 \mid X_1, \ldots, X_n] \geq E[(Y - Y^*)^2 \mid X_1, \ldots, X_n].$$

The result is that the conditional expectation has an expected mean square error that is less than or equal to the best linear estimator.

The reader might wonder why linear mean square estimators are stressed, when we have just shown that the conditional expectation is a better estimator.

The answer is that the conditional expectation is not known unless the joint distribution functions are known, a situation that is very rare in practice. On the other hand a linear best estimate is a function of only first and second moments, which may be known or can be estimated in the manner shown in the next section.

However, in one important special case the minimum mean square estimator is a linear estimator.

Jointly Normal Random Variables

In Section 5.8 it was shown that if X_1 and X_2 are jointly normal, then (5-27)

$$E[X_1 \mid X_2] = \mu_1 + r\frac{\sigma_1}{\sigma_2}(X_2 - \mu_2)$$

or
$$E[X_1 \mid X_2] = \left(\mu_1 - \mu_2 r\frac{\sigma_1}{\sigma_2}\right) + r\frac{\sigma_1}{\sigma_2}X_2 \qquad (9\text{-}34)$$

where $r = \rho$ is the correlation coefficient.

Note that the conditional expected value of jointly normal random variables is a linear estimate, that is, $E[X_1 \mid X_2] = a + bX_2$. Moreover comparison of (9-34) with (9-21) and (9-22) shows that the conditional expectation is exactly the same as the linear estimator. This last statement should be expected, since if the minimum mean square error estimator is linear, then it must agree with the best linear estimator.

Also we see from (5-26) that the conditional variance of X_1 given X_2 is $\sigma_1^2(1 - r^2)$, which is in agreement with (9-27).

9-10 CURVE FITTING OR REGRESSION

In this section we consider the case where laboratory data are available. We would like to find the curve of a specified type (e.g. straight line) that best fits the data. The solution uses the theory of linear estimators developed in Section 9-8 with sample moments substituted for the needed moments. First fitting a straight line is considered.

Suppose that we have data as represented in Figure 9-1. From these data we would like to fit a straight line $\hat{y} = a + bx$ to the data such that the mean square error is minimized. This is often called the *regression* problem. The solution is to use the data to arrive at estimates of μ_Y, μ_X, σ_X^2, and σ_{XY} and then use (9-21) and (9-22). The estimates of the unknown moments can be derived either from the Bayesian approach or from the empirical distribution function approach.

235

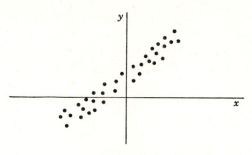

Figure 9-1

If we use the empirical distribution function, then

$$\hat{\mu}_Y = \frac{\sum_{i=1}^{n} y_i}{n},$$

$$\hat{\mu}_X = \frac{\sum_{i=1}^{n} x_i}{n},$$

$$\hat{\sigma}_X^{\ 2} = \frac{\sum_{i=1}^{n} (x_i - \hat{\mu}_X)^2}{n},$$

$$\hat{\sigma}_{XY} = \frac{\sum_{i=1}^{n} (x_i - \hat{\mu}_X)(y_i - \hat{\mu}_Y)}{n}.$$

Use of these results in (9-21) and (9-22) results in

$$b = \frac{\sum_{i=1}^{n} \left[\left(x_i - \frac{1}{n}\sum_{j=1}^{n} x_j \right)\left(y_i - \frac{1}{n}\sum_{k=1}^{n} y_k \right) \right]}{\sum_{i=1}^{n} \left(x_i - \frac{1}{n}\sum_{j=1}^{n} x_j \right)^2}, \qquad (9\text{-}35)$$

$$a = \frac{1}{n}\sum_{i=1}^{n} y_i - \frac{b}{n}\sum_{i=1}^{n} x_i. \qquad (9\text{-}36)$$

As is given in a problem it can be shown that if one pictures the problem as is shown in Figure 9-2, and minimizes the sum of the squared errors where the errors are indicated by ϵ_i on the figure, then (9-35) and (9-36) are obtained.

If one wants to fit curves of the form

$$\hat{Y} = a_0 + a_1 X + a_2 X^2 + a_3 X^3$$

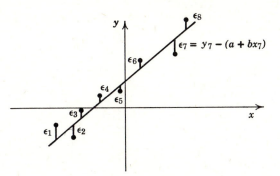

Figure 9-2

then as when moments were assumed to be known, these models can be easily fit by redefining variables until the model is of the form

$$\hat{Y} = \sum_{i=1}^{n} a_i X_i + a_0.$$

The solution in the form of simultaneous equations is given by (9-30) and (9-31) where $\sigma_{X_i X_j}$ is estimated by

$$\hat{\sigma}_{X_i X_j} = \frac{\sum_{k} (x_{ik} - \hat{\mu}_{Xi})(x_{jk} - \hat{\mu}_{Xj})}{n}$$

where x_{ik} is the kth sample of the random variable X_i.

Practically, programs that provide the optimum coefficients and the mean square error before and after fitting, as well as various other useful information, are available for digital computers. Thus the mechanics of solving the problem are not so essential as an understanding of the assumptions involved and an ability to interpret the results.

EXAMPLE 9-9

Fit a straight line to the points $(-1, -1)$, $(0, 0)$, and $(1, 2)$ such that the mean square error is minimized.

Using (9-35)

$$b = \frac{(-1)(-\frac{4}{3}) + (0)(-\frac{1}{3}) + (1)(\frac{5}{3})}{(1)^2 + (0)^2 + (1)^2} = \frac{3}{2}.$$

Using (9-36)

$$a = \frac{1}{3}(-1 + 0 + 2) - \frac{3}{2(3)}(-1 + 0 + 1)$$

$$= \tfrac{1}{3}.$$

Thus $\hat{y} = \tfrac{1}{3} + \tfrac{3}{2}x$ is the best fitting straight line.

9-11 SUMMARY

The first part of this chapter dealt with estimating parameters of a distribution function of a random variable when samples of the random variable are available. Unbiased and minimum variance estimators were defined. Estimates for various properties of a random variable were derived from the empirical distribution function and from the Bayesian distribution of an unknown parameter. Maximum likelihood estimates were treated as a special case of Bayesian estimates. Estimating probabilities and parameters of a normal distribution were discussed. The sample moments from the empirical distribution function were suggested for the moments of the unknown distribution function. Mean and variance were emphasized.

The second part of the chapter considered a different estimation problem. In this case it was assumed that there was a sample of one random variable and the best estimate of the value of a second random variable was desired. Minimum mean square error was used as the criterion for best, and linear estimators were stressed. The problem was formulated with the assumption that the second moments of the joint distribution are known. If these joint moments are not known, then they can be estimated from data, and the estimates are used for the true moments. The problems introduced by linear estimates were discussed, and it was shown that the conditional expectation is the minimum mean square error estimate. The linear estimate is used because it is simpler and more practical.

Additional reading may be found in F2, G2, H1, and P2.

9-12 PROBLEMS

1. If there are measurements with error of an unknown parameter, give examples where the mean of the measurements would not be a good estimator of the parameter.

2. The median m is defined by

$$F_X(m) = .5.$$

Find $\hat{m}$ the estimator of the median derived from the empirical distribution function.

3. Show that the median m minimizes $E[|X - m|]$ by showing for $a < m$

$$E\{|X - a|\} = E\{|X - m|\} + 2 \int_a^m (x - a) f_X(x)\, dx,$$

and for $a > m$

$$E\{|X - a|\} = E\{|X - m|\} + 2 \int_m^a (a - x) f_X(x)\, dx.$$

4. If K is the number of times an event A occurs in n independent trials, show that K/n is an unbiased estimator of $P(A)$.

5. Estimate the mean and variance of the diameter of the rivets described in Problem 22 in Chapter VIII.

6. Find the empirical estimator of the third central moment.

7. Use Chebyshev's inequality to show for all $\epsilon > 0$

$$\lim_{n \to \infty} P[|\bar{X} - \mu| > \epsilon] = 0.$$

8. Why do you average "laboratory readings" of one parameter?

9. What is necessary in order to have *independent* samples?

10. If your original estimate of $P(A)$ is that it has a mean of .9 and a standard deviation of .05, then 150 samples are taken with A occurring 140 times, what is $\hat{P}_{150}$? What is $\hat{\hat{P}}_{150}$? Hint: Assuming the prior density is beta;

$$.9 = \mu = \left(\frac{\gamma + 1}{\eta + 2}\right), \qquad (.05)^2 = \sigma^2 = \frac{(\gamma + 2)(\eta - \gamma + 1)}{(\eta + 2)^2(\eta + 3)}.$$

Solve for η and γ.

11. See Problem 16 in Chapter 8. Using the results of that problem, find $\hat{A} = E[A \mid X_1, X_2, \ldots X_n]$ and compare with (9-18). Also find $\hat{\hat{A}}$ and compare with (9-19).

12. X is Poisson distributed given Λ, that is,

$$P(X = K \mid \Lambda = \lambda) = e^{-\lambda}\frac{\lambda^K}{K!}, \qquad K = 0, 1, 2, \ldots.$$

The prior density of Λ is

$$f_\Lambda(\lambda) = \frac{a^{b+1}\lambda^b}{\Gamma(b + 1)}e^{-a\lambda}, \qquad \lambda \geq 0.$$

$$= 0, \qquad \lambda < 0.$$

Given X_1, X_2, X_3, find the estimator of the value of λ based upon the mode of the posterior density (see Problem 8-15).

13. $$X = S + N$$

where X is an observation of a constant signal plus noise. Assume that the prior density of signal S is

$$f_S(s) = \frac{1}{\sqrt{2\pi}}\exp\left[\frac{-(s - 1)^2}{2}\right].$$

We take an observation and observe $X = 2$. Use Bayes' rule to find the new estimate of S. Assume S and N are statistically independent and that

$$f_N(n) = \frac{1}{\sqrt{2\pi}}e^{-n^2/2}.$$

Hint: First find $f_{X|S=s}(x)$ and use Bayes' rule to find $f_{S|X}$. Then use the mean as the estimate.

239

14. The gain G of a solid state amplifier varies with the temperature of the material. The gain seems to be normal with a mean of 20 and a variance of 1. The measurement M of the gain is modeled by

$$M = g + N$$

where g is the present value of gain and N is a normal random variable with mean 0 and variance 2. Assume the measurement of gain is 19.5. Estimate the present gain.

15. We observe a signal S_1 at time one and we want the best linear estimator of the signal S_2 at a later time. Assume $E[S_1] = E[S_2] = 0$. Find the best linear estimator in terms of moments.

16. Same situation as in Problem 15, but now the derivative S_1' of the signal at time one can also be observed. $E[S_1'] = 0$. Find the best estimator of the form

$$\hat{S}_2 = a_1 S_1 + a_2 S_1'.$$

17. We observe a state variable X of a control system. But the observation Y is actually

$$Y = mX + N$$

where m is some known constant and N is a normal random variable which is independent of X. Find the best linear estimator of X

$$\hat{X} = a + bY,$$

in terms of m and the moments of X and N.

18. Show that if Y and X are jointly normal, then

$$E[Y \mid X = x] = a + bx$$

and $$E\{[Y - (a + bX)]^2\} = \sigma_Y^2(1 - r^2).$$

Note that a and b are as prescribed for the best linear estimate.

19. Refer to Figure 9-2. Find a and b such that

$$\sum_{i=1}^{n} [y_i - (a + bx_i)]^2$$

is minimized. Compare your result to (9-35) and (9-36).

20. $$Y = X^2,$$
$$f_X(x) = \tfrac{1}{2}, \qquad -1 \leq x \leq 1,$$
$$= 0, \qquad \text{elsewhere.}$$

Find a_0, a_1, and a_2 in the model

$$\hat{Y} = a_0 + a_1 X + a_2 X^2$$

such that the mean square error is minimized. Find the minimum mean square error.

21. The following data are available on capacitor diameter Y versus average foil thickness X. Find the best linear fit.

X(.00001 in.)	Y(.001 in.)
58	521
60	530
56	537
65	545
67	554
68	562
71	570
75	577
76	585

CHAPTER X

....

ENGINEERING DECISIONS

10-1 EXPECTED GAIN OF A DECISION

Suppose there is a specified problem given to an engineer, and that there are certain specified actions which he may take to solve the problem. For example the problem may be to design a control system for a certain plant, and there are three possible types of control system that could be used. The purpose of this chapter is to present a method for deciding which of the possible actions is best.

The basic idea is that a gain (or a loss) is assigned to each possible outcome. Then the decision that maximizes the expected gain (minimizes the expected loss) is chosen as best. In general the outcome will be determined by the action and by the state of nature. In the problems of interest the state of nature will not be known, but the probability distribution of the possible states of nature will be known or learned. We now illustrate such decision problems by examples.

10-2 CHOOSING AMONG ALTERNATIVE DESIGNS

EXAMPLE 10-1

A designer can choose design a_1, a_2, or a_3. These are systems to convert wind into electrical energy. The designer (or his company) will operate the chosen system and sell the electricity. The gain is measured in \$ profit/day and the gain of any of the systems will depend on the daily average wind speed. We assume for simplification that there are four possible values of average wind speed W:

$$w_1 = 0 \leq W < 10,$$
$$w_2 = 10 \leq W < 20,$$
$$w_3 = 20 \leq W < 30,$$
$$w_4 = W \geq 30.$$

242

The profit/day for the three systems was determined by tests and is given in Table 10-1. Note that the entry in each block in Table 10-1 represents the gain assigned to a possible outcome.

Table 10-1

System	State of nature			
	w_1	w_2	w_3	w_4
a_1	−10	0	+20	+30
a_2	0	0	0	+30
a_3	+10	+10	+10	+10

Now assume that the probabilities of the various wind speeds are

$$P(w_1) = \tfrac{1}{12}, P(w_2) = \tfrac{3}{12}, P(w_3) = \tfrac{2}{12}, P(w_4) = \tfrac{6}{12}.$$

Note that by stating these probabilities with no mention of which system is chosen, we are implicitly assuming that the choice of system has no effect on the state of nature, that is, the average wind speed.

The expected value $E[V]_{a_1}$ associated with system a_1 is

$$E[V]_{a_1} = -10(\tfrac{1}{12}) + 0(\tfrac{3}{12}) + 20(\tfrac{2}{12}) + 30(\tfrac{6}{12}) = \tfrac{210}{12}.$$

Similarly

$$E[V]_{a_2} = 30(\tfrac{6}{12}) = \tfrac{180}{12},$$
$$E[V]_{a_3} = 10 = \tfrac{120}{12}.$$

Thus a_1 is the best system for the probabilities given. Different probability distributions of wind speeds could make either of the other systems best.

EXAMPLE 10-2

A control system is to position a shaft at a reference point which we will call 0°. The loss associated with positioning the shaft at any given position is assumed to be proportional to the square of the error. This loss versus position curve, or loss function, is shown in Figure 10-1.

Two control systems are proposed. The way either positions the shaft depends on the environmental conditions. A series of tests in the various environments has lead to the following probability density for position X using the first control system,

$$f_1(x) = \frac{1}{\sqrt{2\pi}\,.1} \exp\left[-\frac{x^2}{.02}\right]$$

For the second control system the position is described by

$$f_2(x) = \frac{1}{\sqrt{2\pi}\,.05} \exp\left[-\frac{(x-.01)^2}{.005}\right]$$

243

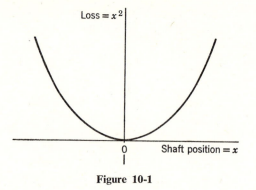

Figure 10-1

The expected loss for system 1 is

$$E[L]_1 = \int_{-\infty}^{\infty} x^2 \frac{1}{\sqrt{2\pi}\,.1} \exp\left[-\frac{x^2}{.02}\right] dx = .01.$$

The integral was readily evaluated because it represents the mean square error or the variance of a normal random variable.

The expected loss for system 2 is

$$E[L]_2 = \int_{-\infty}^{\infty} x^2 \frac{1}{\sqrt{2\pi}(.05)} \exp\left[-\frac{(x-.01)^2}{.005}\right] dx$$

$$= (.01)^2 + (.05)^2 = .0026.$$

Thus the second system is best.

Comments on Examples

The two simple examples given above illustrate decision problems when there are a small number of possible actions. In the first case the environment or the state of nature is modeled to have only four possible states. In the second there is assumed to be an uncountable number of possible states of nature. Thus in the first example the discrete form of expected value is used, while in the second example the continuous form of expected value is used. In the first example a gain is assigned while in the second example a loss is assigned.

The example of the next section is described in more detail because there is the additional complication of different possible observations and because there are important practical implications in the example.

244

10-3 DETECTION OF A SIGNAL IN NOISE

EXAMPLE 10-3

A binary communication system transmits one of two signals, say A or B. The communication system adds noise to the transmitted signal. The received signal (transmitted signal plus noise) is then operated on by the receiver (e.g., matched filter) and a voltage is the output. If there were no noise we assume the output would be 0 if A were transmitted, and the output would be 1 if B were transmitted. However the noise may cause the voltage to be any value. We must decide what action to take given the observed voltage output. We assume in this example that the only possible actions are say A or say B. We also assume that we will say A if the received signal is below y and we will say B if the received signal is above y. (Later we show that with reasonable assumptions this is indeed the optimum form of the decision.) Note that the decision is indexed by y, and y is assumed to be continuous.

We now model the problem more precisely. The probability that A is transmitted is a and the probability that B is transmitted is $1 - a$. The receiver's output voltage Y is the output with no noise plus the noise which is assumed to be normal with mean 0 and variance σ^2. Since the no noise output is zero when A is transmitted

$$f_{Y|A}(y) = \frac{1}{\sqrt{2\pi}\,\sigma} \exp\left[-\frac{y^2}{2\sigma^2}\right] \qquad (10\text{-}1)$$

The no noise output is one when B is transmitted. Thus

$$f_{Y|B}(y) = \frac{1}{\sqrt{2\pi}\,\sigma} \exp\left[-\frac{(y-1)^2}{2\sigma^2}\right]. \qquad (10\text{-}2)$$

The losses are given in Table 10-2.

Table 10-2

	State of Nature	
Action	A transmitted	B transmitted
say A	0	L_{AB}
say B	L_{BA}	0

For a fixed y the expected loss is

$$E[L]_y = L_{AB}P(\text{say } A \text{ and } B \text{ sent}) + L_{BA}P(\text{say } B \text{ and } A \text{ sent})$$

$$= L_{AB}P(\text{say } A \mid B \text{ sent})(1 - a) + L_{BA}P(\text{say } B \mid A \text{ sent})a \quad (10\text{-}3)$$

$$= L_{AB}(1 - a)\int_{-\infty}^{y} \frac{1}{\sqrt{2\pi}\,\sigma} \exp\left[-\frac{(\lambda - 1)^2}{2\sigma^2}\right] d\lambda$$

$$+ L_{BA}a \int_{y}^{\infty} \frac{1}{\sqrt{2\pi}\,\sigma} \exp\left[-\frac{\lambda^2}{2\sigma^2}\right] d\lambda$$

$$= L_{AB}(1 - a)\int_{-\infty}^{y-1} \frac{1}{\sqrt{2\pi}\,\sigma} \exp\left[-\frac{z^2}{2\sigma^2}\right] dz$$

$$+ L_{BA}a\left[1 - \int_{-\infty}^{y} \frac{1}{\sqrt{2\pi}\,\sigma} \exp\left[-\frac{\lambda^2}{2\sigma^2}\right] d\lambda\right],$$

$$E[L]_y = L_{AB}(1 - a)F_N(y - 1) + L_{BA}a[1 - F_N(y)], \quad (10\text{-}4)$$

where F_N is the distribution function of a normal random variable with mean 0 and variance σ^2.

To find the optimum value of y, $E[L]_y$ is differentiated with respect to y and set equal to zero. The solution y^* to this equation results in a minimum loss. Before performing the calculations, what is your guess of y^*? Is it $1/2$?

$$\frac{dE[L]_y}{dy} = L_{AB}(1 - a)\frac{1}{\sqrt{2\pi}\,\sigma} \exp\left[-\frac{(y - 1)^2}{2\sigma^2}\right]$$

$$- L_{BA}a \frac{1}{\sqrt{2\pi}\,\sigma} \exp\left[-\frac{y^2}{2\sigma^2}\right].$$

Setting the derivative equal to zero and solving for y^*

$$L_{AB}(1 - a) \exp\left[-\frac{(y^* - 1)^2}{2\sigma^2}\right] = L_{BA}a \exp\left[-\frac{y^{*2}}{2\sigma^2}\right],$$

$$\frac{y^{*2} - (y^* - 1)^2}{2\sigma^2} = \ln\left(\frac{L_{BA}a}{L_{AB}(1 - a)}\right),$$

$$2y^* - 1 = 2\sigma^2 \ln\left(\frac{L_{BA}a}{L_{AB}(1 - a)}\right),$$

$$y^* = \tfrac{1}{2} + \sigma^2 \ln\left(\frac{L_{BA}a}{L_{AB}(1 - a)}\right). \quad (10\text{-}5)$$

Note that if $L_{BA} = L_{AB}$ and $a = 1 - a$ then $y^* = 1/2$, as predicted. However the dividing line moves up, meaning we are more likely to say A, as

L_{BA}, the loss associated with saying B when A was sent, and a, the probability of A, increase. Similarly the dividing line moves down as L_{AB} and $(1 - a)$ increase. The movement is proportional to the variance of the noise. A little thought will make all of these seem very reasonable and natural.

Derivation of Optimum Form of Decision

We now assume that the optimum decision is simply to say A when the voltage is in some region R and to say B when the voltage is in a region $\bar{R}$. We show that R should be from $-\infty$ to y^*. Using (10-3)

$$E[L] = L_{AB}(1 - a)\int_R \frac{1}{\sqrt{2\pi}\,\sigma} \exp\left[-\frac{(y - 1)^2}{2\sigma^2}\right] dy$$

$$+ L_{BA}a\int_{\bar{R}} \frac{1}{\sqrt{2\pi}\,\sigma} \exp\left[-\frac{(y)^2}{2\sigma^2}\right] dy$$

$$= L_{AB}(1 - a)\int_R \frac{1}{\sqrt{2\pi}\,\sigma} \exp\left[-\frac{(y - 1)^2}{2\sigma^2}\right] dy$$

$$+ L_{BA}a\left\{1 - \int_R \frac{1}{\sqrt{2\pi}\,\sigma} \exp\left[-\frac{(y)^2}{2\sigma^2}\right] dy\right\}$$

$$= L_{BA}a + \int_R\left\{L_{AB}(1 - a)\frac{1}{\sqrt{2\pi}\,\sigma} \exp\left[-\frac{(y - 1)^2}{2\sigma^2}\right]\right.$$

$$\left. - L_{BA}a\frac{1}{\sqrt{2\pi}\,\sigma} \exp\left[-\frac{y^2}{2\sigma^2}\right] dy.\right\}$$

To minimize $E[L]$, R should be chosen to be the region where the integrand is negative, that is,

$$L_{BA}a\frac{1}{\sqrt{2\pi}\,\sigma} \exp\left[-\frac{y^2}{2\sigma^2}\right] > L_{AB}(1 - a)\frac{1}{\sqrt{2\pi}\,\sigma} \exp\left[-\frac{(y - 1)^2}{2\sigma^2}\right]. \quad (10\text{-}6)$$

The inequality (10-6) can be reduced to

$$\frac{L_{BA}a}{L_{AB}(1 - a)} > \exp\left[\frac{(2y - 1)}{2\sigma^2}\right],$$

or

$$\ln\left(\frac{L_{BA}a}{L_{AB}(1 - a)}\right) > \frac{2y - 1}{2\sigma^2},$$

or

$$y < \tfrac{1}{2} + \sigma^2 \, \ln\left(\frac{L_{BA}a}{L_{AB}(1 - a)}\right)$$

which shows R is the region from $-\infty$ to

$$y^* = \tfrac{1}{2} + \sigma^2 \ \ln \ \left(\frac{L_{BA}a}{L_{AB}(1-a)}\right).$$

10-4 SUMMARY OF PROBABILISTIC DECISION THEORY

The previous two sections gave examples of typical engineering decision problems. Here we emphasize the important constituents of a decision problem which are applicable whenever a "best" decision must be made and a probabilistic model is available. Basically there are three necessary components when a problem requiring a decision has been defined:

1. The possible actions,
2. The possible states of nature,
3. A criterion for judging which is the *best* of the possible outcomes.

While describing the possible actions and/or states of nature is often a difficult modeling task, no general description of how to accomplish this is attempted here. However some discussion of a criterion by which outcomes are judged is in order.

A gain or utility may be assigned to each outcome. In this case the best action is the one that maximizes the expected value of utility or gain.

Alternatively a loss may be assigned to each outcome. In this case the best action is the one that minimizes the expected value of the loss.

Basically the two approaches are the same,* with a loss function being (except for a constant which has no effect on the choice of decision) the negative of a gain function.

The preceding examples illustrated different gain and loss functions. Here we discuss two loss functions commonly used in engineering. The first is the squared error loss function. That is, the loss is the square of the difference between the actual outcome and the desired outcome. For example assume that a power supply is expected to produce 10 V; if the actual output is y volts, then the loss assuming a squared error loss function is

$$L(y) = (10 - y)^2.$$

This loss function is plotted in Figure 10-2.

The second common loss function is the 0-1 loss function. In this case it is assumed that the specifications of a system provide a range of values of system performance which are acceptable. Anything within this range has zero loss, and any value outside this range has a loss of one. For example, a

* Problem 9 asks for a proof.

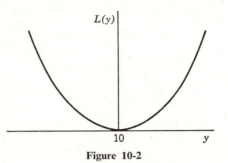

Figure 10-2

zero-one loss function for a circuit designed to produce 10 ± 1 V is shown in Figure 10-3.

The two loss functions given above are very commonly used and are very reasonable measures. However they do not consider cost, appearance, quality, weight, and other factors which may be important. Simply restated, the two example loss functions are good measures for comparing designs if the other factors are equal.

Note that when a loss or gain function is chosen, it must be used throughout the problem. For example, two actions cannot be compared by evaluating one with a squared error loss function and the other with a zero-one loss function.

Normal Form

A decision problem is often displayed as a table or matrix where the possible states of nature define columns and the possible actions define rows (e.g., see Table 10-1 or 10-2). The intersection of a column and a row defines an outcome. The square representing the outcome is usually filled with the gain or loss assigned to that outcome.

Extensive Form

The extensive form of a decision problem is very similar to a tree diagram. An example is shown in Figure 10-4. The diagram uses crosses at the points

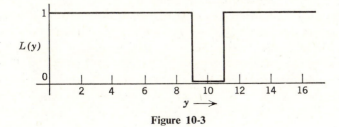

Figure 10-3

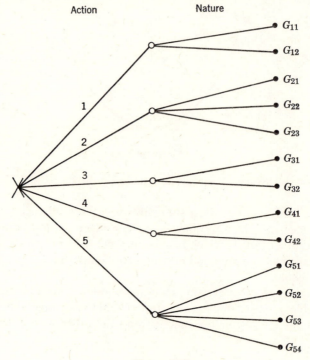

Figure 10-4

where an action is chosen by the decision maker, and circles at points where nature chooses one of its possible states. The two choices result in an outcome which is assigned a gain G_{ij}.

In Figure 10-4 five possible actions are open to the decision maker. If he chooses action 1 then one of two states of nature is possible; if he chooses action 2, then one of three states of nature is possible; etc.

Just as a tree diagram is useful in describing a sequence of experiments, the extensive form of a decision problem is particularly useful in describing a problem when a sequence of decisions needs to be made. However the extensive form is useful for single decisions also. We illustrate by discussing the extensive form of the example of detecting a signal in noise.

The problem is shown in extensive form in Figure 10-5. This is not a very useful diagram from the decision maker's point of view, because at the place (×) where the action is chosen, the decision maker has not observed what was sent and what noise was added. Indeed if he had, there would be no problem. We redraw the tree in Figure 10-6 so that it portrays the problem from the decision maker's point of view. That is, in Figure 10-6 the things to the left of (×) are the things he knows when the decision is to be made. Note that the

250

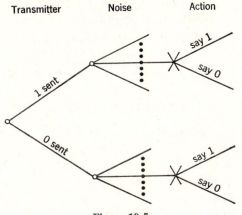

Figure 10-5

decision maker observes the received signal and then makes a decision without knowing what has been sent.

Figure 10-6 emphasizes the reason why the decision problem of Section 10-3 was more complicated than the earlier examples. In order to solve the complete problem we had to specify what we would do for each possible received signal. That is, there is a separate decision problem for each value of received signal, and we must solve each one. Fortunately, in this example, the problem can be solved in parametric form.

Comment on Decisions Theory

Note that decision theory is not a panacea. The engineer must define the problem, list the possible actions and states of nature, choose the gain (or loss) function, and assign the probability distribution of the states of nature. Problems at the end of this chapter illustrate this.

However, decision theory does provide a logical method by which "best" actions may be chosen.

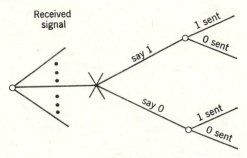

Figure 10-6

The "best" action using this logical method may not be a good one as viewed by hindsight because the state of nature chosen in a particular trial may result in an outcome that has a high loss. The "best" action was best on the average; however, if the state of nature were known, a different action might be better.

10-5 UNKNOWN PROBABILITY DISTRIBUTIONS

In this section we extend probabilistic decision theory to include the case in which we assume that there is a probability distribution of the various states of nature, but the probability distribution is (at least partially) unknown.

What do we do in this case? The general answer to this very important question is to perform experiments, take data, and use the data to estimate the probability distribution. Because data or statistics are involved, the associated theory is called statistical decision theory.

Before considering using data, we consider what can be done if it is not possible to take data. One approach is to guess the probabilities. The guessed probabilities will work quite well in the theory presented. In fact decision theory* does not distinguish known probabilities from guessed probabilities.

But suppose the decision maker knows nothing about the state of nature. It is obvious that the decision maker is not in an enviable position. However certain actions called inadmissible actions can be eliminated. The following example will be used for illustration.

Example 10-4

There are four possible control systems that can be placed in a satellite. Their performance is sensitive to the temperature in the satellite. The temperature is not known but it will be between 50° and 100°. The squared error of the control system is used as a measure of the loss. The conditional squared error given the temperature has been determined in a bench test for each system and is plotted in Figure 10-7.

Admissible Actions

An action a_j is inadmissible if it is dominated by any other action for all possible states of nature. An action a_j is dominated by a_k if the gain $G_{a_j e_i}$ assigned to action a_j and state of nature e_i is such that

$$G_{a_j e_i} \leq G_{a_k e_i} \quad \text{for all } i,$$

and
$$G_{a_j e_i} < G_{a_k e_i} \quad \text{for some } i.$$

* This is consistent with other engineering and physical models. Any number assigned to a parameter in such models is treated as correct. The models do not take into account uncertainty in the parameters due to measurement errors, parameter variation, etc.

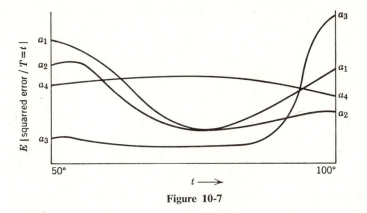

Figure 10-7

This means we need not consider a_j since a_k is never worse and sometimes better. Thus a_j is called inadmissible.

An action is admissible if it is not inadmissible.

Refer to Example 10-4. Is there any action that is dominated by another action? Since loss functions are shown rather than gains, then we must ask if there is an a_j such that the loss $L_{a_j e_i}$ assigned to action a_j and state of nature e_i satisfies

$$L_{a_j e_i} \geq L_{a_k e_i} \quad \text{for all } i$$

and

$$L_{a_j e_i} > L_{a_k e_i} \quad \text{for some } i.$$

From Figure 10-7, a_1 is dominated by a_2, thus a_1 is inadmissible. Note that a_4 has a loss that is always greater than that assigned to some a_j but it is still admissible because its loss is not always above the loss of the same a_j.

Minimax Solution

A minimax solution is one method that has been popular in theoretical studies of decision making. The minimax solution is the action that minimizes the maximum loss. That is, a_j is the minimax solution if

$$\max_{e_i} L_{a_j e_i} \leq \max_{e_i} L_{a_k e_i} \quad \text{for all } k.$$

In normal form $\max_{e_i} L_{a_j e_i}$ is the largest loss in the row corresponding to action a_j. In Example 10-4 a_4 is the minimax solution.

The minimax solution seems very reasonable if one endows nature with a cruel personality and the power to change environment to thwart whatever action the decision maker chooses. This might be the case if noise were the major component of the environment and the noise was caused by a jammer. However in the case where nature is not governed by an antagonist, an

253

engineer who bases his decisions on the worse case is likely to be less successful than his less pessimistic competitors.

We mention briefly that one could also define a minimin criterion that selected the action with the minimum of the minimum losses. This very optimistic criterion is almost never used, either theoretically or practically.

Equally Likely

It has been proposed that if nothing is known about the states of nature, then all will be assumed to be equally likely. Thus the action with the highest average gain or lowest average loss will be chosen.

In Example 10-4, a_3 would be chosen if all values of temperature were equally likely.

The assumption that all states of nature are equally likely simply because we do not know the probabilities is not satisfying. However, as was stated earlier, it is reasonable to expect that we will not find a satisfying means for choosing the best action when we do not know the probabilities. The remainder of the chapter considers using learned probabilities to make decisions.

10-6 STATISTICAL DECISION THEORY

We continue consideration of decision problems except that we now consider a "statistical" description of nature as opposed to a probabilistic description of nature as discussed earlier. In a statistical description of nature, it is assumed that there is some randomness, that is, a probabilistic model is called for but the probability distribution is unknown. Tests and the results of tests are used via the methods of Chapter VIII to learn the probability distributions.

Note that a statistical description of nature, that is, one in which data must be taken before a decision is made, is the almost universal situation in engineering. We always want to test—during research, during development, during production, and during use.

When tests are possible and the results of these tests influence decisions, we must add to our list of the constituents of a decision problem. To the original list

1. The possible actions,
2. The possible states of nature,
3. A criterion for "best;"

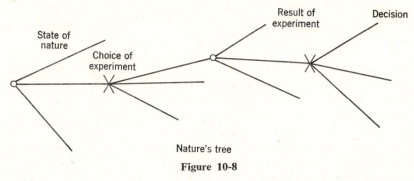

Nature's tree

Figure 10-8

we add

4. The possible tests,
5. The possible test results.

In addition, where necessary it is assumed that the gain function is extended to cover the cost of the tests, and that the test results have some (known) relation to the state of nature.

Extensive Form of Statistical Problems

The extensive form of a decision problem in which experiments and their results are useful is shown in Figure 10-8.

Logically, the state of nature must be determined before experiments are conducted to help determine the state of nature. However from the point of view of the decision maker, the tree is as shown in Figure 10-9.

To the decision maker the state of nature is not known when he chooses the experiment. However the result of the experiment, which is an indication of the state of nature, is known when the decision is made.

We now consider decisions when experiments have been performed to help determine the state of nature. Later in the chapter we consider the choice of experiment.

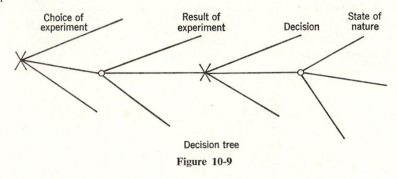

Decision tree

Figure 10-9

10-7 EMPIRICAL DECISION THEORY

In empirical decision theory, the conditional expected gain $E[G \mid \theta]_{a_i}$ (or conditional expected loss) given the random variable θ, of an action a_i, is computed by techniques discussed in previous chapters.

The unknown θ represents the state of nature. In this book θ will be a single random variable. However there is no reason that θ cannot be a random vector. In the context of design θ may describe some part of the environment with which the proposed design must react, or θ may describe some unknown input to the system. We assume that nothing is known about θ; that is, we know neither a likelihood function nor a prior distribution. However independent samples $\theta_1, \theta_2, \ldots, \theta_n$, which have the same distribution as θ, are assumed to be available. Using the method of generating an empirical distribution function presented in Chapter VIII, the expected gain G of an action given the n samples, may be computed by

$$E[G] = \frac{1}{n} \sum_{\text{all } i} E[G \mid \theta = \theta_i]. \tag{10-7}$$

This expected gain (or expected loss) is then used to compare proposed actions. Two examples illustrate.

EXAMPLE 10-5

A system is to control the rate R at which a certain catalyst is fed into a chemical process. The rate of profit G made by the plant is assumed to be

$$G = -40 + 1000R \quad \text{if} \quad R < .05$$
$$= 60 - 1000R \quad \text{if} \quad R \geq .05.$$

Thus the desired rate is .05 and there is a significant penalty for deviations.

The present control system has been in operation for some time and records show that the expected rate of profit is 8. A new control system is proposed and the salesman claims it will increase your profit. You are allowed to test it; the samples of the rate at which it allows catalyst to be fed into the process are

.051, .045, .055, 0.48, .049,

.049, .050, .052, .050, .049.

Is the new control system better?

Using (10-7) to evaluate the expected gain, we find

$$E(G) = \tfrac{1}{10}[(60 - 51) + (-40 + 45) + (60 - 55)$$
$$+ (-40 + 48) + 2(-40 + 49) + (60 - 50)$$
$$+ (60 - 52) + (60 - 50) + (-40 + 49)]$$
$$= \tfrac{1}{10}(9 + 5 + 5 + 8 + 18 + 10 + 8 + 10 + 9) = 8.2.$$

Thus the new control system is better and in fact is worth .2 more than the old control system. If the system can be rented for this figure it should be tried. However it must be pointed out that a different set of samples might lead to a different conclusion.

EXAMPLE 10-6

We return to the example of saying A or saying B after a signal is received. This is the same problem as discussed in Example 10-3; however we now assume that the noise N is unknown; that is, we do not know the distribution function of N. If the distribution function of N is not known, then it is impossible to specify the form of the best detector; that is, we cannot show that it is best to say A if the voltage is below some y^*, and say B if the voltage is above y^*. However in this example we will look for the best decision of the form: call any y below y^* an A, and call any y above y^* a B.

To simplify this example we assume that

$$L_{AB}(1 - a) = L_{BA}a. \tag{10-8}$$

With these assumptions the expected loss is [see (10.4)]

$$E[L]_y = L_{AB}(1 - a)F_N(y - 1) + L_{BA}a[1 - F_N(y)].$$

We wish, by choice of y^*, to minimize $E[L]_y$. Note that F_N is unknown but n samples of the noise can be used to generate an empirical distribution function $\hat{F}_{N,n}$ where

$$\hat{F}_{N,n}(x) = \frac{\text{No. of samples } N_1, N_2, \ldots, N_n, \text{ that are no greater than } x}{n}.$$

Thus with $\hat{F}_{N,n}$ substituted for F_N in (10-4)

$$E[L]_y = \frac{1}{n} L_{AB}(1 - a)[\text{No. of samples} \leq y - 1]$$

$$+ \frac{1}{n} L_{BA}(a)[\text{No. of samples} > y].$$

Because of the assumption (10-8)

$$E[L]_y = \frac{1}{n} L_{AB}(1 - a)[n - \text{No. of samples between } y - 1 \text{ and } y].$$

To find the value y^* that minimizes $E[L]_y$ we should choose y^* to maximize the number of samples between $y - 1$ and y.

For instance if the samples of noise are .21, $-.47$, .59, $-.15$, $-.29$, .17, $-.37$, $-.43$, .28, .10, .06, $-.10$, $-.71$, $+32$, $-.42$, $-.51$, .09, $-.18$, $+.15$, $-.38$, then Table 10-3 shows the results. With these data y^* should be chosen to be somewhere close to .4.

Table 10-3

y	No. of samples between $y - 1$ and y
.7	13
.6	15
.5	17
.4	18
.3	17
.2	16

This example illustrates that with the small amount of data, the empirical distribution function calls for an answer that is quite different from 1/2. Perhaps one might wish to question this result, but the cause for question results from a prior judgment which is not included in empirical solutions. The remainder of this chapter discusses Bayesian decision theory which does use prior judgment.

10-8 BAYESIAN DECISION THEORY

The basic idea of Bayesian decision theory is that the conditional (upon θ) expected gain of a proposed action is found; the Bayesian posterior distribution of θ is found from the results of tests; and the posterior distribution of θ is used to average the conditional expected gain.

An example will illustrate. In the example θ represents the probability of a good unit.

EXAMPLE 10-7

The company can produce a gadget that costs $10 per unit to produce. The probability a produced unit is good is θ, and we assume that the production can be described by Bernoulli trials. The probability θ of a good unit is assumed to be a continuous random variable, equally likely to be anywhere between 0 and 1.

To be specific, assume there are 10 buyers who will buy good gadgets at $20 per unit. We must start 10 units through the production line all at once; that is, it is not possible to start say 10, see how many are good, then start some more, test them, etc. The decision to be made is whether to start 10 units or none at all.

To help determine θ, one unit will be run through the production process and tested. However we assume that the test is destructive; that is, the tested unit, if good, cannot be sold. The only value of the test is to help determine θ.

The conditional probability that there are k good units when 10 are started is

$$P(k \text{ good units} \mid \theta) = \binom{10}{k} \theta^k (1 - \theta)^{10-k}.$$

Before solving the problem as given, we note that if no units were tested, the expected value of not starting any units is 0. Also the expected gain associated with starting 10 units can be easily calculated as follows:

$$E[G \mid \theta] = E[20k - 100 \mid \theta] = 20E[k \mid \theta] - 100.$$

Using the expected value of a binomial random variable

$$E[k \mid \theta] = 10\theta.$$

Thus

$$E[G \mid \theta] = 200\theta - 100.$$

Using the theorem that

$$E[G] = E\{E[G \mid \theta]\},$$

and the assumed uniform density of θ, we have

$$\begin{aligned} E[G] &= E[200\theta - 100] \\ &= 200E[\theta] - 100 \\ &= (200)\tfrac{1}{2} - 100 = 0. \end{aligned}$$

Thus with no testing the expected value is 0 if 10 units are started.

We return to the original problem where one unit is tested. The problem in extensive form is shown in Figure 10-10.

In this problem we have assumed that one unit will be tested. (A later continuation of this example determines how many to test.) From Figure 10-10 it is obvious that we have two decisions to be made: (a) what to do if the tested unit is good; (b) What to do if the tested unit is bad.

We first consider the case where the tested unit is good. Then (see Chapter VIII)

$$f_{\theta \mid 1 \text{ good}}(\lambda) = \frac{\lambda 1}{\displaystyle\int_0^1 x \, dx} = 2\lambda, \qquad 0 \le \lambda \le 1,$$

$$= 0, \qquad\qquad\qquad \text{elsewhere.}$$

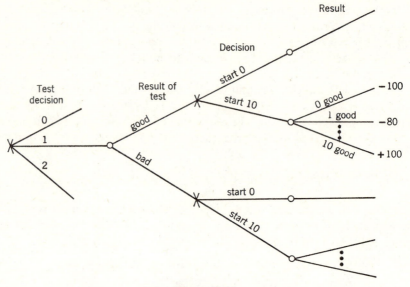

Figure 10-10

Also
$$E[\theta \mid 1 \text{ good}] = \int_0^1 \lambda \, 2\lambda \, d\lambda = 2 \left.\frac{\lambda^3}{3}\right|_0^1 = \tfrac{2}{3}.$$

Thus if 10 units are produced

$$E[G] = E\{E[G \mid \theta]\}$$
$$= E[200\theta - 100]$$
$$= (200)\tfrac{2}{3} - 100 \simeq 33.$$

As before, the expected value associated with producing 0 units is 0, and the correct decision is to start the 10 units into production if the tested unit is good.

We now consider the case where the tested unit is bad. Then

$$f_{\theta \mid 1 \text{ bad}}(\lambda) = \frac{(1-\lambda)1}{\displaystyle\int_0^1 (1-x)\, dx} = 2(1-\lambda), \qquad 0 \leq \lambda \leq 1,$$

$$= 0, \qquad\qquad\qquad\qquad \text{elsewhere.}$$

$$E[\theta \mid 1 \text{ bad}] = \int_0^1 \lambda 2(1-\lambda)\, d\lambda = \left. \lambda^2 - \tfrac{2}{3}\lambda^3 \right|_0^1 = \tfrac{1}{3}.$$

Thus, if 10 units are produced,

$$E[V] = E[200\theta - 100] = (200)\tfrac{1}{3} - 100 \simeq -33.$$

Thus if the tested unit is bad, one should not produce the units.

The above results can be summarized into the following rather natural decision rule. If the tested unit is good, produce; if not good, don't produce.

EXAMPLE 10-8

This is simply a continuation of the previous example except that we now consider that 2 units rather than 1 unit have been tested.

1. If both tested units are good

$$f_{\theta|2 \text{ good}}(\lambda) = \frac{\lambda^2 1}{\displaystyle\int_0^1 x^2 \, dx} = 3\lambda^2, \qquad 0 \le \lambda \le 1,$$

$$= 0, \qquad\qquad\qquad \text{elsewhere.}$$

$$E[\theta] = \int_0^1 \lambda 3\lambda^2 \, d\lambda = \tfrac{3}{4}.$$

If 10 units are started

$$E[G] = (200)\tfrac{3}{4} - 100 = 50.$$

2. If 1 is good and 1 is bad

$$f_{\theta|1 \text{ good and 1 bad}}(\lambda) = \frac{\lambda(1 - \lambda)1}{\displaystyle\int_0^1 x(1 - x) \, dx} = 6\lambda(1 - \lambda), \qquad 0 \le \lambda \le 1,$$

$$= 0, \qquad\qquad\qquad\qquad \text{elsewhere.}$$

$$E[\theta] = \int_0^1 \lambda 6\lambda(1 - \lambda) \, d\lambda = \tfrac{1}{2},$$

$$E[G] = (200)\tfrac{1}{2} - 100 = 0.$$

3. If both are bad,

$$f_{\theta|2 \text{ bad}}(\lambda) = 3(1 - \lambda)^2, \qquad 0 \le \lambda \le 1,$$

$$= 0, \qquad\qquad \text{elsewhere.}$$

$$E[\theta] = \tfrac{1}{4}$$

$$E[G] = (200)\tfrac{1}{4} - 100 = -50.$$

Thus if both tested units are good, produce; if both bad, don't produce; and if one is good and one is bad, it makes no difference which decision is made.

Comments on Preceding Example

Other assumptions about the same problem are discussed in problems at the end of the chapter. The obvious continuation of this example is to now decide whether it is better to test 1 unit or 2 units, or perhaps none or more than two. This is discussed in the last section on testing decisions.

We can also note that in the preceding calculations, we could have used a result from Chapter IX, that the expected value of θ given a uniform prior and m good units in n tests is

$$E[\theta \mid m \quad \text{in} \quad n] = \frac{m+1}{n+2}.$$

We emphasize that because of the loss functions of this simple example, only the mean of the posterior distribution of θ was needed. Obviously different loss functions would require different information from the posterior distribution.

EXAMPLE 10-9

Refer to Example 10-2 and assume that the mean of the control system's output has been adjusted to zero. In this case the conditional expected loss given σ^2 is simply σ^2, that is

$$E[L \mid \sigma^2] = \sigma^2.$$

Assume that m independent samples $X_1, X_2, \ldots, X_m$, of the output X are available. What is the expected loss, given the data and assuming that the prior distribution of σ^2 is

$$f_{\sigma^2}(\lambda) = \frac{1}{\lambda^2} e^{-1/\lambda}, \qquad \lambda > 0,$$

$$= 0, \qquad \lambda \le 0.$$

Assuming X is normal with mean zero and variance σ_1^2, from problem 19 in chapter 8

$$f_{\sigma^2 \mid X_1, \ldots, X_m}(\lambda \mid x_1, \ldots, x_m)$$

$$= \frac{\left(\dfrac{\displaystyle\sum_{i=1}^{m} x_i^2 + 2}{2} \right)^{(m/2)+1}}{(\lambda)^{(m/2)+2} \Gamma\left(\dfrac{m}{2} + 1 \right)} \exp\left[-\left(\frac{\displaystyle\sum_{i=1}^{m} x_i^2 + 2}{2\lambda} \right) \right], \qquad \lambda > 0,$$

$$= 0, \qquad \lambda \le 0.$$

Now
$$E[L] = E\{E[L \mid \sigma^2]\} = E[\sigma^2].$$

From $f_{\sigma^2 | X_1, \ldots, X_m}$ the expected value of σ^2 given $X_1, \ldots, X_m$ is

$$E[\sigma^2 \mid X_1 = x_1, \ldots, X_m = x_m] = \int_0^\infty \lambda f_{\sigma^2 | X_1, \ldots, X_m}(\lambda \mid x_1, \ldots, x_m) \, d\lambda$$

$$= \frac{\left(\dfrac{\sum\limits_{i=1}^{m} x_i^2 + 2}{2} \right)^{(m/2)+1}}{\Gamma\left(\dfrac{m}{2} + 1 \right)} \int_0^\infty \frac{1}{\lambda^{\frac{m}{2}+1}} \exp\left(-\frac{\sum\limits_{i=1}^{m} x_i^2 + 2}{2\lambda} \right) d\lambda$$

$$= \frac{\left(\dfrac{\sum\limits_{i=1}^{m} x_i^2 + 2}{2} \right)^{(m/2)+1}}{\Gamma\left(\dfrac{m}{2} + 1 \right)} \frac{\Gamma(m/2)}{\left(\dfrac{\sum\limits_{i=1}^{m} x_i^2 + 2}{2} \right)^{m/2}}$$

$$= \frac{\sum\limits_{i=1}^{m} x_i^2 + 2}{2} \frac{1}{m/2} = \frac{\sum\limits_{i=1}^{m} x_i^2 + 2}{m} \, .$$

Thus the Bayesian estimate of σ^2 is very nearly the mean square value of the readings. The only difference is the factor $2/m$, which is the influence (bias) of the prior distribution which was assumed for this problem.

Summary of Bayesian Decision Theory

The basic idea is to find the expected value of the conditional gain of each action, where the expected value is taken with respect to the Bayesian distribution of θ. If θ is continuous

$$E[G] = \int_{-\infty}^{\infty} E[G \mid \phi] f_{\theta | \text{data}}(\phi) \, d\phi. \tag{10-9}$$

If θ is discrete

$$E[G] = \sum_{i=1}^{\infty} E[G \mid \theta_i] P(\theta = \theta_i \mid \text{data}). \tag{10-10}$$

Note that as in case of probabilistic decisions, there may be a finite number or an infinite number of actions. All of the examples considered a small number of possible actions. If an infinite number of actions are possible, then the minimization of risk or maximization of expected gain becomes a more difficult mathematical problem.

To this point we have assumed that the samples are already available. In the next two sections we consider the testing decision, or how to choose the best test.

10-9 MAXIMUM AMOUNT TO BE SPENT ON TESTING

We assume that the state of nature is described by θ and that there is a prior distribution function F_θ. The best action can then be chosen by methods already described. Call this best action a^*. We can then compute

$$E[G]_{a^*}.$$

Moreover we also presumably know $E[G \mid \theta=\phi]_{a^*}$ for all possible values ϕ that θ might take on. Given $\theta = \phi$, some other action a_i might have a conditional expected gain greater than a^*, that is, it may be that

$$E[G \mid \theta = \phi]_{a_i} - E[G \mid \theta = \phi]_{a^*} = G(a_i, \phi) > 0.$$

If the state of nature were known to be ϕ, we would pick the action that maximizes $E[G \mid \theta = \phi]$. Call $a(\phi)$ the action that maximizes $E[G \mid \theta = \phi]$. The difference

$$C(\phi) = E[G \mid \theta = \phi]_{a(\phi)} - E[G \mid \theta = \phi]_{a^*}.$$

is called the cost of uncertainty. Note that $C(\phi) \geq 0$.

What does it cost us not to know exactly what state of nature exists? That is, on the average how much could we increase the gain if a clairvoyant could foretell the exact state of nature? This is of interest because it distinguishes the expected gain of a best decision under the probabilistic assumption from the expected gain of a best decision if everything were known, and thus tells how much value should be attached to the clairvoyant's information, or, more realistically, how much is the maximum gain that could be attained by tests designed to determine the state of nature.

If the clairvoyant says that the true state of nature is ϕ, then to be fair we should pay the clairvoyant $C(\phi)$, because he has saved us this much. Now how much should we contract to pay the clairvoyant before knowing what value of θ he will say? We should contract to pay the clairvoyant $E[C(\theta)]$, which is the expected value of uncertainty. If θ is discrete

$$E[C(\theta)] = \sum_{\text{all } i} C(\phi_i)P(\theta = \phi_i)$$

$$= \sum_{\text{all } i} (E[G \mid \theta = \phi_i]_{a(\phi_i)} - E[G \mid \theta = \phi_i]_{a^*})P(\theta = \phi_i), \quad (10\text{-}11)$$

$$E[C(\theta)] = \sum_{\text{all } i} E[G \mid \theta = \phi_i]_{a(\phi_i)}P(\theta = \phi_i) - E[G]_{a^*}. \quad (10\text{-}12)$$

If θ is continuous

$$E[C(\theta)] = \int_{-\infty}^{\infty} C(\phi) f_\theta(\phi)\, d\phi \tag{10-13}$$

$$= \int_{-\infty}^{\infty} E[G \mid \theta = \phi]_{a(\phi)} f_\theta(\phi)\, d\phi - E[G]_{a^*}. \tag{10-14}$$

A very important interpretation of the expected cost of uncertainty is that it is the maximum amount that should be spent on testing. Because any experimental program can provide no more information than our hypothetical clairvoyant, no more should be invested in testing than would be paid the clairvoyant.

EXAMPLE 10-10

Refer to Example 10-1. What is the maximum amount to be spent on testing? From Table 10-1 the best decision and its gain are given in Table 10-4.

Table 10-4

	State of nature			
	w_1	w_2	w_3	w_4
Best system	a_3	a_3	a_1	a_1 or a_2
Conditional gain of best system	10	10	20	30
$C(w)$	20	10	0	0

Thus the expected cost of uncertainty is, using (10-11),

$$E[C(w)] = 20\tfrac{1}{12} + 10\tfrac{3}{12} + 0\tfrac{2}{12} + 0\tfrac{6}{12} = \tfrac{50}{12}.$$

Alternately from (10-12)

$$E[C(w)] = 10\tfrac{1}{12} + 10\tfrac{3}{12} + 20\tfrac{2}{12} + 30\tfrac{6}{12} - \tfrac{210}{12} = \tfrac{260}{12} - \tfrac{210}{12} = \tfrac{50}{12}.$$

EXAMPLE 10-11

Refer to Example 10-7. What is the maximum amount to be spent on testing? With no testing it was shown that either action was equally good. We assume we chose the action "start no units." In this case if the clairvoyant tells us the value of θ is below $1/2$, we do not change the action. However if we are told that the value of θ is above $1/2$, we change the action and $C(\phi) = 200\phi - 100$. $C(\phi)$ is shown in Figure 10-11.

$$E[C(\phi)] = \int_{1/2}^{1} (200\phi - 100)\, d\phi = 100\phi^2 - 100\phi \Big|_{1/2}^{1}$$

$$= 100 - 100 - 25 + 50 = 25.$$

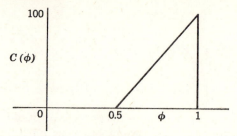

Figure 10-11

Note that if we had taken the action "start 10 units"

$$E[C(\phi)] = \int_0^{1/2} 0 - (200\phi - 100)\, d\phi = 25$$

which agrees with the earlier answer, as it must.

10-10 THE TEST DECISION

We have discussed the optimum decision given the test data. We now consider the best test. Assuming that the best action has been specified for all results of all possible experiments, the expected gain of the best decision can be associated with every result of the experiment. Then Figure 10-9 can be redrawn as shown in Figure 10-12.

In Figure 10-12 $E[G]_{a_i^*}$ is the conditional expected gain (or expected loss could be used) of the best action given the result of the experiment. Note the best action may depend on the result of the experiment.

At this stage we have simply another decision problem of exactly the type solved before. The action in this case is the choice of experiment, and the state of nature is replaced by the result of the experiment.

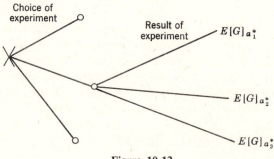

Figure 10-12

We point out that the idea presented here is a fundamental idea in solving sequential decision problems. One starts at the end of the tree and works backward to the beginning. This is often called "backward induction." Backward induction provides a logical method of solving multistage or sequential decision problems. However the amount of calculations required is usually prohibitive. Dynamic programming and a digital computer are essential tools.

Because the evaluation of all possible experiments, results, actions, and outcomes is sometimes too difficult, the expected cost of uncertainty is useful. If one finds the expected cost of uncertainty, then this is an upper bound on the amount to be spent on testing and thus limits the possible tests.

We now illustrate finding the best test by continuing an earlier example.

EXAMPLE 10-12

This is a continuation of Examples 10-7 and 10-8. The problem now is to determine how many units to test. Assume that the cost of a test is $10.

The extensive form of the problem showing the optimum decisions and the associated expected gain is shown in Figure 10-13. The expected gains are results of examples 10-7 and 10-8. Note that the expected gains shown do not include the cost of testing.

We now evaluate the possible test decisions.

1. If we test 0

$$E[G] = 0.$$

2. If we test 1

$$E[G \mid \text{test } 1] = P(\text{good})33 + P(\text{bad})0 - 10$$

$$\text{where } P(\text{good}) = E[\theta] \quad \text{and}$$

$$P(\text{bad}) = E[1 - \theta]$$

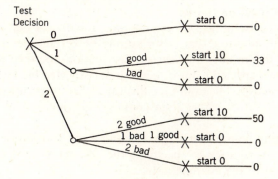

Figure 10-13

267

and the expected values are taken with respect to the prior distribution of θ. Thus since $E[\theta] = 1/2$

$$E[G \mid \text{test } 1] = \tfrac{1}{2}(33) + \tfrac{1}{2}0 - 10 \simeq +6.7.$$

3. If we test 2

$$E[G \mid \text{test } 2] = P(2 \text{ good}) \, 50 + P(1 \text{ good and 1 bad}) \, 0 + P(2 \text{ bad}) \, 0 - 20,$$

$$P(2 \text{ good}) = \int_0^1 \theta^2 \, d\theta = 1/3.$$

Thus $$E[G \mid \text{test } 2] = 1/3 \, (50) - 20 \simeq -3.$$

Thus the best testing decision is to test one unit.

Perhaps testing three units might be even better. We answer this question by recalling that the expected cost of uncertainty is 25 (see Example 10-11). Because \$25 is the maximum amount to be spent in testing, there is no need to consider testing three or more units because this will always result in a smaller expected gain.

10-11 SUMMARY

Probabilistic decision theory is the study of the results of actions and states of nature. The results are quantified by assigning gains or losses to each outcome. The gain used as an example was money while squared error and zero-one loss functions were mentioned as frequently used loss functions. The best decision rule is the one that chooses the action that maximizes the expected gain. The expected gain is usually conditional upon the available information concerning the state of nature. The "best" action is best on the average; however if the state of nature were known, a different action might be chosen. The average cost of uncertainty measures the risk introduced by uncertainty, and there is no way of eliminating this risk so long as there is uncertainty. The average cost of uncertainty provides an upper limit on the amount to be spent on testing.

Statistical decision theory is the same as probabilistic decision theory with the important addition that distributions of certain unknowns must be learned from test results. The learned distributions are then used to average the conditional expected gains or conditional expected losses associated with each possible action. The examples illustrated that in specific problems sometimes expected values rather than complete distributions of the unknowns suffice.

If a Bayesian point of view is taken and prior distribution of unknowns are available, in addition to specifying the best action, it is also possible to specify the best of the possible tests.

Additional reading may be found in C1, H1, and W1.

10-12 PROBLEMS

1. A voltage generator is supposed to produce 100 V. The loss function of voltage output is $(100 - V)^2$ where V is the actual output. There are two proposed designs A and B for the voltage generator. For design A the density of output is given by

$$f_V{}^A(v) = \frac{1}{\sqrt{2\pi}} \exp\left[-\frac{(v - 100)^2}{2} \right].$$

For design B the density of output is

$$f_V{}^B(v) = \frac{1}{\sqrt{2\pi \frac{1}{2}}} \exp\left[-\frac{(v - 99)^2}{1/2} \right].$$

Evaluate the expected loss and choose the best design.

2. Same situation as in Problem 1 except we are now given specifications which say a voltage between 98 and 102 is acceptable; any other output is unacceptable (zero-one loss function). Evaluate the expected loss and choose the best design.

3. Assume a binary communication system like the one in Section 10-3. The following detection system is proposed. Say B if $Y \geq y_2$, say A if $Y \leq y_1$, say *unknown* if $y_1 < Y < y_2$. Set up the decision problem for choosing y_1 and y_2. Assume that the required losses are known. Determine y_1 and y_2.

4. Continuing the binary communication problem, now assume that the signal is present long enough for two independent observations Y_1, Y_2, that is

$$f_{Y_1, Y_2 | A}(y_1, y_2) = \frac{1}{2\pi\sigma^2} \exp\left\{ -\left[\frac{y_1{}^2 + y_2{}^2}{2\sigma^2} \right] \right\},$$

$$f_{Y_1, Y_2 | B}(y_1, y_2) = \frac{1}{2\pi\sigma^2} \exp\left\{ -\frac{(y_1 - 1)^2 + (y_2 - 1)^2}{2\sigma^2} \right\}.$$

Find the optimum decision rule.

5. A man gets up and observes that it is cloudy. He needs to decide whether to carry an umbrella to work. Describe the decision problem in extensive and in normal form. What values do you assign to each of the four possible outcomes?

6. Assume that you are a member of a jury. Describe the decision problem in extensive and normal form and discuss the values or losses.

7. Assume that to produce a new product a company has to decide which of three ideas from the research department to turn over to development; evaluate the results of development; decide whether or not to go into production; observe production cost and quality; set sales price; and observe sales volume and profit. Describe this problem by a tree diagram. Is this a single-state decision problem?

269

8. You have responsibility for a number of radar sets and must decide whether to take all the tubes from the sets and check them to see if they are getting "weak" so that they may be replaced before they cause radar failure at some critical time. The reasons for not doing so is that it might be that a "weak" tube is better than a new tube, and it also takes time to perform the "preventive maintenance." Describe this problem in normal form.

9. In a probabilistic decision problem assume that there are n possible actions $a_1, a_2, \ldots, a_n$ and there are k possible states of nature, $e_1, e_2, \ldots, e_k$, with probabilities $p(e_j) = p_j$, $j = 1, 2, \ldots, k$. The gain of $a_i e_j$ is G_{ij}. Call G the maximum of G_{ij}. Define the loss l_{ij} to be

$$l_{ij} = G - G_{ij}.$$

Show that if a_i maximizes the expected gain then a_i minimizes the expected loss and conversely.

10. One of five systems can be designed. The environment is such that it is assumed to take on four different values. The output for each system in each environment is shown in the table below. Assume that the desired output is 100 and a squared error loss function.

System	Environment			
	e_1	e_2	e_3	e_4
A	103	100	102	97
B	103	100	101	98
C	98	102	97	97
D	100	100	100	95
E	101	101	99	99

(a) Which systems correspond to admissible actions?
(b) Assume that all environments are equally likely and find the best system.

11. The profit U_1 per batch made by a manufacturer depends on the impurity S in the raw material by

$$U_1 = 10 - 100S.$$

S is modeled as a normal random variable with mean .05 and standard deviation .02.

At a cost of $2 the refinery can make the impurity content .04 independent of the original impurity (unreasonable but simplifying). Thus the profit per batch U_2, assuming purification is

$$U_2 = 10 - 100(.04) - 2$$
$$= 4.$$

(a) Should the plant purify?
(b) What is the maximum amount to be spent on testing to determine the impurity content of a particular batch?

12. A major guidance component on a satellite is sensitive to the satellite temperature. Engineering laboratory tests have determined that the expected mean

square error of three proposed modes of operation of the guidance systems is a function of the average satellite temperature $\bar{T}$. The average mean square error, that is, conditional loss, is shown in the following table for three possible values of $\bar{T}$.

System mode of operation	$\bar{T}$		
	60°	62°	64°
a_1	1.7	.9	1.1
a_2	.8	1.0	1.6
a_3	2.0	1.3	.5

By listing only three possible values of $\bar{T}$, we are assuming that a continuous range of temperatures is not needed in this problem. Such a model might arise when experience has shown that similar satellites have had an average temperature of one of these values depending on the orbit of the satellite, or because the engineer is simply willing to assume that this approximation for average temperature is adequate for his purposes.

During the first day of operation, the temperature is sampled at four different times, which we assume are independent samples, with readings 60°, 64°, 62°, and 62°. Which mode of operation of the guidance system should be chosen?

13. Refer to Examples 10-7 and 10-8. Assume that at most five units can be sold and any number of units can be started into production, but the chosen number must be started immediately and others cannot be started later. Additional good units beyond five are worth $0. How many should be started? What is the expected gain?

14. Same as Problem 13 except that units can be started one at a time, the results observed and then another started, etc., until five good ones are obtained. What is the expected gain?

15. In order to determine if there is oil in a certain location, there are three different kinds of tests, A, B, C. The tests have the following characteristics:

Test	Cost	P(positive test/oil)	P(positive test/no oil)
A	10,000	.6	.4
B	30,000	.8	.4
C	50,000	.8	.1

The cost of actually drilling is $200,000. Assume that if there is oil, drilling will always be successful, and conversely. The return if there is no oil is $0, and the return if there is oil is $400,000. Assume that the prior probability of oil is .5.

(a) Draw an extensive (tree) diagram of the decision problem.
(b) Find the best decision given each test.
(c) Find the best test. (It is not possible to perform more than one test.)

16. In the detection problem of Example 10-3, assume that the noise is known to be normal with a mean of 0 and a variance of 1/2. Also $L_{AB} = L_{BA} = 1$.

   (a) Find and plot the best decision boundary y^* versus a.
   (b) Assume a to be a random variable with a density.

$$f_a(x) = 1, \quad 0 \le x \le \tfrac{1}{2},$$
$$= \tfrac{1}{2}, \quad \tfrac{1}{2} < x \le \tfrac{3}{4},$$
$$= \tfrac{3}{2}, \quad \tfrac{3}{4} < x \le 1,$$
$$= 0, \quad \text{elsewhere.}$$

   Find the best boundary.
   (c) How might one take data to modify the distribution of a?

17. A probabilistic decision problem is described by the matrix, where the entries in the matrix are gains.

	e_1	e_2	e_3	e_4	e_5
a_1	10	8	4	-2	-5
a_2	4	3	3	0	-1
a_3	1	1	1	1	1
a_4	10	8	3	-2	-4

   (a) Assume that $P(e_i) = 1/5$, $i = 1, 2, 3, 4, 5$, and find the best action and its expected value. Find the expected cost of uncertainty.
   (b) Assume $P(e_1) = P(e_2) = 0$, $P(e_3) = 1/2$, $P(e_4) = 1/10$, $P(e_5) = 4/10$. Find the best action and its expected gain. Find the expected cost of uncertainty.

18. We return to the binary communication problem discussed in Example 10-3. Recall that the best decision in that case was if the received signal was above $\tfrac{1}{2} + \sigma^2\{ \ln \ [L_{BA}a]/[L_{AB}(1 - a)]\}$, then we said signal B was sent; otherwise we said A was sent. We now consider the same problem except we now assume that the variance σ^2 of the noise is not known; however samples of the noise are available. The losses L_1 and L_2 and the probability a of A being sent are known. The noise is assumed to be normal with a mean of 0 and an unknown variance.
   (a) Use the approximation for expected values that

$$E[g(x)] \simeq g(E[x])$$

and arrive at the conclusion

$$y^* \simeq \tfrac{1}{2} + E[\sigma^2] \ \ln \ \frac{L_{BA}a}{L_{AB}(1 - a)}.$$

   (b) What is the expected value of σ^2?
   (c) Discuss the approximation used in (a), and what is involved in solving the problem exactly. Hint: see Chapter VI.

···

CHAPTER XI

INTRODUCTION TO RANDOM PROCESSES

11-1 GENERAL DESCRIPTION

In the study of random variables we are interested in a probabilistic description of numbers. That is, to each outcome s of an experiment, a number $X(s)$ is assigned by the random variable X. In this beginning study of random processes we are interested in a probabilistic description of functions of time, that is, in methods of describing random signals.

Such a description is useful when considering the time varying response of a system when the input is random (noise) or when there is random variation of some constituents of the system. Random processes are used to model phenomena such as statistical mechanics, electron emission, weather, noise, population growth, queues, inventory control, and economic changes (e.g., the stock market).

A random process is often called a stochastic process; these terms are interchangeable in this text. A random process is defined as a family of random variables, $\{X(t), t \in \mathcal{T}\}$. $\mathcal{T}$ represents some set of values of time (although more general application is possible). $\mathcal{T}$ may be an interval, in which case the random process is called a continuous parameter random process, or $\mathcal{T}$ may be a set of discrete points, in which case the random process is called a discrete parameter random process. In this book, unless otherwise stated, $\mathcal{T}$ will be the entire time axis. Thus a random process is, like a random variable, a function and the domain of the function is, like a random variable, the sample space. The difference between a random variable and a random process is that a random variable maps an outcome into a number, while a random process maps an outcome in the sample space into a function of time. Alternatively, a random process can be viewed as a function of two variables, $t \in \mathcal{T}$ and $s \in S$.

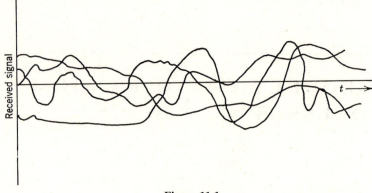

Figure 11-1

We will use the notation $X(t)$, $Y(t)$ or in general some capital letter with the argument t to represent a random process. It must be understood that there is another variable which represents the probability space which is suppressed in this notation. If the point in the probability space is fixed, we will use the notation $X_r(t)$, where r is an index of the outcome in the sample space.

EXAMPLE 11-1

A radio receiver is tuned to a frequency at which there is no transmitter operating. The received signal is called noise. Each time the receiver is turned on a different signal is received. We assume the choice of turn-on time is the choice of outcome and measure time from the turn-on time. Four received functions of time are shown in Figure 11-1. These are viewed as mappings of the four outcomes to functions of time.

EXAMPLE 11-2

One of three voltage generators A, B, C is to be connected to a load. The voltage outputs are described in Figure 11-2.

In this example we assume that the selection of a particular generator is determined by chance and that

$$P(A) = \tfrac{1}{3},$$
$$P(B) = \tfrac{1}{2},$$
$$P(C) = \tfrac{1}{6}.$$

Note that in this example $\mathscr{T}$ is an interval and is approximately 0 to 4. In this simple example the probability space is described by the probabilities of each event.

274

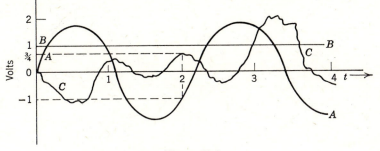

Figure 11-2

Fixed Point in the Sample Space

If the point in the sample space is fixed, then a single function of time $X_r(t)$ results. This function of time is called a sample function or a realization. For instance in Example 11-2 if the point in the sample space corresponds to A, then the sample function is the one that looks like the sine wave. A sample function can be an ordinary function of time such as is studied in, say, transient circuit analysis.

Note that if the set $\mathscr{T}$ is a series of disjoint points (discrete), then a sample function is a series of values and corresponds to what is usually called sample data. Figure 11-3 illustrates a sample function of a random process when $\mathscr{T} = \{1, 2, 3, 4, \ldots, 14\}$.

Time Fixed

If the parameter t is fixed at some value, say t_i, and the point in sample space is not fixed, then $X(t_i)$ is simply a random variable. For instance in Example 11-2 $X(2)$ is a random variable which is described by

$$P[X(2) = -1] = \tfrac{1}{3},$$
$$P[X(2) = \tfrac{3}{4}] = \tfrac{1}{6},$$
$$P[X(2) = 1] = \tfrac{1}{2}.$$

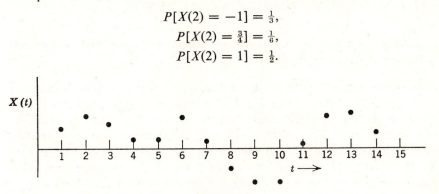

Figure 11-3

275

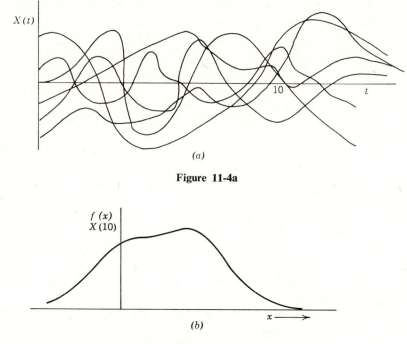

$X(t)$

10

t

(a)

Figure 11-4a

$f(x)$
$X(10)$

$x \longrightarrow$

(b)

Figure 11-4b

Depending on the sample space, for a fixed value of time the resulting random variable may be discrete, continuous, or mixed.

A random process is illustrated by drawing some of the sample functions in Figure 11-4a. When t is fixed at 10 the probability density function of $X(10)$ is shown in Figure 11-4b. Note that the density function implies that there are an uncountable number of sample functions.

Relation Between an n-dimensional Random Vector and a Discrete Parameter Random Process

In an earlier part of the book we discussed random vectors. In the case where the set $\mathcal{T}$ is a set of n values of time, then the random process is exactly the same as the n-dimensional random vector. For example is $\mathcal{T} = \{1, 2, 3, 4, 5\}$, then $X(1)$, $X(2)$, $X(3)$, $X(4)$, $X(5)$ are five random variables which completely describe the random process. The random vector $Y = (X_1, X_2, X_3, X_4, X_5)$, where $X_i = X(i)$, for $i = 1, 2, \ldots, 5$, is the same as the random process in the sense that a description of Y (e.g., a joint density function) is the same as a description of $X(t)$.

276

Methods of Description

In the previous two examples and in the examples of the next section random processes are described by referring to a physical situation and modeling that situation. In addition, random processes can also be described by joint distribution functions and as analytic functions of random variables.

JOINT DISTRIBUTION FUNCTION. Subject to very reasonable symmetry and compatibility conditions, a random process can be described by its joint distribution function:

$$F_{X(t_1),\ldots,X(t_n)} \tag{11-1}$$

for all n, t_1, t_2, ..., $t_n \in \mathcal{T}$. An example is discussed in Section 11.5.

ANALYTIC DESCRIPTION. It is possible to define random processes as functions of random variables, that is

$$X(t) = g(X_1, X_2, \ldots, X_n, t), \tag{11-2}$$

where $X(t)$ is a random process, X_i, $i = 1, \ldots, n$ are random variables, g is an ordinary function.

EXAMPLE 11-3. $X(t) = A \cos(\omega t + \phi)$

This is an example of a random process described by an analytic description. A and ω are assumed to be fixed constants, while ϕ is a random variable uniformly distributed between $-\pi$ and $+\pi$. Portions of two sample functions are shown in Figure 11-5.

EXAMPLE 11-4

$X(t) = A + Bt$ where A and B are uncorrelated normal random variables with means μ_a and μ_b and variances σ_a^2 and σ_b^2. Sample functions are shown in Figure 11-6.

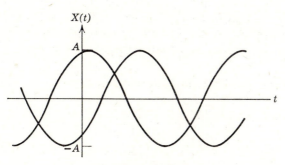

Figure 11-5

277

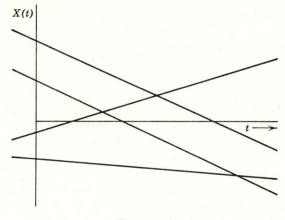

Figure 11-6

11-2 EXAMPLES OF SOME SPECIAL PROCESSES

This section will describe two special continuous parameter random processes, each of which is an example of an independent increments process. A continuous parameter random process, $X(t)$, $-\infty < t < \infty$, is an independent increments process if, for all choices of $t_0 < t_1 < \cdots < t_n \in \mathcal{T}$, the n random variables

$$X(t_1) - X(t_0), \; X(t_2) - X(t_1), \ldots, X(t_n) - X(t_{n-1})$$

are independent.

Random Walk

Assume that a coin tossing experiment with $P(\text{heads}) = p$, $P(\text{tails}) = q$, and that a toss occurs every second beginning with $t = 1$. Note $\mathcal{T} = \{t : 0 \leq t\}$. $X(0) = 0$, $X(t)$ increases by one if a head appears and $X(t)$ decreases by one if a tail appears. A sample function corresponding to $h, h, t, h, h, h, t, h, t, t$ is shown in Figure 11-7. Mathematically this process is described by

$$X(i) = \sum_{j=0}^{i} Y_j \qquad \text{where the } Y_j\text{'s are independent} \qquad i = 0, 1, \cdots$$

$$Y_0 = 0,$$
$$P(Y_j = 1) = p,$$
$$P(Y_j = -1) = q = 1 - p. \qquad j = 1, 2, \cdots$$
$$X(t) = X(i), \qquad i \leq t < (i + 1).$$

278

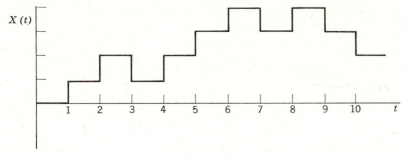

Figure 11-7

Thus

$$P[X(i) = n] = P[Y_j = 1 \ n \text{ more times than } Y_j = -1]$$

$$= P\left[Y_j = 1 \ \frac{i + n}{2} \text{ times, and } Y_j = -1 \ \frac{i - n}{2} \text{ times} \right]$$

$$= \binom{i}{(i + n)/2} p^{\frac{i+n}{2}} q^{\frac{i-n}{2}},$$

and this expression is meaningful only when $-i \leq n \leq i$ and $i + n$ is even.

The Poisson Process

The Poisson process arises in many applied problems. One of the important cases is the random emission of electrons from the filament of a vacuum tube. Others are time to failure of a system and time of demand for services.

The usual concept that leads to the Poisson process is that the number of occurrences of some event (e.g., emission of an electron) is random in time or space. In the following paragraphs we more carefully define random occurrences and derive the Poisson process from basic assumptions via a differential equation.

1. The occurrences in any period of time are independent of the occurrences at all other nonoverlapping time periods (independent increments).
2. As the time period Δt approaches 0; for any starting time:

   (a) $P(0 \text{ occurrences in } \Delta t) = 1 - \mu \, \Delta t$.
   (b) $P(2 \text{ or more occurrences in } \Delta t) = 0$.
   (c) $P(1 \text{ occurrence in } \Delta t) = \mu \, \Delta t$.

Note that μ is a parameter which is assumed to be known.

279

We denote the probability that there are k occurrences in time interval T by $P(k, T)$ and show that this results in the Poisson distribution by relating $P(k, t + dt)$ to $P(k, t)$ and then solving the resulting differential equation.

The probability of 0 occurrences in time $t + \Delta t$ is the product of the independent (by assumption 1) probabilities of no occurrences in t and no occurrences between t and $t + \Delta t$, or

$$P(0, t + \Delta t) = P(0, t)P(0, \Delta t). \tag{11-3}$$

Using assumption 2a

$$P(0, t + \Delta t) = P(0, t)[1 - \mu \Delta t]$$

or

$$\frac{P(0, t + \Delta t) - P(0, t)}{\Delta t} = -\mu P(0, t).$$

Taking the limit as $\Delta t \to 0$

$$\frac{dP(0, t)}{dt} = -\mu P(0, t).$$

This simple differential equation has the solution

$$P(0, t) = Ae^{-\mu t}, \tag{11-4}$$

and since $P(0, 0) = 1$, $A = 1$, or

$$P(0, t) = e^{-\mu t}.$$

Note that $P(0$ occurrences from T to $T + t) = e^{-\mu t}$ does not depend on T. For $k > 0$:

$$P(k, t + \Delta t) = P(k, t)P(0, \Delta t) + P(k - 1, t)P(1, \Delta t).$$

This elementary difference equation simply states that the probability of k occurrences in time $t + \Delta t$ is equal to the probability of k in time t and none in time Δt plus the probability of $k - 1$ in time t and one in time Δt. Note that the assumption of independence is used in order to multiply the probabilities and assumption 2b restricts to two the number of probabilities summed. Now using assumptions 2a and 2c

$$P(k, t + \Delta t) = P(k, t)(1 - \mu \Delta t) + P(k - 1, t)\mu \Delta t$$

or

$$\frac{P(k, t + \Delta t) - P(k, t)}{\Delta t} = [-P(k, t) + P(k - 1, t)]\mu.$$

Letting $\Delta t \to 0$,

$$\frac{dP(k, t)}{dt} + \mu P(k, t) = \mu P(k - 1, t).$$

This differential equation has the solution

$$P(k, t) = \mu e^{-\mu t} \int_0^t e^{\mu \tau} P(k - 1, \tau) \, d\tau.$$

For $k = 1$, using $P(0, t) = e^{-\mu t}$:

$$P(1, t) = \mu e^{-\mu t} \int_0^t e^{\mu \tau} e^{-\mu \tau} \, d\tau, \tag{11-5}$$

$$P(1, t) = \mu t e^{-\mu t}.$$

For $k = 2$

$$P(2, t) = \mu e^{-\mu t} \int_0^t \mu \tau \, d\tau,$$

$$P(2, t) = \frac{(\mu t)^2}{2} e^{-\mu t}. \tag{11-6}$$

In general

$$P(k, t) = \mu e^{-\mu t} \int_0^t \frac{(\mu \tau)^{k-1}}{(k - 1)!} \, d\tau,$$

$$P(k, t) = \frac{(\mu t)^k}{k!} e^{-\mu t}. \tag{11-7}$$

Letting $X(t)$ equal the number of events that have occurred in time t

$$X(0) = 0,$$

$$P\{[X(t + T) - X(t)] = k\} = P\{X(T) = k\} = P(k, T) = \frac{e^{-\mu T}(\mu T)^k}{k!}. \tag{11-8}$$

From Example 5-5 the mean is

$$E[X(t)] = \mu t,$$

and μ is called the mean rate, or average rate.

EXAMPLE 11-5

A radar set fails according to a Poisson process at the average rate of one failure per month. If we have one spare, what is the probability that there will not be an operating radar one month from now? Let $X(t)$ be the number of failures. Then $\mu = 1/\text{month}$. Thus

$$P(X(t) \geq 2) = 1 - P[X(t) = 0] - P[X(t) = 1],$$
$$P[X(1) \geq 2] = 1 - e^{-1} - e^{-1}$$
$$= 1 - 2e^{-1} \simeq .264.$$

EXAMPLE 11-6

A certain radioactive substance emits alpha particles at a mean rate of one every 4 sec. Let $X(t)$ represent the number of alpha particles counted by a counter. Observations have indicated that a Poisson process is a reasonable model for $X(t)$. Find

(a) $P[X(10) < 3]$,

(b) $P[X(5) > 3]$.

(a) $P[X(10) < 3] = \sum\limits_{i=0}^{2} P[X(10) = i]$

$$= e^{-10/4}\left[1 + \frac{10}{4} + \frac{(10/4)^2}{2}\right].$$

(b) $P[X(5) > 3] = 1 - [e^{-5/4}]\left[1 + \frac{5}{4} + \frac{(5/4)^2}{2} + \frac{(5/4)^3}{6}\right].$

11-3 MOMENTS

Just as with random variables, moments of random processes are defined, however the moments of random processes are functions of time. The first and second moments are emphasized.

Mean

$$\mu(t) = E[X(t)] = \int_{-\infty}^{\infty} x f_{X(t)}(x)\, dx, \tag{11-9}$$

where $f_{X(t)}$ is the probability density function of the random variable $X(t)$. If $X(t)$ is discrete

$$\mu(t) = \sum_{\text{all } i} x_i P[X(t) = x_i].$$

Note that in general $\mu(t)$ will vary with t, time, because $f_{X(t)}$ will change with t. The mean and some sample functions are shown in Figure 11-8.

Other moments, for example, the variance, can be defined as functions of time by a simple extension of the ideas for single random variables. For the kth central moment μ_k

$$\mu_k(t) = E\{[X(t) - \mu(t)]^k\}. \tag{11-10}$$

Although more and more moments better describe the distribution of $X(t)$ for fixed t, such moments do not describe the time variation. The two

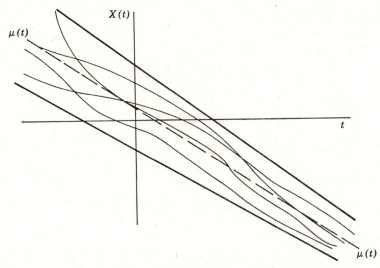

Figure 11-8

stochastic processes shown in Figures 11-8 and 11-9 might have identical distribution functions at any t. However it is obvious that they are entirely different processes.

Thus an important second moment including two values of time is also widely used to partially describe the relation between the values of the random process at two instants of time.

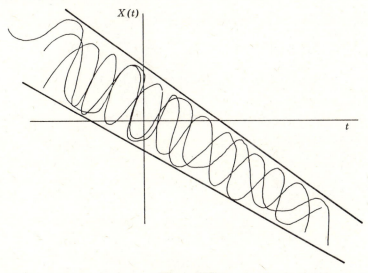

Figure 11-9

283

Autocorrelation Function

The autocorrelation function, $R_{XX}(t_1, t_2)$ is defined by

$$R_{XX}(t_1, t_2) = E[X(t_1)X(t_2)].$$

In the case where $X(t_1)$ and $X(t_2)$ have a joint density function

$$R_{XX}(t_1, t_2) = E[X(t_1)X(t_2)] = \int\int_{-\infty}^{\infty} x_1 x_2 f_{X(t_1), X(t_2)}(x_1, x_2)\, dx_1\, dx_2. \quad (11\text{-}11)$$

The autocovariance function is defined by

$$C_{XX}(t_1, t_2) = E\{[X(t_1) - \mu(t_1)][X(t_2) - \mu(t_2)]\}. \quad (11\text{-}12)$$

It is easy to show that

$$C_{XX}(t_1, t_2) = R_{XX}(t_1, t_2) - \mu(t_1)\mu(t_2). \quad (11\text{-}13)$$

Note that the autocorrelation function $R_{XX}(t_1, t_2)$ is similar to the correlation coefficient of two random variables. In fact if $\mu(t_1) = 0 = \mu(t_2)$ and if $E[X^2(t_1)] = E[X^2(t_2)] = 1$, then $R_{XX}(t_1, t_2)$ is the correlation coefficient between $X(t_1)$ and $X(t_2)$. See Problem 8.

EXAMPLE 11-7

Find the mean and autocorrelation function for the random walk.

In the random walk, for i a positive integer, $X(i) = \sum_{j=1}^{i} Y_j$ where $Y_j = +1$ if the jth toss is a head and $Y_j = -1$ if the jth toss is a tail.

$$\mu(i) = E[X(i)] = E\left[\sum_{j=1}^{i} Y_j\right] = iE[Y_j] = i(p - q),$$

$$\mu(t) = E[X(t)] = i(p - q), \qquad i \leq t < (i + 1).$$

If $p = q = 1/2$ (fair coin), then
$$\mu(t) = 0.$$

Let $t_i = i, t_j = j, j \geq i$. Then

$$E[X(t_i)X(t_j)] = R_{XX}(i, j)$$

$$= E\left[\sum_{k=1}^{i} Y_k \sum_{m=1}^{j} Y_m\right]$$

$$= E\left[\sum_{k=1}^{i} Y_k \sum_{m=1}^{i} Y_m\right] + E\left[\sum_{k=1}^{i} Y_k \sum_{m=i+1}^{j} Y_m\right].$$

Now $E[Y_k Y_m] = E[Y_k]E[Y_m]$ if $k \neq m$ due to independence. Thus

$$R_{XX}(i,j) = iE[Y_k^2] + (i^2 - i)[E(Y_k)]^2 + iE[Y_k](j - i)E[Y_m]$$
$$= i + (i^2 - i)(p - q)^2 + i(j - i)(p - q)^2$$
$$= i + (ij - i)(p - q)^2, \quad i \leq j.$$

If $j \leq i$ then

$$R_{XX}(i,j) = j + (ij - j)(p - q)^2, \quad j \leq i.$$

If $p = q$ (fair coin) then

$$R_{XX}(i,j) = i, \quad i \leq j.$$

The autocorrelation function for the case where t_i and/or t_j are not integers is easily derived from the above results.

EXAMPLE 11-8

Find the mean and autocorrelation of the random process given in Example 11-3. The mean $\mu(t)$ is

$$\mu(t) = E[A \cos (\omega t + \phi)]$$
$$= \int_{-\pi}^{\pi} A \cos (\omega t + \theta) \frac{1}{2\pi} d\theta$$
$$= \frac{A}{2\pi} [\sin (\omega t + \theta)] \Big|_{-\pi}^{\pi}$$
$$= 0.$$

The autocorrelation function is

$$R_{XX}(t_1, t_2) = E[A \cos (\omega t_1 + \phi)A \cos (\omega t_2 + \phi)]$$
$$= \int_{-\pi}^{\pi} A^2 \cos (\omega t_1 + \theta) \cos (\omega t_2 + \theta) \frac{1}{2\pi} d\theta$$
$$= \frac{A^2}{2\pi} \int_{-\pi}^{\pi} \{\tfrac{1}{2} \cos (\omega t_1 + \omega t_2 + 2\theta) + \tfrac{1}{2} \cos [\omega(t_1 - t_2)]\} d\theta$$
$$= 0 + \frac{A^2}{2} \cos \omega(t_1 - t_2).$$

EXAMPLE 11-9

Find the mean and autocorrelation function for the Poisson process. The expected value of $X(t)$ is

$$E[X(t)] = \mu t.$$

285

This mean is a direct result of the fact that

$$E[X(t)] = \sum_{i=0}^{\infty} iP[X(t) = i],$$

and that

$$P[X(t) = i] = \frac{e^{-\mu t}(\mu t)^i}{i!},$$

and that μt is the mean of this Poisson random variable.

The autocorrelation function $R_{XX}(t_1, t_2)$ is (for $t_2 > t_1$)

$$\begin{aligned} R_{XX}(t_1, t_2) &= EX[(t_1)X(t_2)] \\ &= E\{X(t_1)[X(t_1) + X(t_2) - X(t_1)]\} \\ &= E[X^2(t_1)] + E[X(t_1)]E[X(t_2) - X(t_1)]. \end{aligned}$$

The expected value of $X^2(t_1)$ is found in Example 5-5 to be $(\mu t_1)^2 + \mu t_1$. Thus

$$\begin{aligned} R_{XX}(t_1, t_2) &= (\mu t_1)^2 + \mu t_1 + \mu t_1[\mu(t_2 - t_1)], \\ R_{XX}(t_1, t_2) &= \mu t_1(1 + \mu t_2), \qquad t_1 \leq t_2, \\ &= \mu t_2(1 + \mu t_1), \qquad t_1 \geq t_2. \end{aligned}$$

EXAMPLE 11-10. BINARY TRANSMISSION

The random process consists of sequences of pulses with the following properties.

1. Each pulse is of duration L.
2. Pulses are equally likely to be $+1$ or -1.
3. All pulses are statistically independent.
4. The sequence of pulses is not synchronized, that is, the starting time of the first pulse following $t = 0$ is equally likely to be any value between 0 and L.

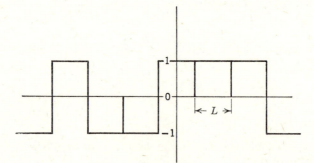

Figure 11-10

286

A typical sample function is shown in Figure 11-10. Find the mean and autocorrelation function. The mean is

$$E[X(t)] = 0$$

because of assumption 2.

To simplify the calculation of R_{XX}, define $S(t_1, t_2)$ as the event that t_1 and t_2 occur during the same pulse. Assumption 4 implies that S depends only on $|t_1 - t_2|$, not on the absolute location of t_1 or t_2. Then

$$R_{XX}(t_1, t_2) = E[X(t_1)X(t_2) \mid S(t_1, t_2)]P[S(t_1, t_2)]$$
$$+ E[X(t_1)X(t_2) \mid \overline{S(t_1, t_2)}]\{1 - P[S(t_1, t_2)]\}.$$

These terms are relatively easy to evaluate.

$$E[X(t_1)X(t_2) \mid S(t_1, t_2)] = \tfrac{1}{2}(1)^2 + \tfrac{1}{2}(-1)^2 = 1,$$
$$E[X(t_1)X(t_2) \mid \overline{S(t_1, t_2)}] = E[X(t_1)]E[X(t_2)] = 0,$$
$$P[S(t_1, t_2)] = 0 \quad \text{if} \quad |t_1 - t_2| > L.$$

If $|t_1 - t_2| \leq L$, then because of assumption 4, for $|t_1 - t_2| < L$, $P[t_1$ and t_2 occur on different pulses] is simply the ratio of $|t_1 - t_2|$ to L. Thus

$$P[S(t_1, t_2)] = 1 - \frac{|t_1 - t_2|}{L}.$$

Thus

$$R_{XX}(t_1, t_2) = 1 - \frac{|t_1 - t_2|}{L} \qquad \text{if} \qquad |t_1 - t_2| \leq L,$$
$$= 0 \qquad \text{if} \qquad |t_1 - t_2| > L.$$

11-4 STATIONARY PROCESSES

Stationary random processes can be described as those processes which are independent of the choice of zero on the time axis. Another idea of stationarity is that a time translation of a sample function results in another sample function (with the same probability) of the random process. To be more precise, stationarity is defined in terms of the joint distribution function as follows.

A random process is strictly stationary if, for all $t_1, t_2, \ldots, t_n$ and $\tau \in \mathcal{T}$

$$F_{X(t_1), \ldots, X(t_n)} = F_{X(t_1+\tau), \ldots, X(t_n+\tau)}. \qquad (11\text{-}14)$$

This definition is useful from a mathematical viewpoint, but is difficult to check experimentally. If, as is often the case, one is willing to use the mean and autocorrelation function to describe a random process, then it seems only reasonable that if the mean and autocorrelation function do not vary with a change in the time origin, for practical purposes one could call the process stationary.

Thus processes which satisfy the earlier definition of stationarity are called strict sense stationary, and the following definition of wide sense stationarity is given.

A process is wide sense stationary if

$$E[X(t_1)] = K, \qquad \forall t_1 \in \mathcal{T}, \tag{11-15}$$

$$R_{XX}(t_1, t_2) = R_{XX}(t_1 + t, t_2 + t) \, \forall t_1, t_2, t_1 + t, t_2 + t \in \mathcal{T}. \tag{11-16}$$

If the mean and autocorrelation function are available then it is relatively easy to decide if a random process is stationary in the wide sense. For example, in Example 11-8 the random process is wide sense stationary because $\mu(t) = 0$ and

$$R_{XX}(t_1, t_2) = \frac{A^2}{2} \cos \omega(t_1 - t_2),$$

$$R_{XX}(t_1 + t, t_2 + t) = \frac{A^2}{2} \cos \omega(t_1 - t_2).$$

The random walk of Example 11-7 is not wide sense stationary, because although the mean is a constant, the autocorrelation function is not invariant under a shift in the time origin.

If the mean and autocorrelation function are not available, then collecting data may seem to be the answer but, practically, the amount of data necessary is staggering. In this case the practical answer is analyzing the problem and reasoning about stationarity with available data used as a guide. If the mechanism generating the random process depends on the starting time, then stationarity is not a good assumption. If there is no obvious change, then there is hope that the assumption of stationarity may result in an adequate model.

Relation Between Strict Sense and Wide Sense Stationarity

It is easy to show that if a random process is strictly stationary, then it is wide sense stationary. See Problem 13. However wide sense stationarity does not imply strict sense stationarity except in a special case, because wide sense stationarity involves only first- and second-order moments.

288

Autocorrelation Function of Stationary Random Processes

If a random process is wide sense stationary, then by the definition of stationarity

$$R_{XX}(t_1 + \tau, t_1) = R_{XX}(t_1 + t + \tau, t_1 + t) = R_{XX}(t_2 + \tau, t_2). \quad \forall t_1, t_2.$$

$$(11\text{-}17)$$

Thus R_{XX} for stationary processes depends only on τ, the time difference, and we will write $R_{XX}(t_1 + \tau, t_1)$ as $R_{XX}(\tau)$. We now show some of the properties of $R_{XX}(\tau)$.

$$R_{XX}(\tau) = R_{XX}(-\tau). \tag{11-18}$$

This follows from letting $t_2 = t_1 - \tau$ in (11-17) and recalling that $R_{XX}(t_1, t_2) = E[X(t_1)X(t_2)] = E[X(t_2)X(t_1)] = R_{XX}(t_2, t_1)$.

From (11-17) with $\tau = 0$

$$R_{XX}(0) = E[X^2(t)]. \tag{11-19}$$

Thus $$R_{XX}(0) \geq 0.$$

Also $$R_{XX}(0) \geq |R_{XX}(\tau)|. \tag{11-20}$$

This last result follows from the same argument made earlier (Chapter V) to show that the correlation coefficient was bounded by ± 1.

If $X(t_1)$ and $X(t_1 + \tau)$ are independent

$$R_{XX}(\tau) = \mu^2. \tag{11-21}$$

In many practical cases $X(t_1)$ and $X(t_1 + \tau)$ will be independent for large values of τ.

EXAMPLE 11-11

Why are the functions shown in Figure 11-11 not autocorrelation functions?

(a) Is not symmetrical and thus violates (11-18).
(b) Violates (11-20).

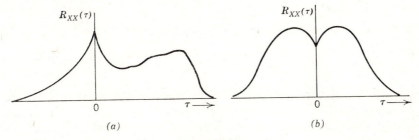

(a) (b)

Figure 11-11

289

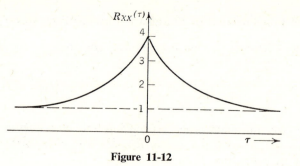

Figure 11-12

EXAMPLE 11-12

Assume that $X(t)$ and $X(t + \tau)$ are independent for large τ. The auto-correlation function is given in Figure 11-12. What is the mean and variance of $X(t)$?

Because the autocorrelation function is given as a function of only one variable, we assume that the process is wide sense stationary. Therefore the mean is a constant and for large values of τ

$$1 = R(\tau) = E[X(t)X(t + \tau)] = E[X(t)]E[X(t + \tau)] = \mu^2.$$

Thus $\mu = \pm 1$. From (11-19)

$$4 = R_{XX}(0) = E[X^2] = \sigma^2 + \mu^2.$$

Thus

$$\sigma^2 = 4 - 1 = 3.$$

11-5 NORMAL PROCESSES

It was stated earlier that a random process can be described by the joint distribution function of $X(t_1), \ldots, X(t_n)$ for all n and all $t_i \in \mathcal{T}$. One important special case is described in this section.

We say a random process is a normal process if for $n \geq 1$, any n points within $\mathcal{T}$, $t_1, t_2, \ldots, t_n$, have a joint normal distribution. That is (see Equation 4-41)

$$f_{X(t_1), X(t_2), \ldots, X(t_n)}(x_1, x_2, \ldots, x_n) = \frac{1}{|C|^{1/2}(2\pi)^{n/2}} \exp \left\{ -\tfrac{1}{2} \sum_{i=1}^{n} \sum_{j=1}^{n} D_{ij} \right.$$
$$\left. \times [(x_i - \mu(t_i)][(x_j - \mu(t_j)] \right\} \quad (11\text{-}22)$$

where $C_{ij} = C_{XX}(t_i, t_j)$; $C = $ matrix $[C_{ij}]$; $|C| = $ determinant of C; $D = C^{-1}$, that is, inverse of C; $D_{ij} = ij$th element of D.

To complete the description of this process one need only specify $\mu(t)$ and $C_{XX}(t_i, t_j)$ for all $t_i, t_j \in \mathcal{T}$. If $\mathcal{T}$ is some finite set then this stochastic process is just the multivariate normal random variable studied in Section 4-6.

We consider the case where $\mathscr{T}$ is continuous and

$$\mu(t) = 0,$$
$$R_{XX}(t_1, t_2) = R_{XX}(\tau); \quad \tau = t_1 - t_2.$$

That is, we consider a wide sense stationary process with zero mean. In this case the one-dimensional distribution is

$$f_{X(t)}(x) = \frac{1}{\sqrt{2\pi R(0)}} \exp\left[-\frac{x^2}{2R(0)}\right]. \tag{11-23}$$

The joint distribution of $X(t)$ and $X(t + \tau)$ is given by

$$f_{X(t), X(t+\tau)}(x_1, x_2) = \frac{1}{2\pi R(0)(1 - \rho^2)^{1/2}} \exp\left\{-\frac{1}{2(1 - \rho^2)R(0)}\right.$$

$$\left. \times [x_1^2 - 2\rho x_1 x_2 + x_2^2]\right\} \tag{11-24}$$

where
$$\rho = \frac{E[X(t)X(t + \tau)]}{\sqrt{E\{[X(t)]^2\}E\{[X(t + \tau)]^2\}}} = \frac{R_{XX}(\tau)}{R_{XX}(0)}.$$

Note the significant fact that in the case of a normal random processes with zero mean, the process is completely described by the autocorrelation function. Because of this fact, wide sense stationarity and strict sense stationarity are equivalent for normal random processes.

EXAMPLE 11-13

A stationary normal random process with zero mean and

$$R_{XX}(\tau) = \sigma^2 e^{-\alpha|\tau|} \tag{11-25}$$

is often used to represent "low pass noise" with σ^2 and α estimated from data. Such a process is called the Ornstein-Uhlenbeck process.

11-6 TIME AVERAGES AND ERGODICITY

The time average from a to b of a sample function of a random process is called $\eta(a, b)$, where in the case of a continuous parameter set $\mathscr{T}$

$$\eta(a, b) = \frac{1}{b - a} \int_a^b X(t) \, dt, \tag{11-26}$$

or in the case of a discrete parameter set,

$$\eta(a, b) = \frac{1}{n(a, b)} \sum_{i=n_a}^{n_b} X(i) \tag{11-27}$$

where n_a is the first index following time a, n_b is the last index before time b, and $n(a, b)$ is the number of measurements between a and b.

For a fixed point in the sample space and thus a fixed sample function, $\eta(a, b)$ is a number. As different outcomes in the sample space are chosen, $\eta(a, b)$ will take on different numbers; thus $\eta(a, b)$ is a random variable. We assume that the second moment of $\eta(a, b)$ exists.

We now seek the mean and second moment of $\eta(a, b)$ in terms of the mean $\mu(t)$ and autocorrelation function $R_{XX}(t)$ of $X(t)$. The continuous form of $\eta(a, b)$, (11-26), will be used. The mean and variance in the discrete case are requested in a problem. We assume that the linear operations of integration and expectation can be interchanged. Thus

$$E[\eta(a, b)] = E\left[\frac{1}{b - a} \int_a^b X(t)\, dt\right]$$

$$= \frac{1}{b - a} \int_a^b E[X(t)]\, dt,$$

$$E[\eta(a, b)] = \frac{1}{b - a} \int_a^b \mu(t)\, dt. \tag{11-28}$$

$$E[\eta^2(a, b)] = E\left[\frac{1}{b - a} \int_a^b X(t)\, dt\, \frac{1}{b - a} \int_a^b X(t')\, dt'\right]$$

$$= \frac{1}{(b - a)^2} \int_a^b\!\!\int E[X(t)X(t')]\, dt\, dt'. \tag{11-29}$$

If $X(t)$ is a stationary random process,

$$E[\eta(a, b)] = \frac{1}{b - a} \mu(b - a) = \mu \tag{11-30}$$

and

$$E[\eta^2(a, b)] = \frac{1}{(b - a)^2} \int_a^b\!\!\int R_{XX}(t - t')\, dt\, dt'.$$

Making the change of variable,

$$x = t',$$

$$\tau = t - t',$$

with the aid of Figure 11-13.

$$E[\eta^2(a, b)] = \frac{1}{(b - a)^2}$$

$$\times \left\{\int_{a-b}^0 \left[\int_{a-\tau}^b R_{XX}(\tau)\, dx\right] d\tau + \int_0^{b-a} \left[\int_a^{b-\tau} R_{XX}(\tau)\, dx\right] d\tau\right\},$$

$$E[\eta^2(a, b)] = \frac{1}{(b - a)^2}$$

$$\times \left[\int_{a-b}^0 (b - a + \tau)R_{XX}(\tau)\, d\tau + \int_0^{b-a} (b - a - \tau)R_{XX}(\tau)\, d\tau\right]. \tag{11-31}$$

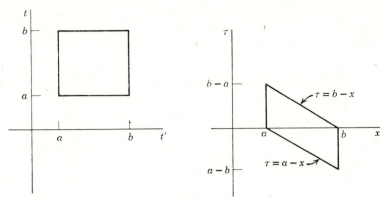

Figure 11-13

If we average from 0 to T, that is, $a = 0$, $b = T$, then

$$E[\eta^2(0, T)] = \frac{1}{T^2}\left[\int_{-T}^{0}(T + \tau)R_{XX}(\tau)\,d\tau + \int_{0}^{T}(T - \tau)R_{XX}(\tau)\,d\tau\right],$$

$$E[\eta^2(0, T)] = \frac{1}{T^2}\int_{-T}^{T}(T - |\tau|)R_{XX}(\tau)\,d\tau$$

$$= \frac{1}{T}\int_{-T}^{T}\left[1 - \frac{|\tau|}{T}\right]R_{XX}(\tau)\,d\tau. \tag{11-32}$$

EXAMPLE 11-14

Find $E[\eta(0, T)]$ and $E[\eta^2(0, T)]$ for the process given in Examples 11-3 and 11-8. Using (11-28)

$$E[\eta(0, T)] = \frac{1}{T}\int_{0}^{T}0\,dt = 0.$$

Using (11-32)

$$E[\eta^2(0, T)] = \frac{1}{T^2}\int_{-T}^{T}[T - |\tau|]\frac{A^2}{2}\cos \omega\tau\,d\tau = \frac{A^2}{\omega^2 T^2}[1 - \cos \omega T].$$

EXAMPLE 11-15. THE EFFECT OF AVERAGING

Let $X(t) = s + N(t)$ where s is an unknown constant which is to be determined, and $N(t)$ is a stationary Gaussian process independent of s with mean 0 and $R_{NN}(\tau) = e^{-a|\tau|}$. Now assume that we want to learn the value of s, and we average $X(t)$, that is,

$$\eta(T) = \frac{1}{T}\int_{0}^{T}X(t)\,dt.$$

$$E[\eta(T)] = \frac{1}{T}\int_{0}^{T}(s + 0)\,dt = s$$

293

and

$$E[\eta^2(T)] = \frac{1}{T^2}\left[\int_{-T}^{0}(T+\tau)[s^2 + e^{+a\tau}]\,d\tau\right] + \frac{1}{T^2}\left[\int_{0}^{T}(T-\tau)[s^2 + e^{-a\tau}]\,d\tau\right]$$

$$= \frac{2}{T^2}\int_{0}^{T}(T-\tau)[s^2 + e^{-a\tau}]\,d\tau$$

$$= \frac{2}{T^2}\left[s^2\left(T\tau - \frac{\tau^2}{2}\right) - \frac{T}{a}e^{-a\tau} + \frac{e^{-a\tau}}{a^2}(+a\tau+1)\right]_{0}^{T}$$

$$= \frac{2}{T^2}\left[s^2\left(T^2 - \frac{T^2}{2}\right) - \frac{T}{a}e^{-aT} + \frac{e^{-aT}}{a^2}(+aT+1)\right] - \frac{2}{T^2}\left[-\frac{T}{a} + \frac{1}{a^2}\right]$$

$$= s^2 + \frac{2}{aT} - \frac{2}{a^2T^2}(1 - e^{-aT}).$$

Note that the mean is s and the variance is

$$\sigma^2_{\eta(T)} = E[\eta^2(T)] - \{E[\eta(T)]\}^2$$

$$= \frac{2}{aT} - \frac{2}{a^2T^2}(1 - e^{-aT}).$$

This is plotted in Figure 11-14. Note that averaging produces the same mean and reduces the variance.

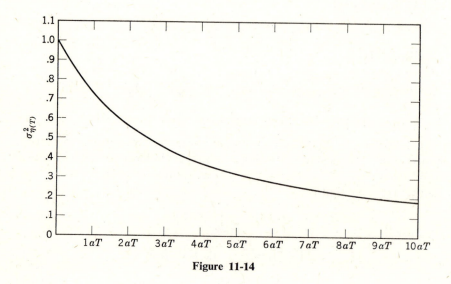

Figure 11-14

Ergodicity

The basic idea of ergodicity is that expected values can be replaced by time averages. This concept seems natural when "noise" is considered. In this case randomness is present not only in the selection of the sample function, but also in how the function behaves in time. The thermal noise of a resistor is random not because of the choice of a resistor, but because of the uncertainty associated with the collisions of the "free" electrons in the metal. The randomness of shot noise similarly is due to the random behavior of the electrons in the filament and one might expect that time averages of the output noise of one tube would be the same as averaging over the instantaneous output of many tubes of the same type operating under equivalent conditions.

Such ideas have led to a loose definition of ergodicity. A random process is ergodic if (with probability one)

$$E\{g[X(t)]\} = \lim_{T \to \infty} \frac{1}{2T} \int_{-T}^{T} g[X(t)] \, dt \qquad (11\text{-}33)$$

for all functions g.

For our purposes the most interesting functions g are

$$g[X(t)] = X(t)$$

and
$$g[X(t)] = X(t)X(\tau + t).$$

We now examine what is necessary in order for the expected value μ to equal the time average η.

$$\eta = \lim_{T \to \infty} \frac{1}{2T} \int_{-T}^{T} X(t) \, dt.$$

Note that if the time average η exists, this average must not be a function of time. Also note that in general η will depend on which sample function is chosen.

Thus if the expected value μ is to equal η then μ must be constant, that is, $X(t)$ must be stationary in the mean. Moreover μ is not a function of the sample space, that is, we have averaged over the sample space. Thus η must not depend on the sample space, that is, η must be the same (with probability one) for all choices of sample functions.

If $\mu = \eta$ then we say that $X(t)$ is ergodic in the mean. We have seen that stationarity of the mean is necessary for ergodicity in the mean. We have also seen that η must be a constant over the sample space. Thus if η is a constant this will imply that $\sigma_\eta^2 = 0$. Similarly if $\sigma_\eta^2 = 0$, then by Tchebycheff's inequality η is a constant. Thus a necessary and sufficient condition for stationarity of the mean is that $\sigma_\eta^2 = 0$.

The same ideas are useful in determining if a random process is ergodic in the autocorrelation function. The process must be wide sense stationary. In addition the average over the sample space $R_{XX}(\tau)$ must equal the time average autocorrelation function $\mathscr{R}_{XX}(\tau)$ where

$$\mathscr{R}_{XX}(\tau) = \lim_{T \to \infty} \mathscr{R}_T(\tau) = \lim_{T \to \infty} \frac{1}{2T} \int_{-T}^{T} X(t)X(t + \tau) \, dt.$$

In order for $R_{XX}(\tau)$ to be the same as $\mathscr{R}_{XX}(\tau)$ it is, in addition to wide sense stationarity, necessary that (with probability one) $\mathscr{R}_{XX}(\tau)$ does not depend on which sample function is chosen.

In practice it is usually impossible to decide on the basis of data if a random process is ergodic. One has to decide based on reasoning about the situation. In order to be ergodic a random process must be stationary, and "randomness" must be evident in the time variation as well as in the selection of a sample function. In addition the time average must not depend on which sample function is selected.

EXAMPLE 11-16

Is the process of Examples 11-3 and 11-8 ergodic in the mean and in the autocorrelation function?

We see from Example 11-14 that $\lim_{T \to \infty} \eta(0, T) = 0$. Thus the mean (see Example 11-8) and the time average agree. We now find the time average of $X(t)X(t + \tau)$.

$$\mathscr{R}_T(\tau) = \frac{1}{2T} \int_{-T}^{T} A \cos(\omega t + \phi)A \cos(\omega t + \omega\tau + \phi) \, dt$$

$$= \frac{A^2}{4T} \int_{-T}^{T} [\cos(2\omega t + \tau + 2\phi) + \cos(\omega\tau)] \, dt$$

$$= \frac{A^2}{4T} \left[\frac{1}{2\omega} \sin(2\omega T + \tau + 2\phi) \right.$$

$$\left. - \frac{1}{2\omega} \sin(-2\omega T + \tau + 2\phi) \right] + \frac{A^2}{2} \cos \omega\tau.$$

Now $\lim_{T \to \infty} \mathscr{R}_T(\tau) = (A^2/2) \cos \omega\tau$. Note that as $T \to \infty$, the time average does not depend on the sample function selected, that is, ϕ is not in the final answer, and the time average autocorrelation function agrees with $R_{XX}(\tau)$. Thus this example is ergodic in mean and in autocorrelation.

11-7 POWER DENSITY SPECTRUM

If the autocorrelation function $R(\tau)$ of a stationary random process is such that

$$\int_{-\infty}^{\infty} |R(\tau)| \, d\tau$$

exists, then its Fourier transform

$$\Phi(\omega) = \int_{-\infty}^{\infty} e^{-j\omega\tau} R(\tau)\, d\tau, \qquad (11\text{-}34)$$

is called the power density spectrum.

The usual inversion formula for Fourier transforms produces

$$R(\tau) = \frac{1}{2\pi} \int_{-\infty}^{\infty} e^{j\omega\tau} \Phi(\omega)\, d\omega \qquad (11\text{-}35)$$

and $R(\tau)$ is uniquely (except for discontinuities) reproduced by finding the power density spectrum and then finding $R(\tau)$ from $\Phi(\omega)$.

The Name "Power Density"

From the definition (11-35)

$$R(0) = \frac{1}{2\pi} \int_{-\infty}^{\infty} e^{0} \Phi(\omega)\, d\omega$$

or

$$E[X^2(t)] = \frac{1}{2\pi} \int_{-\infty}^{\infty} \Phi(\omega)\, d\omega,$$

and letting $2\pi f = \omega$

$$E[X^2(t)] = \int_{-\infty}^{\infty} \Phi(2\pi f)\, df. \qquad (11\text{-}36)$$

If $X(t)$ is considered the current through a 1-ohm resistance, then $E[X^2(t)]$ is the expected instantaneous power dissipated by the resistance. Thus the integral of $\Phi(2\pi f)$ is the expected power and Φ is naturally called the power density spectrum in the sense that its integral is expected instantaneous power. In Chapter XII, we will show that the integral of Φ between two frequencies represents the power between those frequencies.

If $X(t)$ is any voltage or any current, then $E[X^2(t)]$ represents the power (times a constant factor) that would be dissipated in a resistance. Thus we justify the use of the name power density spectrum for the transform of the autocorrelation function of stationary processes.

If $X(t)$ is ergodic then

$$E[X^2(t)] = \int_{-\infty}^{\infty} \Phi(2\pi f)\, df = \lim_{T \to \infty} \frac{1}{2T} \int_{-T}^{T} X^2(t)\, dt$$

is the (time) average power dissipated in a 1-ohm resistance.

Fourier Transform Review

Table 11-1 is presented as a review of Fourier transforms. Small letters will be used for functions of time and corresponding capital letters for the Fourier transform, for example, $f(t)$ has the Fourier transform $F(\omega)$. Note that some

297

Table 11-1

Time function	Fourier Transform	Comments
1. $af(t) + bg(t)$	$aF(\omega) + bG(\omega)$	Superposition applicable in both domains
2. $f(t - t_0)$	$e^{-j\omega t_0}F(\omega)$	Time delay causes a phase shift
3. $f(Kt)$	$\dfrac{1}{|K|}F\left(\dfrac{\omega}{K}\right)$	Time-bandwidth invariance
4. $\displaystyle\int_{-\infty}^{\infty} f(\tau)g(t - \tau)\, d\tau$	$F(\omega)G(\omega)$	Convolution in the time domain corresponds to multiplication in frequency domain
5. $f(t)g(t)$	$\dfrac{1}{2\pi}\displaystyle\int_{-\infty}^{\infty} F(\lambda)G(\omega - \lambda)\, d\lambda$	Convolution in the frequency domain corresponds to multiplication in time domain
6. $f(t) = e^{-at}$, $\quad t > 0$ $ = 0$, $\qquad t < 0$	$F(\omega) = \dfrac{1}{a + j\omega}$	$a > 0$ Note correspondence with Laplace transform
7. $f(t) = e^{at}$, $\quad t < 0$ $ = 0$, $\qquad t > 0$	$F(\omega) = \dfrac{1}{a - j\omega}$	$a > 0$ The mirror image in the time domain has the complex conjugate as a transform
8. $f(t) = e^{-a|t|}$	$F(\omega) = \dfrac{2a}{a^2 + \omega^2}$	$a > 0$ Sum of $f(t)$ of 6 and 7 produces sum of $F(\omega)$ of 6 and 7
9. $f(t) = 1$, $\quad |t| < T$ $ = 0$, $\quad |t| > T$	$F(\omega) = 2\dfrac{\sin \omega T}{\omega}$	Time limited function
10. $f(t) = \dfrac{\sin \omega_0 t}{t}$	$F(\omega) = \pi$, $|\omega| < \omega_0$ $ = 0$, $|\omega| > \omega_0$	Band limited function
11. $f(t) = 1 - \dfrac{|t|}{T}$, $|t| < T$ $ = 0$, $\qquad |t| > T$	$F(\omega) = \dfrac{T \sin^2 (\omega T/2)}{(\omega T/2)^2}$	
12. $f(t) = \delta(t)$	$F(\omega) = 1$	Transform of Dirac delta function
13. $f(t) = 1$	$F(\omega) = 2\pi\delta(\omega)$	Transform of DC
14. $f(t) = \cos \omega_0 t$	$F(\omega) = \pi[\delta(\omega + \omega_0) + \delta(\omega - \omega_0)]$	All of the signal is concentrated at $\pm\omega_0$.

of the transforms do not exist as defined above. However by using generalized functions (e.g., the Dirac delta function) useful transforms are defined.

As with Laplace transforms, finding inverse Fourier transforms is often difficult, and inverse transforms are usually found by partial fraction expansion and use of tables.

We now find the power density spectrum of some of the example random processes.

EXAMPLE 11-17

In Example 11-8 we found that $R_{XX}(\tau) = (A^2/2) \cos \omega_0\tau$. Using 1 and 14 from Table 11-1

$$\Phi_{XX}(\omega) = \frac{\pi A^2}{2} [\delta(\omega + \omega_0) + \delta(\omega - \omega_0)].$$

This is shown in Figure 11-15. Note that the figure shows that all of the power is concentrated at $\pm\omega_0$. That certainly is reasonable since each sample function is a sinusoidal wave with frequency ω_0.

It is easier (using the properties of the delta function) to show that the inverse transform of $\Phi_{XX}(\omega)$ produces the correct answer.

$$R_{XX}(\tau) = \frac{1}{2\pi} \int_{-\infty}^{\infty} \frac{\pi A^2}{2} [\delta(\omega + \omega_0) + \delta(\omega - \omega_0)]e^{j\omega\tau} \, d\omega$$

$$= \frac{A^2}{4} [e^{-j\omega_0\tau} + e^{+j\omega_0\tau}] = \frac{A^2}{2} \cos \omega_0\tau.$$

EXAMPLE 11-18

In Example 11-13 we found that $R_{XX}(\tau) = \sigma^2 e^{-\alpha|\tau|}$. It is easy to find

$$\Phi_{XX}(\omega) = \int_{-\infty}^{0} \sigma^2 e^{\alpha\tau} e^{-j\omega\tau} \, d\tau + \int_{0}^{\infty} \sigma^2 e^{-\alpha\tau} e^{-j\omega\tau} \, d\tau$$

$$= \sigma^2 \frac{1}{\alpha - j\omega} e^{(\alpha - j\omega)\tau} \Big|_{-\infty}^{0} + \frac{-\sigma^2}{\alpha + j\omega} e^{-(\alpha + j\omega)\tau} \Big|_{0}^{\infty}$$

$$= \frac{\sigma^2}{\alpha - j\omega} + \frac{\sigma^2}{\alpha + j\omega}$$

$$= \frac{2\alpha\sigma^2}{\alpha^2 + \omega^2}.$$

This result is essentially found in 8 in Table 11-1.

299

Figure 11-15

Properties of $\Phi_{XX}(\omega)$

From the fact that $R_{XX}(\tau)$ is even we can note that

$$\Phi_{XX}(\omega) = \int_{-\infty}^{\infty} R_{XX}(\tau)[\cos \omega\tau + j \sin \omega\tau] \, d\tau,$$

$$\Phi_{XX}(\omega) = \int_{-\infty}^{\infty} R_{XX}(\tau) \cos \omega\tau \, d\tau \qquad (11\text{-}37)$$

because $[R_{XX}(\tau) \sin \omega\tau]$ is an odd function and its integral over symmetric limits is zero.

Thus we note that

$$\Phi_{XX}(\omega) \text{ is a real function.}$$

That is, there is no phase information in $\Phi_{XX}(\omega)$.

Also

$$\Phi_{XX}(-\omega) = \int_{-\infty}^{\infty} R_{XX}(\tau) \cos (-\omega\tau) \, d\tau$$

$$= \int_{-\infty}^{\infty} R_{XX}(\tau) \cos \omega\tau \, d\tau = \Phi_{XX}(\omega). \qquad (11\text{-}38)$$

Thus $\Phi_{XX}(\omega)$ is an even function of ω.

Note that since $\Phi_{XX}(\omega)$ is the transform of $R_{XX}(\tau)$, $\Phi_{XX}(\omega)$ only contains information concerning the second moment of a stationary random process. Thus the power density spectrum, although a very useful measure, does not describe a random process. Stated another way, there are many different random processes which have the same power density spectrum.

Example 11-19

In the example of binary transmission (11-10) we found that

$$R_{XX}(\tau) = 1 - \frac{|\tau|}{L} \qquad \text{if} \quad |\tau| < L,$$

$$= 0 \qquad \text{if} \quad |\tau| > L.$$

The power density spectrum is

$$\Phi_{XX}(\omega) = \int_{-L}^{L} \left(1 - \frac{|\tau|}{L}\right) e^{-j\omega\tau} \, d\tau.$$

Using (11.37)

$$\Phi_{XX}(\omega) = \int_{-L}^{L} \left(1 - \frac{|\tau|}{L}\right) \cos \omega\tau \, d\tau$$

$$= 2 \int_{0}^{L} \left(1 - \frac{\tau}{L}\right) \cos \omega\tau \, d\tau$$

$$= 2 \int_{0}^{L} \cos \omega\tau \, d\tau - \frac{2}{L} \int_{0}^{L} \tau \cos \omega\tau \, d\tau.$$

$$\Phi_{XX}(\omega) = 2 \left.\frac{\sin \omega\tau}{\omega}\right|_{0}^{L} - \frac{2}{L}\left[\frac{\tau}{\omega} \sin \omega\tau + \frac{\cos \omega\tau}{\omega^2}\right]_{0}^{L}$$

$$= 2 \left\{ \frac{\sin \omega L}{\omega} - \frac{1}{L}\left[\frac{L \sin \omega L}{\omega} + \frac{\cos \omega L - 1}{\omega^2}\right]\right\}$$

$$= \frac{2}{L}\frac{1 - \cos \omega L}{\omega^2}$$

$$= L \left(\frac{\sin \dfrac{\omega L}{2}}{\dfrac{\omega L}{2}}\right)^2.$$

This is essentially result 11 from Table 11-1.

White Noise

White noise is defined as any stationary random process with zero mean whose power density spectrum is constant, that is, not dependent on frequency. Physically and mathematically there are difficulties. Physically white noise is not possible because this demands infinite power. Mathematically the inverse transform does not exist by the usual criterion of existence of an integral. Nevertheless it is useful and possible to use white noise as a model of the physical world and to extend the definition of an integral such that the result is that given in 12 in Table 11-1.

Note that white noise implies that the autocorrelation function is 0 for $\tau \neq 0$. Thus $X(t)$ is uncorrelated with $X(t + \tau)$ for $\tau \neq 0$.

301

11-8 TWO RANDOM PROCESSES

We will often be interested in the joint properties of two random processes (e.g., the input and the output of a system).

Two random processes $X(t)$ and $Y(t)$ can be described by their joint distribution function:

$$F_{X(t_1), X(t_2), \ldots, X(t_n), Y(t_{n+1}), \ldots, Y(t_{n+m})}$$

for all $n, m, t_1, \ldots, t_{n+m} \in \mathcal{T}$.

Cross-Correlation

The problems with using the joint distribution function mentioned earlier with respect to one random process are simply magnified, and the joint properties of two random processes are often described by the cross-correlation function $R_{XY}(t_1, t_2)$ where

$$R_{XY}(t_1, t_2) = E[X(t_1) Y(t_2)]. \tag{11-39}$$

The cross-covariance is defined by

$$C_{XY}(t_1, t_2) = E\{[X(t_1) - \mu_X(t_1)][Y(t_2) - \mu_Y(t_2)]\} \tag{11-40}$$

and it is easy to show that

$$R_{XY}(t_1, t_2) = C_{XY}(t_1, t_2) + \mu_X(t_1)\mu_Y(t_2). \tag{11-41}$$

Also

$$R_{XY}(t_1, t_2) = E[X(t_1) Y(t_2)] = E[Y(t_2)X(t_1)] = R_{YX}(t_2, t_1). \tag{11-42}$$

Stationarity

Two random processes are called jointly strict sense stationary if

$$F_{X(t_1), \ldots, X(t_n), Y(t_{n+1}), \ldots, Y(t_{n+m})} = F_{X(t_1+\tau), \ldots, X(t_n+\tau), Y(t_{n+1+\tau}), \ldots, Y(t_{n+m+\tau})}$$

for all $n, m, t_1, \ldots, t_{n+m}, t_1 + \tau, \ldots, t_{n+m} + \tau \in \mathcal{T}$.

Two random processes are called jointly wide sense stationary if for all t_1, t_2, τ

$$R_{XY}(t_1, t_1 - \tau) = R_{XY}(t_1 + t_2, t_2 + t_1 - \tau) = R_{XY}(\tau). \tag{11-43}$$

Note that for jointly stationary processes

$$R_{XY}(\tau) = E[X(t + \tau) Y(t)] = E[X(t) Y(t - \tau)]. \tag{11-44}$$

If the processes are jointly stationary then (11-42) implies that

$$R_{XY}(-\tau) = R_{YX}(\tau). \tag{11-45}$$

By the same type of argument used before we can show that

$$R_{XY}^2(\tau) \leq R_{XX}(0)R_{YY}(0). \tag{11-46}$$

We briefly review this argument.

$$E\{[X(t + \tau) + aY(t)]^2\} = R_{XX}(0) + a^2R_{YY}(0) + 2aR_{XY}(\tau) \geq 0. \tag{11-47}$$

The discriminant of the quadratic (in a) is nonpositive, or

$$R_{XY}^2(\tau) \leq R_{XX}(0)R_{YY}(0). \tag{11-48}$$

Also setting $a = \pm 1$ in the right-hand side of (11-47) results in

$$2\,|R_{XY}(\tau)| \leq R_{XX}(0) + R_{YY}(0). \tag{11-49}$$

Independence

Two random processes are independent if

$$F_{X(t_1),\ldots,X(t_n),Y(t_{n+1}),\ldots,Y(t_{n+m})} = F_{X(t_1),\ldots,X(t_n)}F_{Y(t_{n+1}),\ldots,Y(t_{n+m})}$$

for all $n, m, t_1, \ldots, t_n$ and $t_{n+1}, \ldots, t_{n+m} \in \mathcal{T}$.
Two random processes are uncorrelated if

$$R_{XY}(t_1, t_2) = \mu_X(t_1)\mu_Y(t_2), \qquad \forall t_1, t_2 \in \mathcal{T}.$$

As with random variables, if two random processes are independent then they are uncorrelated, but being uncorrelated does not imply independence except in the special case of jointly normal random processes.

Cross-Power Density Spectrum

If $X(t)$ and $Y(t)$ are wide sense stationary, the Fourier transform of $R_{XY}(\tau)$,

$$\Phi_{XY}(\omega) = \int_{-\infty}^{\infty} R_{XY}(\tau)e^{-j\omega\tau}\,d\tau, \tag{11-50}$$

is called the cross-power density spectrum. We have the usual inverse transformation

$$R_{XY}(\tau) = \frac{1}{2\pi}\int_{-\infty}^{\infty} \Phi_{XY}(\omega)e^{+j\omega\tau}\,d\omega. \tag{11-51}$$

If $X(t)$ is voltage across a two-terminal device and $Y(t)$ is the current through the same device, then the expected power

$$E[X(t)Y(t)] = R_{XY}(0) = \int_{-\infty}^{\infty} \Phi_{XY}(\omega)\,d\omega$$

justifies the name power density spectrum.

303

From (11-45) we see that

$$\Phi_{XY}(\omega) = \int_{-\infty}^{\infty} R_{XY}(\tau)e^{-j\omega\tau}\,d\tau = \int_{-\infty}^{\infty} R_{YX}(-\tau)e^{-j\omega\tau}\,d\tau$$

$$= \int_{-\infty}^{\infty} R_{YX}(t)e^{j\omega t}\,dt = \Phi_{YX}(-\omega). \tag{11-52}$$

11-9 SUMMARY

The major purpose of this chapter was to introduce the concept of a random process. Random processes were described by a differential equation (Poisson process), by an analytic description [$A \cos (\omega t + \phi)$], or by a joint distribution function (normal noise).

The mean and autocorrelation function were stressed as measures of a random process. Stationary random processes were defined, and for this class of processes the idea of ergodicity was introduced. An ergodic process is basically one in which time averages and expected values are interchangeable.

The power density spectrum of a stationary random process was defined as the Fourier transform of the autocorrelation function. The power density spectrum of a real stationary process was shown to be an even real function.

The cross-correlation function and the cross-power density spectrum were emphasized as useful measures of two random processes. Independence and stationarity of two random processes were defined.

Additional reading may be found in Ref. B1, D4, H2, H3, P1, P2, P4, R1, T1, and S1.

11-10 PROBLEMS

1. If both time and the point in the sample space are fixed, what is $X_r(t_i)$?

2. Let $\tau = \{1, 2\}$. If $X(1)$ and $X(2)$ are jointly normally distributed each with mean 0 and variance 1, and with $r = 1/2$, find

(a) $P[X(2) < 0]$;

(b) $P[X(2) < 0 \mid X(1) = 0]$;

(c) $P[X(2) < 0 \mid X(1) = -2]$;

(d) $E[X(2) \mid X(1) = x]$.

Hint: See (5-26) and (5-27).

3. Let $X(t) = A \cos (\omega t + \phi)$.

(a) Let ω and ϕ be constants and A be a random variable normally distributed with mean 1 and variance 2. Draw three sample functions. Are any of the sample functions (180°) out of phase?

(b) Describe the density function of $X(2)$. Let $\omega = 1$ and $\phi = 0$.

4. Consider the random walk described in Section 11-2. Find

(a) $P[-5 < X(10) < 5]$;

(b) $P[X(1) = 0]$;

(c) $P[X(8) = 0]$;

(d) $P[X(8) = 0 \mid X(6) = 2]$;

(e) $P[X(8) = 0 \mid X(6) = 4]$;

(f) $E[X(10)]$;

(g) $E[X(10) \mid X(9) = 1]$;

(h) $E[X(10) \mid X(4) = 4]$.

5. A radioactive source emits particles at the average rate of 2/msec. If the emitted particles can be modeled by a Poisson random process, find the probability of observing:

(a) More than two particles in 6 msec.

(b) More than two particles in 3 msec.

(c) More than two particles in two nonoverlapping time periods of 3 msec each.

6. The number of electrons emitted from the cathode of a thermionic vacuum tube is modeled as a Poisson process. Data have shown that for a certain tube

$$\mu = 10^{15} \text{ electrons/sec}$$

Find an expression for the probability that the number of electrons emitted within 1 sec is within 1% of 10^{15}. (To evaluate the expression, an approximation is suggested. Assume that the number of electrons emitted is a continuous random variable which has a normal distribution with the same mean and variance as the Poisson random variable.)

7. If $X(t) = A + Bt$ with A and B independent random variables.

$$\mu_A = 1, \qquad \sigma_A^2 = 9,$$

$$\mu_B = 2, \qquad \sigma_B^2 = 4.$$

Find $\mu_X(t)$ and $\sigma_X^2(t)$.

8. Show that if $\mu(t_1) = 0 = \mu(t_2)$ and $E[X^2(t_1)] = 1 = E[X^2(t_2)]$, then

$$\rho_{X(t_1)X(t_2)} = R_{XX}(t_1, t_2).$$

9. Show that

$$C_{XX}(t_1, t_2) = R_{XX}(t_1, t_2) - \mu(t_1)\mu(t_2).$$

305

10. Find the mean and autocorrelation function of $X(t) = A \cos (\omega t + \phi)$.

(a) When ω and ϕ are fixed and A is a random variable uniformly distributed between 1 and 3.

(b) When ω is fixed and A and ϕ are independent random variables, A normally distributed with a mean 0 and variance σ^2 and ϕ is uniformly distributed between $-\pi$ and π.

(c) A and ϕ are fixed and ω has a density

$$f_\omega(\lambda) = e^{-\lambda}, \qquad \lambda > 0,$$
$$= 0, \qquad \lambda < 0.$$

11. Is the Poisson process stationary?

12. Refer to Problem 10. Which of the processes are wide sense stationary?

13. Show that strict sense stationarity implies wide sense stationarity.

14. For a stationary normal random process with zero mean and autocorrelation function $R(\tau) = e^{-|\tau|}$ find

(a) $f_{X(t)|X(t-1)}(x \mid \tfrac{1}{2})$,

(b) $f_{X(t)|X(t-1)}(x \mid \lambda)$;

(c) $E[X(t) \mid X(t - 3) = +2]$;

(d) $E[(X(t) - 2)^2 \mid X(t - 3) = +2]$;

(e) $E[X(t) \mid X(t - \tfrac{1}{10}) = +2]$;

(f) $E[(X(t) - 2)^2 \mid X(t - \tfrac{1}{10}) = 2]$.

Hint: $X(t)$ and $X(t + \tau)$ are normally distributed. Use the results on conditional normal random variables.

15. $R_{XX}(\tau) = 7 - |\tau|, \qquad |\tau| \leq 3,$

$\qquad\qquad = 4, \qquad\qquad |\tau| > 3.$

$\qquad \mu_X = 2,$

$$Y = \int_0^2 X(t) \, dt.$$

Find the mean and variance of Y.

16. $R_{XX}(0) = \mu^2 + \sigma^2; \qquad R_{XX}(j) = 0, \qquad j \neq 0, \qquad \mu_X(t) = \mu,$

$$Y = \frac{1}{n} \sum_{j=1}^{n} X_j. \text{ Find } E[Y], \sigma_Y^2.$$

17. Refer to Example 11-15. Let $T = 10/a$.

(a) Use Chebycheffs inequality to find p where $P[|\eta(T) - s| < 1/2] \geq p$.

(b) Assuming $\eta(T)$ is normally distributed, find $P[|\eta(T) - s| < 1/2]$.

18. Let $X(t) = A \cos (t + \theta)$. A and θ are independent random variables with

$$f_A(a) = \frac{1}{\sqrt{2\pi}} e^{-(a^2)/2}.$$

$$f_\theta(\phi) = \frac{1}{2\pi}, \qquad -\pi < \phi < \pi,$$

$$= 0, \qquad \text{elsewhere.}$$

(a) Show that $X(t)$ is wide sense stationary.
(b) Compute $E[X(t)]$ and $R_{XX}(\tau)$,

$$\lim_{T \to \infty} \frac{1}{2T} \int_{-T}^{T} X(t)\, dt, \quad \text{and} \quad \lim_{T \to \infty} \frac{1}{2T} \int_{-T}^{T} X(t)X(t + \tau)\, dt.$$

(c) Is $X(t)$ ergodic in the mean?
(d) Is $X(t)$ ergodic in autocorrelation function?

19. Which of the following are possible power density spectrum of random processes?

(a) $\dfrac{\omega^2}{\omega^2 + 1}$

(b) $\dfrac{\omega}{\omega^2 + 1}$

(c) $\dfrac{(3 + j\omega)}{\omega^2 + 1}$

(d) $\delta(\omega + 377) + \delta(\omega - 377)$

20. Can you find the power density spectrum of the Poisson random process?

21. Find the power density spectrum if

$$R_{XX}(\tau) = 1 + 3e^{-2|\tau|}.$$

22. Find the autocorrelation function if

$$\Phi_{XX}(\omega) = \frac{2\omega^2 + 5}{(\omega^2 + 1)(\omega^2 + 4)}.$$

23. Show that if $X(t)$ and $Y(t)$ are jointly stationary

$$R_{XY}(\tau) = C_{XY}(\tau) + \mu_X \mu_Y$$

where $\tau = t_1 - t_2$.

24. If $X(t)$ and $Y(t)$ are independent random processes, what is $R_{XY}(t_1, t_2)$?

25. If $X(t) = \cos (t + \phi)$ and $Y(t) = \cos (t + \theta)$ where ϕ and θ are random variables, find $R_{XY}(t_1, t_2)$

(a) If ϕ and θ are independent and each is uniformly distributed between $-\pi$ and $+\pi$;
(b) If $\phi = \theta$ and ϕ is uniformly distributed between $-\pi$ and π.

26. Is $\Phi_{XY}(\omega)$ real or complex?

307

CHAPTER XII

SYSTEMS AND RANDOM SIGNALS

In this chapter we are primarily concerned with the problem of calculating some description of the output $Y(t)$ of a system when the input is $X(t)$, a random process. This is pictured in Figure 12-1.

12-1 OPERATIONS ON RANDOM PROCESSES

If one or more of the inputs to a system are random processes, then it seems reasonable that the output would also be a random process. In this section we consider this problem, define more exactly what is the problem, and then consider some special types of systems which can be treated by methods studied before.

A system such as pictured in Figure 12-1 will be the basic model. When the system is fixed, then the input $X(t)$ produces the output $Y(t)$. This is the usual deterministic model and is the basis for our analysis. We assume that the system itself is fixed, that is, there is no randomness. However we wish to let one or more inputs be random processes. In this case, when the point in the sample space is fixed, there is a specific realization $X_r(t)$. This realization then is acted upon by the system to produce the output $Y_r(t)$, where given the

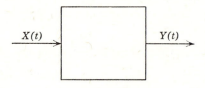

Figure 12-1

system and $X_r(t)$, $Y_r(t)$ is calculated by the usual deterministic procedures. Probability enters only in the selection of the input sample function and thus the output sample function.

The problem of describing the output in the general case is complicated, but some special cases are relatively easy. Two such special cases are treated in this section.

One Input Instantaneous Systems

We consider systems which can be described by

$$Y(t_0) = g[X(t_0)]. \tag{12-1}$$

That is, the output at time t_0 depends only on the input at time t_0, not on inputs at earlier times.

Systems which are modeled by equations of the form of (12-1) include resistive circuits; square law detectors, that is

$$Y = X^2;$$

voltage limiters, for example

$$Y = \begin{cases} -10, & X \le -10, \\ X, & -10 < X < 10, \\ +10, & X \ge 10; \end{cases}$$

and ideal amplifiers. However circuits with storage devices such as capacitors and inductances are not modeled by equations of the form of (12-1) since the output at t_0 depends on inputs at times before t_0.

Given that a model like (12-1) is to be used, then the distribution of

$$Y = g(X)$$

can be determined from the distribution of X as discussed in Chapter VI. Problem 2 reviews this situation.

Thus if the distribution of $X(t)$ is known, the distribution of $Y(t)$ may be found. However as pointed out in Chapter XI we may want to know the joint distribution of $Y(t_1)$, $Y(t_2)$, ..., $Y(t_n)$. This can become a formidable calculation. If $X(t)$ is such that $X(t_1)$, $X(t_2)$, ..., $X(t_n)$ are independent (e.g., white noise), then the problem is solved since $Y(t_1)$, ..., $Y(t_n)$ will be independent in this case.

In some cases only the mean and autocorrelation function of the output $Y(t)$ are desired. If this is the case

$$E[Y(t)] = \int_{-\infty}^{\infty} g(x) f_{X(t)}(x)\, dx, \tag{12-2}$$

or equivalently, $\qquad E[Y(t)] = \int_{-\infty}^{\infty} y f_{Y(t)}(y)\, dy \tag{12-3}$

where the distribution of $Y(t)$ is found from the distribution of $X(t)$ by methods already discussed.

Also

$$R_{YY}(t_1, t_2) = \int\int_{-\infty}^{\infty} g(x_1)g(x_2)f_{X(t_1), X(t_2)}(x_1, x_2)\, dx_1\, dx_2. \qquad (12\text{-}4)$$

$R_{YY}(t_1, t_2)$ may also be expressed in terms of the joint distribution of $Y(t_1)$, $Y(t_2)$ but the difficulty in finding this joint distribution usually makes it impractical.

An example concludes this discussion.

EXAMPLE 12-1

Let $X(t)$ be a stationary normal process with

$$\mu_X(t) = 0,$$

$$R_{XX}(\tau) = e^{-|\tau|}.$$

$Y(t) = X^2(t)$, that is, square law detection. Find $\mu_Y(t)$ and $R_{YY}(t_1, t_2)$.

$$E[Y(t)] = E[X^2(t)] = \int_{-\infty}^{\infty} x^2 \frac{1}{\sqrt{2\pi}} \exp\left[-(x)^2/2\right] dx.$$

Because $E[X^2(t)] = R_{XX}(0)$,

$$\mu_Y(t) = R_{XX}(0) = 1.$$

$$E[Y(t_1)Y(t_2)] = \int\int_{-\infty}^{\infty} \frac{x_1{}^2 x_2{}^2}{2\pi\sqrt{1 - \exp\{-2\,|t_1 - t_2|\}}}$$

$$\times \exp\left[-\frac{(x_1)^2 - 2x_1x_2 \exp\{-\,|t_1 - t_2|\} + (x_2)^2}{2[1 - \exp\{-2\,|t_1 - t_2|\}]}\right] dx_1\, dx_2.$$

The evaluation of this integral is bypassed by using the relation for normal random variables, which have zero means (see Problem 3).

$$E[X_1{}^2X_2{}^2] = E[X_1{}^2]E[X_2{}^2] + 2E^2[X_1X_2].$$

Thus

$$E[Y(t_1)Y(t_2)] = E[X^2(t_1)X^2(t_2)] = E[X^2(t_1)]E[X^2(t_2)] + 2E^2[X(t_1)X(t_2)]$$

$$= R_{XX}(0)R_{XX}(0) + 2R_{XX}^2(|t_1 - t_2|)$$

$$= 1 + 2[e^{-|t_1-t_2|}]^2 = 1 + 2e^{-2|\tau|}.$$

310

Two Input Instantaneous Systems

We consider systems that can be modeled:

$$Z(t_0) = g[X(t_0), Y(t_0)].$$

For a fixed t_0 the distribution of the random variable $Z(t_0)$ may be found from the distribution of the random variables $X(t_0)$ and $Y(t_0)$ (see Chapter VI).

This last idea suggests that linear combinations of random processes can be handled relatively easily while nonlinear combinations are difficult. For this reason we shall consider only linear combinations of random processes in this introductory treatment. Let

$$Z(t) = aX(t) + bY(t).$$

In this case we seek the mean and autocorrelation function of $Z(t)$, assuming that the means, autocorrelation, and cross-correlation functions of $X(t)$ and $Y(t)$ are known.

$$\mu_Z(t) = E[Z(t)] = a\mu_X(t) + b\mu_Y(t), \tag{12-5}$$

$$\begin{aligned} R_{ZZ}(t_1, t_2) &= E\{[aX(t_1) + bY(t_1)][aX(t_2) + bY(t_2)]\} \\ &= a^2 R_{XX}(t_1, t_2) + ab[R_{XY}(t_1, t_2) \\ &\quad + R_{YX}(t_1, t_2)] + b^2 R_{YY}(t_1, t_2). \end{aligned} \tag{12-6}$$

If $a = b = 1$, then $Z(t) = X(t) + Y(t)$ has an autocorrelation function:

$$R_{ZZ}(t_1, t_2) = R_{XX}(t_1, t_2) + R_{XY}(t_1, t_2) + R_{YX}(t_1, t_2) + R_{YY}(t_1, t_2). \tag{12-7}$$

If in addition X and Y are independent random processes and $\mu_X = 0$ or $\mu_Y = 0$, then

$$R_{ZZ}(t_1, t_2) = R_{XX}(t_1, t_2) + R_{YY}(t_1, t_2). \tag{12-8}$$

If $X(t)$ and $Y(t)$ are jointly wide sense stationary and if

$$Z(t) = X(t) + Y(t)$$

then from (12-7)

$$R_{ZZ}(t_1, t_2) = R_{XX}(\tau) + R_{XY}(\tau) + R_{YX}(\tau) + R_{YY}(\tau) \tag{12-9}$$

where $\tau = t_1 - t_2$.

It may be observed from (12-9) that if $X(t)$ and $Y(t)$ are jointly wide sense stationary, then $Z(t)$ is wide sense stationary.

311

EXAMPLE 12-2

A received signal $X(t)$ is the sum of an intelligence signal $S(t)$ and noise $N(t)$. $S(t)$ and $N(t)$ both have mean zero and they are independent with

$$R_{SS}(\tau) = e^{-5|\tau|},$$

$$R_{NN}(\tau) = \frac{\sin 1000\tau}{\tau}.$$

Find μ_X and $R_{XX}(\tau)$.

By (12-5)

$$\mu_X = E[X(t)] = E[S(t) + N(t)] = 0 + 0 = 0.$$

By (12-8)

$$R_{XX}(\tau) = \exp\{-5|\tau|\} + \frac{\sin 1000\tau}{\tau}.$$

The next example illustrates a linear combination of random processes when the sample space is very simple.

EXAMPLE 12-3

Let $X(t)$ consist of two sample functions, $\sin t$ and t. Let $Y(t)$ also consist of two sample functions t and 1. Let $Z(t) = X(t) - Y(t)$. The joint probabilities are given in the following table.

$X(t)$	$Y(t)$	
	t	1
$\sin t$	1/4	1/8
t	1/2	1/8

Note from the table that $X(t)$ and $Y(t)$ are not independent.

From the given data it is easy to note that there are four sample functions possible for $Z(t)$ and to note the probability of each sample function.

$$P[Z(t) = \sin t - t] = \tfrac{1}{4},$$
$$P[Z(t) = \sin t - 1] = \tfrac{1}{8},$$
$$P[Z(t) = 0] = \tfrac{1}{2},$$
$$P[Z(t) = t - 1] = \tfrac{1}{8}.$$

EXAMPLE 12-4. SUM OF INDEPENDENT AND DEPENDENT PROCESSES

Let $X(t) = A \cos(\omega t + \phi)$ where A and ω are constants and ϕ is a random variable uniformly distributed between $-\pi$ and π. Let $Y(t) = A \sin(\omega t + \theta)$ where A and ω are constants and θ is a random variable uniformly distributed between $-\pi$ and π. The mean and autocorrelation function of $X(t)$ were found in Example 11-8. It is easily shown that $Y(t)$ has the same mean and autocorrelation function as $X(t)$, that is,

$$\mu_Y(t) = 0,$$

$$R_{YY}(\tau) = \frac{A^2}{2} \cos \omega\tau.$$

Now if the processes are independent, then $Z(t) = X(t) + Y(t)$ has the following mean and autocorrelation function.

$$\mu_Z(t) = E[X(t) + Y(t)] = 0,$$

$$R_{ZZ}(\tau) = A^2 \cos \omega\tau.$$

Note that independence in this case means independence of ϕ and θ, the respective phase angles.

Now suppose that ϕ and θ are not independent, but that $\theta = \phi - (\pi/2)$; that is, for every sample function $X(t)$ the sample function $Y(t)$ is strictly determined.

Thus

$$Y(t) = A \sin (\omega t + \phi - \pi/2) = -A \cos (\omega t + \phi),$$

$$Z(t) = X(t) + Y(t) = A \cos (\omega t + \phi) - A \cos (\omega t + \phi) = 0.$$

In this case each sample function of $Z(t)$ is identically zero and thus

$$\mu_Z(t) = 0,$$

$$R_{ZZ}(\tau) = 0.$$

Problem 7 requests that you arrive at this result by finding R_{XY} and R_{YX}.

12-2 DIFFERENTIATION OF A RANDOM PROCESS

One of the most common operations used when modeling systems is the derivative. In this section we discuss the derivative of a random process.

If the stochastic process $X(t)$ is such that each sample function has a derivative, then there is a stochastic process $X'(t)$, which is naturally called the derivative. If $X_r(t)$ is a sample function of $X(t)$ with $P[X_r(t)] = p_r$ then $X_r'(t)$ is a sample function of $X'(t)$ with $P[X_r'(t)] = p_r$.

In addition when there are an infinite number of sample functions it may be possible that some of the sample functions do not have a derivative, but that there exists a random process $L(t)$ such that

$$\lim_{T \to 0} E\left\{\left[\frac{X(t + T) - X(t)}{T} - L(t)\right]^2\right\} = 0. \qquad (12\text{-}10)$$

If this is the case then $L(t)$ is called the mean square derivative of $X(t)$.

313

We now find the mean and autocorrelation function of $X'(t)$ assuming that the derivative exists, that is, we assume that limit and expectation can be interchanged.

$$\mu_{X'}(t) = E[X'(t)] = E\left\{\lim_{T \to 0}\left[\frac{X(t+T)-X(t)}{T}\right]\right\},$$

$$\mu_{X'}(t) = \lim_{T \to 0}\frac{\mu(t+T)-\mu(t)}{T} = \frac{d\mu(t)}{dt}. \tag{12-11}$$

Note that if $X(t)$ is stationary

$$E[X'(t)] = 0.$$

We next find the cross-correlation function $R_{XX'}(t_1, t_2)$.

$$R_{XX'}(t_1, t_2) = E\left\{X(t_1)\lim_{T \to 0}\left[\frac{X(t_2+T)-X(t_2)}{T}\right]\right\}$$

$$= \lim_{T \to 0}\frac{1}{T}\{R_{XX}(t_1, t_2+T) - R_{XX}(t_1, t_2)\},$$

$$R_{XX'}(t_1, t_2) = \frac{\partial R_{XX}(t_1, t_2)}{\partial t_2}. \tag{12-12}$$

If $X(t)$ is stationary

$$R_{XX'}(t_1 - t_2) = R_{XX'}(\tau) = -\frac{dR_{XX}(\tau)}{d\tau}. \tag{12-13}$$

Now

$$R_{X'X'}(t_1, t_2) = E\left\{\lim_{T \to 0}\left[\frac{X(t_1+T)-X(t_1)}{T}\right]X'(t_2)\right\}$$

$$= \lim_{T \to 0}\frac{R_{XX'}(t_1+T, t_2) - R_{XX'}(t_1, t_2)}{T}$$

$$= \frac{\partial R_{XX'}(t_1, t_2)}{\partial t_1},$$

$$R_{X'X'}(t_1, t_2) = \frac{\partial^2 R_{XX}(t_1, t_2)}{\partial t_1 \partial t_2}. \tag{12-14}$$

If $X(t)$ is stationary

$$R_{X'X'}(t_1 - t_2) = R_{X'X'}(\tau) = -\frac{d^2 R(\tau)}{d\tau^2}. \tag{12-15}$$

314

Note that if $X(t)$ is stationary, then

$$R_{XX}(\tau) = R_{XX}(-\tau)$$

and thus

$$\frac{-dR_{XX}(\tau)}{d\tau} = \frac{dR_{XX}(-\tau)}{d\tau}$$

and with $\tau = 0$

$$\frac{-dR_{XX}(0)}{d\tau} = \frac{dR_{XX}(0)}{d\tau}$$

which implies that $R_{XX'}(0) = R_{X'X}(0) = 0$.

EXAMPLE 12-5

The current $i(t)$ through a 1-henry inductance is a stationary random process with

$$\mu_i(t) = 2,$$
$$R_{ii}(\tau) = 4 + \cos 5\tau.$$

Find the mean and autocorrelation function of the voltage across the inductance.

Since

$$V(t) = L\left(\frac{di}{dt}\right) = \left(\frac{di(t)}{dt}\right),$$

$$\mu_{V(t)} = \frac{d2}{dt} = 0,$$

and (see 12-15)

$$R_{VV}(\tau) = -\frac{d^2(4 + \cos 5\tau)}{d\tau^2}$$

$$= -\frac{d(-5 \sin 5\tau)}{d\tau}$$

$$= 25 \cos 5\tau.$$

EXAMPLE 12-6

$$R_{XX}(\tau) = e^{-a|\tau|}.$$

Find $R_{X'X'}(\tau)$. It is obvious that $R_{XX}(\tau)$ does not have a derivative at $\tau = 0$, because

$$\lim_{\substack{\tau \to 0 \\ \tau > 0}} \frac{e^{-0} - e^{-a\tau}}{\tau} \neq \lim_{\substack{\tau \to 0 \\ \tau < 0}} \frac{e^{-0} - e^{a\tau}}{\tau}.$$

315

Therefore we say that the derivative $X'(t)$ does not exist. Although there are ways of changing either definitions or models, for the purposes of this text if the derivative of the autocorrelation function does not exist, then we say $X'(t)$ does not exist.

12-3 REVIEW OF LINEAR SYSTEM ANALYSIS

We now want to consider linear systems when the input is a random process, but first we review linear time-invariant systems assuming that the input is a deterministic function of time.

A linear system is one that if $x_1(t)$ produces an output $y_1(t)$ and if $x_2(t)$ produces an output $y_2(t)$, then the input

$$x_3(t) = a_1 x_1(t) + a_2 x_2(t)$$

produces the output

$$y_3(t) = a_1 y_1(t) + a_2 y_2(t).$$

A time-invariant system is one in which if $x(t)$ produces $y(t)$ then $x(t + T)$ produces $y(t + T)$.

By far the most important class of such systems is that modeled by linear differential equations with constant (time-invariant) coefficients.

Impulse Response

Consider a linear time-invariant system as pictured in Figure 12-1. If $x(t)$ is approximated by a sequence of impulses $\sum_{\text{all } i} x(t_i)\delta(t - t_i) = \hat{x}(t)$ as shown in Figure 12-2, then the approximate output at time t_0 is the sum of the responses of the system due to the various inputs, that is

$$y(t_0) \simeq \sum_{\text{all } i} h_i \hat{x}(t_i). \tag{12-16}$$

Note that the output of the linear system is a linear combination of the inputs. But what is h_i, the multiplying factor or weight of an impulse that

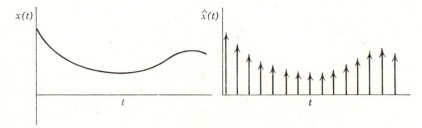

Figure 12-2

occurred $(t_0 - t_i)$ seconds ago? If we have the response of a linear time-invariant system to an impulse plotted as in Figure 12-3, then the response to a unit impulse that occurred 10 sec. ago is 2. In general h_i is the response to an impulse that occurred $t_0 - t_i$ seconds ago. If the input pulse occurred at t_i and we are interested in the output at t_0, then the input pulse occurred $t_0 - t_i$ seconds ago. Thus

$$y(t_0) \simeq \sum_{\text{all } i} h(t_0 - t_i)x(t_i). \qquad (12\text{-}17)$$

Because we are assuming that the system is time invariant, h does not change with t_0, but h depends only on the difference between t_0 and t_i, or how long ago $x(t_i)$ occurred.

The approximation represented by (12-17) has as its limiting continuous form

$$y(t_0) = \int_{-\infty}^{\infty} h(t_0 - t)x(t)\, dt. \qquad (12\text{-}18)$$

This is a convolution integral which was discussed in some detail in Section 6.3. For physical systems

$$h(t) = 0, \qquad t < 0,$$

and thus the upper limit in (12-18) may be changed to t_0.

By making a change of variable $\tau = t_0 - t$, we have an alternate form:

$$y(t_0) = \int_{-\infty}^{\infty} h(\tau)x(t_0 - \tau)\, d\tau, \qquad (12\text{-}19)$$

or with physical systems

$$y(t_0) = \int_{0}^{\infty} h(\tau)x(t_0 - \tau)\, d\tau. \qquad (12\text{-}20)$$

Note that if $x(t) = \delta(t)$ then $y(t_0) = h(t_0)$, which justifies calling $h(t)$ the impulse response.

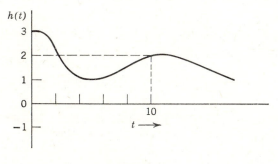

Figure 12-3

Fourier Analysis

Defining the usual Fourier transforms

$$Y(j\omega) = \int_{-\infty}^{\infty} y(t)e^{-j\omega t}\,dt,$$

$$H(j\omega) = \int_{-\infty}^{\infty} h(t)e^{-j\omega t}\,dt,$$

$$X(j\omega) = \int_{-\infty}^{\infty} x(t)e^{-j\omega t}\,dt.$$

Then assuming the integrals converge so that interchanging order of integration is allowable

$$Y(j\omega) = \int_{-\infty}^{\infty}\left[\int_{-\infty}^{\infty} h(\tau)x(t-\tau)\,d\tau\right]e^{-j\omega t}\,dt$$

$$= \int_{-\infty}^{\infty}\left[\int_{-\infty}^{\infty} x(t-\tau)e^{-j\omega(t-\tau)}\,dt\right]e^{-j\omega\tau}h(\tau)\,d\tau$$

$$= \int_{-\infty}^{\infty}\left[\int_{-\infty}^{\infty} x(\lambda)e^{-j\omega\lambda}\,d\lambda\right]e^{-j\omega\tau}h(\tau)\,d\tau$$

$$= \int_{-\infty}^{\infty} X(j\omega)e^{-j\omega\tau}h(\tau)\,d\tau$$

$$= X(j\omega)\int_{-\infty}^{\infty} e^{-j\omega\tau}h(\tau)\,d\tau,$$

$$Y(j\omega) = X(j\omega)H(j\omega), \tag{12-21}$$

or

$$H(j\omega) = \frac{Y(j\omega)}{X(j\omega)}, \tag{12-22}$$

and $H(j\omega)$ is called the system function.

If we have the linear differential equation

$$a_n y^{(n)}(t) + a_{n-1}y^{(n-1)}(t) + \cdots + a_0 y(t) = x(t) \tag{12-23}$$

where a_i, $i = 0, 1, \ldots, n$ are constants and $y^{(i)}$ is the ith derivative with respect to (t) of $y(t)$, then, taking Fourier transforms

$$[a_n(j\omega)^n + a_{n-1}(j\omega)^{n-1} + \cdots + a_0]Y(j\omega) = X(j\omega). \tag{12-24}$$

318

Thus comparing (12-22) and (12-24)

$$H(j\omega) = \frac{1}{\sum_{i=0}^{n} a_i(j\omega)^i} \qquad (12\text{-}25)$$

where

$$(j\omega)^0 = 1.$$

EXAMPLE 12-7

Consider the simple system shown in Figure 12-4. It follows from elementary circuit analysis that

$$Y(j\omega) = \frac{R}{R + j\omega L} X(j\omega)$$

or

$$H(j\omega) = \frac{R}{R + j\omega L}.$$

We show the same result from the differential equation point of view.

$$L\frac{di(t)}{dt} + Ri = x(t) \qquad \text{and} \qquad y(t) = Ri$$

or

$$\frac{L}{R}\frac{dy(t)}{dt} + y(t) = x(t).$$

Using (12-25)

$$H(j\omega) = \frac{1}{\dfrac{L}{R}j\omega + 1} = \frac{R}{j\omega L + R}.$$

The impulse response is

$$h(t) = \frac{R}{L}e^{-Rt/L}, \qquad t \geqslant 0,$$

$$= 0, \qquad t < 0.$$

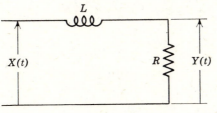

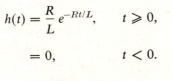

Figure 12-4

319

To show $h(t)$ is correct we show

$$\int_0^\infty h(t)e^{j\omega t}\, dt = H(j\omega) = \frac{R}{j\omega L + R}.$$

$$H(j\omega) = \int_0^\infty \frac{R}{L}\exp\left\{-\frac{Rt}{L}\right\}\exp\left\{-j\omega t\right\}dt = \frac{-R/L}{R/L + j\omega}\exp\left\{-t\left(\frac{R}{L} + j\omega\right)\right\}\Big|_0^\infty$$

$$= \frac{R/L}{R/L + j\omega} = \frac{R}{R + j\omega L}.$$

If the input is

$$x(t) = 2 + e^{-3t}, \qquad t \geqslant 0,$$
$$= 2, \qquad t < 0,$$

and using Table 11-1, $X(j\omega) = 4\pi\delta(\omega) + \dfrac{1}{j\omega + 3}$. The output is (assuming $R/L \neq 3$)

$$Y(j\omega) = H(j\omega)X(j\omega) = \frac{R}{R + j\omega L}\left[4\pi\delta(\omega) + \frac{1}{j\omega + 3}\right]$$

$$= \frac{R}{R + j\omega L}\, 4\pi\delta(\omega) + \frac{R/L}{-3 + R/L}\left[\frac{1}{j\omega + 3} - \frac{1}{j\omega + R/L}\right].$$

Taking inverse transforms

$$y(t) = 2 + \frac{R/L}{-3 + R/L}\,[e^{-3t} - e^{-(R/L)t}], \qquad t \geqslant 0,$$

$$= 2, \qquad t < 0.$$

If we use (12-18)

$$y(t) = \int_{-\infty}^t h(t - \tau)x(\tau)\, d\tau$$

For $t \geq 0$

$$y(t) = \int_{-\infty}^t \frac{R}{L}\exp\left\{-\frac{R}{L}(t - \tau)\right\}2\, d\tau + \int_0^t \frac{R}{L}\exp\left\{-\frac{R}{L}(t - \tau)\right\}e^{-3\tau}\, d\tau$$

$$= 2\exp\left\{-\frac{R}{L}t\right\}\exp\left\{\frac{R}{L}\tau\right\}\Big|_{-\infty}^t + \frac{R/L}{(R/L) - 3}\exp\left\{-\frac{R}{L}t\right\}\exp\left\{\left(\frac{R}{L} - 3\right)\tau\right\}\Big|_0^t$$

$$= 2 + \frac{R/L}{(R/L) - 3}\,[e^{-3t} - e^{-(R/L)t}], \qquad t \geq 0.$$

For $t < 0$:

$$y(t) = \int_{-\infty}^t \frac{R}{L}\exp\left\{\frac{R}{L}(\tau - t)\right\}2\, d\tau = 2, \qquad t < 0.$$

320

12.4 RANDOM PROCESS INPUT TO A LINEAR SYSTEM

We now assume the same type of system described in Figure 12-1 except we now assume that $X(t)$, the input, is a random process. The impulse response is assumed to be deterministic and time invariant, that is, the coefficients in the differential equation are constants.

If there are a finite number of members of the random process, then for each sample function input $X_r(t)$, the corresponding sample function output can be found by techniques described in Section 12.3. Naturally each output would have the same probability as the corresponding input. If $Y_i(t)$ is produced by two different inputs, say X_k and X_j, then the probability of Y_i is

$$P(Y_i) = P(X_j \cup X_k).$$

The case usually encountered is the one where $X(t)$ for any fixed t is a continuous random variable and $X(t)$ is described by its mean and its autocorrelation function. Assuming that this is the case and that $X(t)$ is stationary, we seek the mean of $Y(t)$, the cross-correlation function between $X(t)$ and $Y(t)$, and the autocorrelation function of $Y(t)$. In the developments we assume that integration and expectation can be interchanged.

Mean

$$E[Y(t)] = E\left[\int_{-\infty}^{\infty} X(t - \tau)h(\tau)\,d\tau\right]$$

$$= \int_{-\infty}^{\infty} h(\tau)E[X(t - \tau)]\,d\tau$$

$$= E[X(t)]\int_{-\infty}^{\infty} h(\tau)\,d\tau,$$

$$\mu_Y = \mu_X H(0). \tag{12-26}$$

From the assumption of stationarity, $E[X(t)] = E[X(t - \tau)]$ and does not vary with the argument t or $t - \tau$, thus is a constant and can be taken outside the integral sign.

If $H(0)$ is derived from a differential equation as given in (12-23) then it is easily seen from (12-25) that

$$H(0) = \frac{1}{a_0}. \tag{12-27}$$

Taking the expected value of (12-23) and using (12-26) also results in (12-27) if it is recognized that if $Y(t)$ is stationary then $E[Y^{(i)}(t)] = 0$ for $i \geq 1$.

321

Cross Correlation

$$R_{XY}(\tau) = E[X(t + \tau)Y(t)] = E[X(t + \tau)\int_{-\infty}^{\infty} X(t - \lambda)h(\lambda)\, d\lambda]$$

$$= \int_{-\infty}^{\infty} E[X(t + \tau)X(t - \lambda)]h(\lambda)\, d\lambda,$$

$$R_{XY}(\tau) = \int_{-\infty}^{\infty} R_{XX}(\tau + \lambda)h(\lambda)\, d\lambda. \tag{12-28}$$

Then

$$R_{YX}(\tau) = R_{XY}(-\tau) = \int_{-\infty}^{\infty} R_{XX}(\lambda - \tau)h(\lambda)\, d\lambda = \int_{-\infty}^{\infty} R_{XX}(\tau - \lambda)h(\lambda)\, d\lambda. \tag{12-29}$$

Using the fact that convolution in the time domain corresponds to multiplication in the frequency domain, and taking the Fourier transform of (12-29)

$$\Phi_{YX}(\omega) = \Phi_{XY}(-\omega) = \Phi_{XX}(\omega)H(j\omega). \tag{12-30}$$

Estimation of $H(j\omega)$

Equation (12-30) has an interesting and useful application. If the input is white noise, that is, $R_{XX}(t) = \delta(t)$, then using (12-29)

$$R_{YX}(\tau) = h(\tau).$$

If $X(t)$ and $Y(t)$ are ergodic, then for large T

$$R_{YX}(\tau) \simeq 1/T \int_{0}^{T} Y_r(t + \tau)X_r(t)\, dt$$

or

$$R_{YX}(\tau) \simeq 1/T \int_{\tau}^{\tau+T} Y_r(t')X_r(t' - \tau)\, dt'.$$

Thus $h(\tau)$ can be estimated as shown in Figure 12-5. The correlation

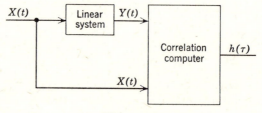

Figure 12-5

computer is essentially a device that delays the input by τ seconds, multiplies the delayed input by the output, and integrates the product.

Practical white noise generators for both electrical and mechanical systems are available. Such a white noise generator must have a power density spectrum that is constant from the lowest to the highest frequency to which the system will significantly respond.

As a practical matter if the system to be modeled is a plant to which the input cannot be shut off, then consider

$$X(t) = i(t) + n(t)$$

where $i(t)$ is the usual known input and $n(t)$ is white noise, which has zero mean and is independent of the input. Then correlate $n(t)$ with the output.

$$R_{Yn}(\tau) = E[Y(t + \tau)n(t)],$$

$$R_{Yn}(\tau) = E\left[n(t) \int_{-\infty}^{\infty} [i(t + \tau - \lambda) + n(t + \tau - \lambda)]h(\lambda)\, d\lambda \right]$$

$$= \int_{-\infty}^{\infty} R_{nn}(\tau - \lambda)h(\lambda)\, d\lambda,$$

$$R_{Yn}(\tau) = \int_{-\infty}^{\infty} \delta(\tau - \lambda)h(\lambda)\, d\lambda = h(\tau).$$

If $x(t)$ is stationary and ergodic but not white noise, then estimates of $\Phi_{XX}(\omega)$ and $\Phi_{YX}(\omega)$ produce an estimate of $H(j\omega)$ via (12-30). That is

$$H(j\omega) = [\Phi_{YX}(\omega)/\Phi_{XX}(\omega)].$$

Autocorrelation of Output

$$R_{YY}(\tau) = E[Y(t + \tau)Y(t)]$$

$$= E\left[\int_{-\infty}^{\infty} Y(t + \tau)X(t - \alpha)h(\alpha)\, d\alpha \right],$$

$$R_{YY}(\tau) = \int_{-\infty}^{\infty} R_{YX}(\tau + \alpha)h(\alpha)\, d\alpha. \qquad (12\text{-}31)$$

From (12-29)

$$R_{YX}(\tau + \alpha) = \int_{-\infty}^{\infty} R_{XX}(\lambda - \tau - \alpha)h(\lambda)\, d\lambda.$$

Thus

$$R_{YY}(\tau) = \iint_{-\infty}^{\infty} R_{XX}(\lambda - \tau - \alpha)h(\lambda)\, d\lambda\, h(\alpha)\, d\alpha. \qquad (12\text{-}32)$$

323

Equation (12-32) is much simpler when viewed in terms of power density spectrum.

$$\Phi_{YY}(\omega) = \iiint\limits_{-\infty}^{\infty} R_{XX}(\lambda - \tau - \alpha)h(\alpha)\, d\alpha\, h(\lambda)\, d\lambda\, e^{-j\omega\tau}\, d\tau$$

$$= \int_{-\infty}^{\infty} h(\alpha) \left[\int_{-\infty}^{\infty} h(\lambda)e^{+j\omega(\alpha-\lambda)} \right.$$

$$\left. \times \left(\int_{-\infty}^{\infty} R_{XX}(\lambda - \tau - \alpha)e^{+j\omega(\lambda-\tau-\alpha)}\, d\tau \right) d\lambda \right] d\alpha$$

$$= \int_{-\infty}^{\infty} h(\alpha)e^{j\omega\alpha}\, d\alpha \int_{-\infty}^{\infty} h(\lambda)e^{-j\omega\lambda}\, d\lambda \Phi_{XX}(-\omega)$$

$$= H(-j\omega)H(j\omega)\Phi_{XX}(-\omega).$$

Since $H(-j\omega)H(j\omega) = |H(j\omega)|^2$ and $\Phi_{XX}(\omega)$ is an even function of ω

$$\Phi_{YY}(\omega) = |H(j\omega)|^2\, \Phi_{XX}(\omega). \qquad (12\text{-}33)$$

Interpretation

Result (12-33) is a further justification of the term power density spectrum. We saw earlier that

$$E[X^2(t)] = \frac{1}{2\pi} \int_{-\infty}^{\infty} \Phi_{XX}(\omega)\, d\omega.$$

Now if $H(j\omega)$ is an ideal band pass filter, that is

$$H(j\omega) = 1, \qquad \omega_1 \le |\omega| \le \omega_2,$$
$$= 0, \qquad \text{elsewhere},$$

then the average power of the output is

$$E[Y^2(t)] = \frac{1}{2\pi} \int_{-\infty}^{\infty} \Phi_{YY}(\omega)\, d\omega = \frac{1}{\pi} \int_{\omega_1}^{\omega_2} \Phi_{XX}(\omega)\, d\omega.$$

Since the average power of the output of the band pass filter is the power density spectrum of the input integrated between ω_1 and ω_2, we say that the power between ω_1 and ω_2 is proportional to the area under the power density spectrum between ω_1 and ω_2.

The above equation also shows, since $E[Y^2(t)] \ge 0$, that $\Phi_{XX}(\omega) \ge 0$ for *all* ω.

EXAMPLE 12-8

$X(t)$ is a stationary random process with $\mu_X = 0$, $R_{XX}(\tau) = e^{-\alpha|\tau|}$. $X(t)$ is the input to a system that has a system function (Example 12-7)

$$H(j\omega) = \frac{R}{R + j\omega L}.$$

Find μ_Y, $\Phi_{YY}(\omega)$, and $R_{YY}(\tau)$.

$$\mu_Y = 0$$

$$\Phi_{XX}(\omega) = \int_{-\infty}^{0} e^{\alpha\tau} e^{-j\omega\tau}\, d\tau + \int_{0}^{\infty} e^{-\alpha\tau} e^{-j\omega\tau}\, d\tau$$

$$= \frac{1}{\alpha - j\omega} + \frac{1}{\alpha + j\omega} = \frac{2\alpha}{\alpha^2 + \omega^2}.$$

$$\Phi_{YY}(\omega) = \frac{2\alpha}{\alpha^2 + \omega^2} \frac{R^2}{R^2 + \omega^2 L^2} = \frac{2\alpha(R^2/L^2)}{(\alpha^2 + \omega^2)[\omega^2 + (R^2/L^2)]}$$

$$= \frac{A}{\alpha^2 + \omega^2} + \frac{B}{\omega^2 + (R^2/L^2)},$$

where

$$A = \frac{2\alpha R^2}{R^2 - \alpha^2 L^2}, \qquad B = \frac{2R^2\alpha}{L^2\alpha^2 - R^2}.$$

Thus taking inverse transforms (see 8 in Table 11-1).

$$R_{YY}(\tau) = \frac{R^2/L^2}{(R^2/L^2) - \alpha^2} e^{-\alpha|\tau|} + \frac{(R/L)\alpha}{\alpha^2 - (R^2/L^2)} e^{-(R/L)|\tau|}.$$

EXAMPLE 12-9

A differentiator is a system for which

$$H(j\omega) = j\omega.$$

Thus if the stationary random process $X(t)$ is differentiated, then the output $X'(t)$ has the power density spectrum

$$\Phi_{X'X'}(\omega) = (j\omega)(-j\omega)\Phi_{XX}(\omega) = \omega^2 \Phi_{XX}(\omega).$$

Thus

$$R_{X'X'}(\tau) = -\frac{d^2 R_{XX}(\tau)}{d\tau^2}.$$

This agrees with the result derived earlier.

If the input $X(t)$ is some deterministic signal $s(t)$, plus noise $n(t)$, which has a mean of zero, then

$$E[X'(t)] = E[s'(t) + n'(t)] = s'(t).$$

325

However we see from $\Phi_{X'X'}$ that the power density spectrum of the noise is multiplied by the factor ω^2 (the high frequency noise is amplified). This provides a theoretical explanation for the practical result that it is impossible to build a differentiator that is not "noisy."

EXAMPLE 12-10

An averaging circuit (for integration period T) has an impulse response

$$h(t) = \frac{1}{T}, \qquad 0 \leq t \leq T,$$

$$= 0, \qquad \text{elsewhere.}$$

Indeed it is called averaging because

$$Y(t) = X(t) * h(t),$$

$$Y(t) = \frac{1}{T} \int_{t-T}^{t} X(\tau)\, d\tau,$$

$$H(j\omega) = \frac{1}{T} \int_{0}^{T} e^{-j\omega t}\, dt = \frac{1}{j\omega T} [1 - e^{-j\omega T}]$$

$$= \frac{1}{j\omega T} e^{-j\omega T/2} [e^{j\omega T/2} - e^{-j\omega T/2}]$$

$$= e^{-j\omega T/2} \frac{\sin(\omega T/2)}{(T\omega/2)}.$$

Thus

$$\Phi_{YY}(\omega) = \Phi_{XX}(\omega) \frac{\sin^2(\omega T/2)}{(T\omega/2)^2}$$

Normal Processes

If the input to a linear system is a normal random process, then (12-17) shows that the output (for the discrete form) is a linear combination of normal random variables. Thus $y(t_1)$, $y(t_2)$, $\ldots$, $y(t_n)$ are jointly normally distributed. Thus the output $y(t)$ is a normal random process. Under suitable conditions this is also true when the output is the limiting case of the sum, that is, the convolution integral (12-18).

Because $y(t)$ is normal, it is completely described by its mean and auto-correlation function. Thus if the input to a linear system is a normal random process the output, which is a normal random process, is completely described by (12-26) and (12-33).

326

12-5 THERMAL NOISE

Thermal agitation of free electrons in a resistor produces a fluctuating voltage across the resistor. In Figure 12-6 the voltmeter registers a time-varying voltage. The input to the amplifier in the absence of any intended signal is called thermal noise.

The noise voltage is the sum of many voltage pulses caused by the individual electrons, so we expect from the central limit theorem that thermal noise is a normal random process. Measurements tend to confirm this hypothesis. It also seems reasonable that there is no preferred direction of motion, and therefore the average value of the voltage should be zero. This is also supported by measurements.

Measurements also show that the power density spectrum is approximately constant, that is

$$\Phi(\omega) = C.$$

The constant C has been found to be proportional to the resistance of the conductor and the temperature, and

$$\Phi(\omega) = 2KTR \frac{\text{volts}^2}{\text{hertz}} \tag{12-34}$$

where T is the temperature (°K), R is the resistance (ohms), K is the Boltzmann constant, $K \simeq 1.38 \times 10^{-23}$ joule/deg.

The thermal noise power $N(\omega_1, \omega_2)$ measured between ω_1 and ω_2 will be

$$N(\omega_1, \omega_2) = \frac{1}{2\pi} \int_{-\omega_2}^{-\omega_1} 2KTR \, d\omega + \frac{1}{2\pi} \int_{\omega_1}^{\omega_2} 2KTR \, d\omega$$

$$= \frac{2(\omega_2 - \omega_1)}{\pi} KTR,$$

$$N(\omega_1, \omega_2) = 4KTR(f_2 - f_1) \text{ volts}^2 \tag{12-35}$$

where $2\pi f = \omega$.

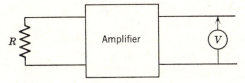

Figure 12-6

327

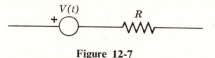

Figure 12-7

Thus we have modeled the noise voltage appearing across a resistor as a stationary normal random process with mean 0 and $\Phi(\omega) = 2KTR$. We now consider circuits made up of noisy resistors. In this case each resistor is modeled as a voltage source as specified above in series with a noiseless resistance R as shown in Figure 12-7.

Series Resistors

Consider two noisy resistors in series as shown in Figure 12-8. This is modeled as shown in Figure 12-9. The output voltage is $V_1(t) + V_2(t)$ and thus
$$E[V(t)] = 0 + 0 = 0.$$

Assuming $V_1(t)$ and $V_2(t)$ are independent
$$E[V(t)V(t + \tau)] = R_{V_1V_1}(\tau) + R_{V_2V_2}(\tau)$$
or
$$\Phi_{VV}(\omega) = \Phi_{V_1V_1}(\omega) + \Phi_{V_2V_2}(\omega)$$
$$= 2KT(R_1 + R_2),$$

assuming the resistors are at the same temperature.

Resistors in Parallel

Two noisy resistors in parallel would be modeled as shown in Figure 12-10. Thus
$$V(t) = \frac{R_2V_1(t) + R_1V_2(t)}{R_1 + R_2}$$
and
$$E[V(t)] = 0.$$

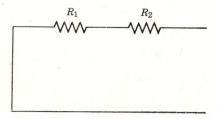

Figure 12-8

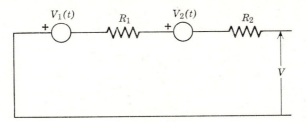

Figure 12-9

Using independence

$$R_{VV}(\tau) = E[V(t)V(T + \tau)] = E\left[\left(\frac{R_2}{R_1 + R_2}\right)^2 V_1(t)V_1(t + \tau)\right]$$

$$+ E\left[\left(\frac{R_1}{R_1 + R_2}\right)^2 V_2(t)V_2(t + \tau)\right]$$

$$= \left(\frac{R_2}{R_1 + R_2}\right)^2 R_{V_1 V_1}(\tau) + \left(\frac{R_1}{R_1 + R_2}\right)^2 R_{V_2 V_2}(\tau).$$

Thus

$$\Phi_{VV}(\omega) = \left(\frac{R_2}{R_1 + R_2}\right)^2 \Phi_{V_1 V_1}(\omega) + \left(\frac{R_1}{R_1 + R_2}\right)^2 \Phi_{V_2 V_2}(\omega)$$

$$= 2KT\left[R_1 \frac{R_2^2}{(R_1 + R_2)^2} + R_2 \frac{R_1^2}{(R_1 + R_2)^2}\right]$$

$$= 2KT \frac{R_1 R_2}{R_1 + R_2}.$$

RLC Circuits

Consider the circuit made up of a noisy resistor and a capacitor modeled as shown in Figure 12-11. Then

$$\frac{V(j\omega)}{V_n(j\omega)} = H(j\omega) = \frac{1/j\omega C}{R + (1/j\omega C)} = \frac{1}{1 + j\omega RC}.$$

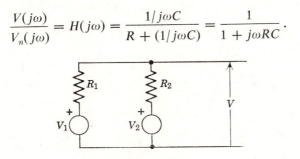

Figure 12-10

329

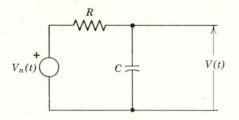

Figure 12-11

Using (12-26)

$$E[V(t)] = 0H(0) = 0,$$

$$\Phi_{VV}(\omega) = \Phi_{V_nV_n}(\omega)\frac{1}{1 + \omega^2R^2C^2}$$

$$= 2KTR\frac{1}{1 + \omega^2R^2C^2}.$$

Notice that the input impedance of the circuit is

$$Z(j\omega) = \frac{R}{1 + j\omega RC}$$

and $Re\, Z(j\omega) = Re\left[\dfrac{R(1 - j\omega RC)}{(1 + j\omega RC)(1 - j\omega RC)}\right] = \dfrac{R}{1 + \omega^2R^2C^2}.$

Thus

$$\Phi_{VV}(\omega) = 2KTRe[Z(j\omega)].$$

It may be shown that in general

$$\Phi_{VV}(\omega) = 2KTRe[Z(j\omega)] \tag{12-36}$$

where $Z(j\omega)$ is the input impedance of a circuit composed of resistances, inductances and capacitances. This result is known as Nyquist's theorem.

12-6 SUMMARY

In the case of instantaneous systems the problem of describing the output when the input is a random process can be reduced to finding the joint distribution of functions of random variables.

In the rest of the chapter the mean of the output, the cross-correlation of the input and the output, and the autocorrelation of the output were the measures of the output random process. Only linear systems were discussed.

330

Differentiation of a random process was defined and it was shown that a simple relation exists between the autocorrelation function of the derivative and the autocorrelation of the random process being differentiated.

When a random process is the input to a linear system, the output is also a random process. If the input is stationary, so is the output. If the input is normal, so is the output. Simple formulas for the mean of the output, the cross-correlation of input and output, and the autocorrelation of the output were derived. Applications of these formulas included a method for determining the system function of an unknown linear system, checking the earlier results on differentiation, and further explaining the useful interpretation of power density spectrum.

Thermal noise was introduced as an example of noise and to illustrate the use of some of the results derived for linear systems.

Additional reading may be found in Ref. B1, D1, H2, H3, L1, L3, P2, T1, and S1.

12-7 PROBLEMS

1. Let $X(t) = A \cos \omega t$ where ω is fixed and A is a normally distributed random variable. If $X(t)$ is the input to the system shown in Figure P12-1 find $Y(t)$ in terms of the parameters ω, R, C and the random variable A.

2. Let $X(t_0)$ have a normal distribution with mean 0 and variance σ^2. Find the density function of $Y(t_0)$ if

   (a) $Y(t_0) = [X(t_0)]^2$,   square law detector;

   (b) $Y(t_0) = |X(t_0)|$,   full wave rectifier;

   (c) $Y(t_0) = \frac{1}{2}[X(t_0) + |X(t_0)|]$,   half-wave rectifier;

   (d) $Y(t_0) = 1$ if $X(t_0) \geq \sigma^2$,   limiter;

             $= X(t_0)$ if $-\sigma^2 < X(t_0) < \sigma^2$

             $= -1$ if $X(t_0) \leq \sigma^2$.

Hint: See Examples 6-2 through 6-5.

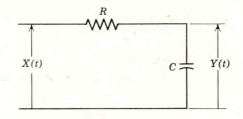

Figure P12-1

3. Show if X_1 and X_2 are jointly normal random variables with zero means

$$E[X_1{}^2X_2{}^2] = E[X_1{}^2]E[X_2{}^2] + 2E^2[X_1X_2].$$

Hint:
$$E[X_1{}^2X_2{}^2] = E\{E[X_1{}^2X_2{}^2 \mid X_2]\}$$
$$= E\{X_2{}^2E[X_1{}^2 \mid X_2]\},$$

see (5-26),
$$= E\left\{X_2{}^2\left[\sigma_1{}^2(1 - r^2) + r^2\frac{\sigma_1{}^2}{\sigma_2{}^2}(X_2{}^2)\right]\right\}$$

$$= \sigma_2{}^2[\sigma_1{}^2(1 - r^2)] + r^2\frac{\sigma_1{}^2}{\sigma_2{}^2}E(X_2{}^4).$$

Use (5-15) to find $E[X_2{}^4] = 3\sigma_2{}^4$.

4. Let $R_{XX}(\tau) = \delta(\tau)$. $X(t)$ is uniformly distributed between -1 and $+1$. $Y(t) = 10X(t)$, i.e. ideal amplifier. Find $\mu_Y(t)$ and $R_{YY}(t_1, t_2)$.

5. Let $X(t)$ be composed of sample functions, each of which is a constant. Let the magnitude of the constant be a normal random variable with mean 3 and variance 1.

   (a) Show that $\mu_X(t) = 3$ and $R_{XX}(\tau) = 10$.
   (b) Let $Y(t)$ be stationary with a power density spectrum

$$\Phi_{YY}(\omega) = \frac{4\sin^2(\omega/2)}{\omega^2}.$$

   Find $R_{YY}(\tau)$.
   (c) Assume that $X(t)$ and $Y(t)$ are independent random processes. What is the autocorrelation function of $X(t) + Y(t)$?

6. Let $X(t)$ be a Poisson process with parameter μ_X and $Y(t)$ be an independent Poisson process with parameter μ_Y. $Z(t) = X(t) + Y(t)$. Show that $Z(t)$ is Poisson with parameter $\mu_X + \mu_Y$.

   The above result can be extended by induction to show that the sum of N Poisson process is a Poisson process with mean the sum of the means.

   If a Poisson process models the number of failures of equipment and if N such equipments are used in a system, the number of failures of the system is Poisson with mean the sum of the means, assuming independence.

7. Refer to Example 12-4. Let $X(t) = A\cos(\omega t + \phi)$, $Y(t) = -A\cos(\omega t + \phi)$ with ϕ uniformly distributed between $-\pi$ and π.
   Find $R_{XY}(\tau)$ and $R_{YX}(\tau)$ and show $R_{ZZ}(\tau) = R_{XX}(\tau) + R_{XY}(\tau) + R_{YX}(\tau) + R_{YY}(\tau) = 0$ where $Z(t) = X(t) + Y(t)$.

8. Suppose $X(t) = f(t) + N(t)$ where $f(t)$ is some deterministic time function and $N(t)$ is noise, with

$$\mu_N = 0,$$
$$R_{NN}(\tau) = e^{-100\tau^2}.$$

If we average $X(t)$ show that

$$\mu_X(t) = E\left[\frac{1}{T}\int_{t-T}^{T} X(\lambda)\, d\lambda\right] = \frac{1}{T}\int_{t-T}^{T} f(\lambda)\, d\lambda$$

and

$$\sigma^2 = \frac{1}{T}\int_{-T}^{T}\left(1 - \frac{|\tau|}{T}\right)e^{-100\tau^2}\, d\tau.$$

Hint: see Section 11-6.

9. Refer to the same situation of a deterministic signal plus noise discussed in problem 8. Suppose that

$$Y(t) = \frac{dX(t)}{dt},$$

Find $E[Y(t)]$ and $\sigma^2_{Y(t)}$.

10. For a discrete time system let the input $x(t)$ be impulses with magnitudes

$$x(0) = 1,$$
$$x(1) = 1,$$
$$x(2) = 3,$$
$$x(3) = 2,$$
$$x(t) = 0, \qquad \text{for } t \neq 0, 1, 2, 3.$$

The impulse respose is

$$h(t) = e^{-t}, \qquad t \geq 0,$$
$$= 0, \qquad t < 0,$$
$$y(t) = x(t) * h(t).$$

Find

$$y(0), y(1), y(2), y(3), y(4), y(5).$$

11. (a) $x(t) = 2, \qquad 0 \leq t \leq 3,$
   $\quad = 0, \qquad$ elsewhere.
   $h(t) = 1, \qquad 0 \leq t \leq 1,$
   $\quad = 0, \qquad$ elsewhere.

   Find $y(t) = x(t) * h(t)$.

   (b) $x(t)$ as above,
   $h(t) = e^{-t}, \qquad t \geq 0,$
   $\quad = 0, \qquad t < 0.$

   Find $y(t)$.

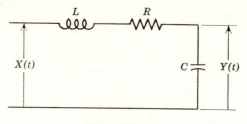

Figure P12-2

12. Let the voltage $V(t)$ across a 1-μfarad capacitor be a random process with

$$\mu_{V(t)} = 10,$$
$$R_{VV}(\tau) = e^{-\tau^2} + 100.$$

Find the mean and autocorrelation function of the current.

13. Let $X(t)$ be a stationary random process with mean 0 and $R_{XX}(\tau) = e^{-\tau^2}$.

(a) Find $\Phi_{XX}(\omega)$.

(b) If $H(j\omega) = \dfrac{a + j\omega}{b + cj\omega + \omega^2}$

find μ_Y and $\Phi_{YY}(\omega)$.

14. White noise is the input to the circuit shown in Figure P12-2. What is the mean and power density spectrum of the output?

15. White noise is applied to an ideal band pass filter.

$$H(j\omega) = 1, \qquad \omega_1 < |\omega| < \omega_2,$$
$$= 0, \qquad \text{elsewhere.}$$

What is the power density and autocorrelation of the output?

16. Find the mean and power density spectrum of the thermal noise voltage of the circuit shown in Figure P12-3.

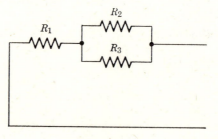

Figure P12-3

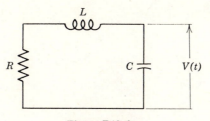

Figure P12-4

17. What is the noise power between 1000 and 100,000 hertz of a 1KΩ resistor at room temperature?

  Consider that a very weak signal is received and amplified. At which stage of the amplifier would the thermal noise be most critical?

18. Find the power density spectrum of the noise voltage $V(t)$ in the circuit shown in Figure P12-4.

REFERENCES

B1. Beckmann, P., *Probability in Communication Engineering*. Harcourt, Brace and World, New York, 1967.

B2. Burr, I. W., *Engineering Statistics and Quality Control*. McGraw-Hill, New York, 1953.

C1. Chernoff, H., and L. Moses, *Elementary Decision Theory*. Wiley, New York, 1959.

C2. Cox, D. R., and H. D. Miller, *The Theory of Stochastic Processes*. Methuen, London, 1964.

C3. *C.R.C. Standard Mathematical Tables*. 12th ed., The Chemical Rubber Co., Cleveland, Ohio.

D1. Davenport, W. B., Jr. and W. L. Root, *Random Signals and Noise*. McGraw-Hill, New York, 1958.

D2. David, F. N., *Games, Gods, and Gambling*. Hafner, New York, 1962.

D3. Drake, A. W., *Fundamentals of Applied Probability Theory*. McGraw-Hill, New York, 1967.

D4. Dubes, R. C., *The Theory of Applied Probability*. Prentice-Hall, Englewood Cliffs, N.J., 1968.

D5. Duncan, A. J., *Quality Control and Industrial Statistics*. Richard D. Irwin, Homewood, Ill., 1959.

F1. Feller, W., *An Introduction to Probability Theory and Its Applications*. Vol. 1, 2nd ed., Wiley, New York, 1957.

F2. Fraser, D. A. S., *Statistics: An Introduction*. Wiley, New York, 1958.

G1. Grant, E. L., *Statistical Quality Control*. 3rd ed., McGraw-Hill, New York, 1964.

G2. Guttman, I., and S. S. Wilks, *Introductory Engineering Statistics*. Wiley, New York, 1965.

H1. Hadley, G., *Introduction to Probability and Statistical Decision Theory*. Holden-Day, San Francisco, 1967.

H2. Hancock, J. C., *An Introduction to the Principles of Communication Theory*. McGraw-Hill, New York, 1961.

H3. Harman, W. W., *Principles of the Statistical Theory of Communication*. McGraw-Hill, New York, 1963.

J1. Jaynes, E. T., *Probability Theory in Science and Engineering*. Socony Mobil Oil Co., Dallas, Texas, 1959.

K1. Kaplan, W., *Operational Methods For Linear Systems*. Addison-Wesley, Reading, Mass., 1962.

337

L1. Laning, J. H. and R. H. Battin, *Random Processes in Automatic Control*. McGraw-Hill, New York, 1956.

L2. Lathi, B. P., *An Introduction to Random Signals and Communication Theory*. International Textbook Co., Scranton, Pa., 1968.

L3. Lee, Y. W., *Statistical Theory of Communication*. Wiley, New York, 1960.

L4. Lindgren, B. W., *Statistical Theory*. Macmillan, New York, 1960.

N1. National Bureau of Standards, "Tables of the Binomial Probability Distribution." *Applied Mathematics Series* 6, U.S. Gov. Printing Office, Washington, D.C., 1949.

P1. Papoulis, A., *The Fourier Integral and Its Applications*. McGraw-Hill, New York, 1962.

P2. Papoulis, A., *Probability, Random Variables, and Stochastic Processes*. McGraw-Hill, New York, 1965.

P3. Parzen, E., *Modern Probability Theory and Its Applications*. Wiley, New York, 1960.

P4. Parzen, E., *Stochastic Processes*. Holden-Day, San Francisco, 1962.

P5. Pfeiffer, P. E., *Concepts of Probability Theory*. McGraw-Hill, New York, 1965.

R1. Reza, F. M., *An Introduction to Information Theory*. McGraw-Hill, New York, 1961.

R2. Rosenblatt, M., *Random Processes*. Oxford University Press, New York, 1962.

S1. Sakrison, David, *Communication Theory: Transmission of Waveforms and Digital Information*. Wiley, New York, 1968.

S2. Savage, L. J., *The Foundations of Statistics*. Wiley, New York, 1954.

S3. Shooman, M. L., *Probabilistic Reliability: An Engineering Approach*. McGraw-Hill, New York, 1968.

T1. Thomas, John B., *An Introduction to Statistical Communication Theory*. Wiley, New York, 1969.

W1. Weiss, Lionel, *Statistical Decision Theory*. McGraw-Hill, New York, 1961.

ANSWERS TO SELECTED
PROBLEMS

2.3   (a) no; (b) yes.

2.4   (a) A.

2.5 $A \cup B \cup C = A \cup \bar{A}B \cup \bar{A}\bar{B}C$.

2.7   (a) $\{1, 2, 3, 4, 5, 6, 7\}$.

2.15   (a) $\frac{1}{2}$; (b) $\frac{1}{20}$; (c) $\frac{1}{10}$; (d) $\frac{11}{20}$.

2.16 $\dfrac{2}{\mu + 2}$.

2.19 $P(N_1) = .3; P(N_2) = .4; P(N_3) = .3;$
$P(M_1) = .6; P(M_2) = .4;$
$P(N_1 \mid M_2) = .5;$
$P(M_1 \mid N_3) \simeq .67;$
$P(N_1 \mid M_1) \simeq .167;$

not independent.

2.21 $P(B_j \mid A_i)P(A_i) = P(A_iB_j)$; (2-32) completes proof.

2.23
Sum	2	3	4	5	6	7	8	9	10	11	12
Prob	$\frac{1}{36}$	$\frac{2}{36}$	$\frac{3}{36}$	$\frac{4}{36}$	$\frac{5}{36}$	$\frac{6}{36}$	$\frac{5}{36}$	$\frac{4}{36}$	$\frac{3}{36}$	$\frac{2}{36}$	$\frac{1}{36}$.

2.25 $P\left(\bigcap_{i=1}^{4} A_i\right) = \prod_{i=1}^{4} P(A_i),$

$P\left(\bigcap_{i=1}^{3} A_i\right) = \prod_{i=1}^{3} P(A_i); P(A_1A_2A_4) = P(A_1)P(A_2)P(A_4);$

$P(A_1A_3A_4) = P(A_1)P(A_3)P(A_4); P\left(\bigcap_{i=2}^{4} A_i\right) = \prod_{i=2}^{4} P(A_i);$

$P(A_1A_2) = P(A_1)P(A_2); P(A_1A_3) = P(A_1)P(A_3);$
$P(A_1A_4) = P(A_1)P(A_4); P(A_2A_3) = P(A_2)P(A_3);$
$P(A_2A_4) = P(A_2)P(A_4); P(A_3A_4) = P(A_3)P(A_4).$

2.27 $P(A_1\bar{A}_2) = P(A_1) - P(A_1A_2)$

$\qquad\qquad = P(A_1) - P(A_1)P(A_2)$

$\qquad\qquad = P(A_1)[1 - P(A_2)] = P(A_1)P(\bar{A}_2).$

2.29 The probability of one or both of the events is zero.

2.31 $\qquad P(B) = P(B \mid A)P(A) + P(B \mid \bar{A})P(\bar{A})$

$\qquad\qquad = (.9)(.5) + (.1)(.5) = .5;$

$\qquad P(A \mid B) = \dfrac{P(B \mid A)P(A)}{P(B)} = \dfrac{(.9)(.5)}{(.5)} = .9;$

$\qquad P(A \mid \bar{B}) = \dfrac{P(\bar{B} \mid A)P(A)}{P(\bar{B})} = \dfrac{(.1)(.5)}{(.5)} = .1.$

2.33 (a) $(.98)^{10}$;

(b) $10(.02)(.98)^9$;

(c) $(.02)(.98)^9$.

2.35 (a) $4\dfrac{13\ 13\ 12\ 11}{52\ 51\ 50\ 49}$;

(b) $12\dfrac{13\ 13\ 13\ 12}{52\ 51\ 50\ 49}$;

(c) $\dfrac{13\ 12}{52\ 51} + (2)\left(\dfrac{13\ 39\ 12}{52\ 51\ 50}\right) + (3)\left(\dfrac{39\ 38\ 13\ 12}{52\ 51\ 50\ 49}\right).$

3.1 (a) $R = p_A p_B p_G[1 - (1 - p_C p_D)(1 - p_E p_F)][1 - q_H q_I q_J];$

$\qquad Q = 1 - R;$

$\qquad Q \simeq q_A + q_B + q_G + (q_C + q_D)(q_E + q_F) + q_H q_I q_J.$

3.2 (a) $3(.01) + (.01)^3 \simeq .03.$

3.3 $R = P(S_A S_B) + P(S_A F_B S_C) + P(F_A S_B S_C)$

$\qquad = (.7)(.8) + (.7)(.2)(.75) + (.3)(.6)(.65).$

3.4 (a) $R = BA_a[\theta_1 R_2 + R_1\theta_2 + R_1 R_2]L.$

3.5 (a) $R = B(X_1 + Y_1 - X_1 Y_1)L.$

3.6 (a) $P(ss \mid E_1) = (.9)^3 = .729;$

$\qquad E(ss \mid E_2) = (.9)(.8)(.7) = .504.$

(b) $P(ss) = (.729)\tfrac{1}{2} + (.504)\tfrac{1}{2} = .6165.$

(c) Assuming conditional independence:

$\qquad P(\text{one of two works} \mid E_1) = 1 - (.271)^2 \simeq .927;$

$\qquad P(\text{one of two works} \mid E_2) = 1 - (.496)^2 \simeq .754;$

$\qquad P(\text{one of two works}) \simeq .927(\tfrac{1}{2}) + .754(\tfrac{1}{2}) \simeq .84.$

$\qquad$ Assuming independence:

$\qquad P(\text{one of two works}) = 1 - (.3835)^2 \simeq .853.$

Conditional independence is more reasonable because environment affects both systems in roughly the same manner.

4.3 $p_X(x_i) \geq 0, \sum\limits_{\text{all } i} p_X(x_i) = 1$

4.5 $p_X(N) = P\{\text{1st tail appears on } N\text{th toss}\};$

$p_X(1) = \frac{1}{2}, p_X(2) = (\frac{1}{2})^2; p_X(3) = (\frac{1}{2})^3;$

$F_X(N) = \sum\limits_{i=1}^{N} (\frac{1}{2})^i; \quad N = 1, 2, \ldots ;$

$F_X(x) = 0, x \leq 1.$

4.7 $p_X(n) = \binom{10}{n}(.01)^n(.99)^{10-n}, \quad n = 0, 1, \ldots, 10;$

$F_X(x) = 0, \quad x < 0;$

$F_X(x) = \sum\limits_{i=0}^{n} \binom{10}{i}(.01)^i(.99)^{10-i}, \quad \begin{matrix} n \leq x < n + 1; \\ 0 \leq x \leq 10. \end{matrix}$

$F_X(x) = 1, \quad x \geq 10.$

4.9 In this case F_X is easier to find.

$F_X(1) = \frac{1}{216} = P(\text{all equal } 1);$

$F_X(2) = \frac{8}{216} = [P(1 \text{ die } \leq 2)]^3;$

$F_X(3) = \frac{27}{216}; F_X(4) = \frac{64}{216}, F_X(5) = \frac{125}{216}, F_X(6) = 1;$

$p_X(1) = \frac{1}{216}, p_X(2) = F_X(2) - F_X(1) = \frac{7}{216};$

$p_X(3) = \frac{19}{216}, p_X(4) = \frac{37}{216}, p_X(5) = \frac{61}{216};$

$p_X(6) = \frac{91}{216}.$

4.11 (a) yes;

(b) yes;

(c) yes;

(d) no.

4.13 .0262.

4.15 $p_X(0) = e^{-\lambda};$

$p_X(1) = \lambda e^{-\lambda};$

$p_X(2) = \frac{\lambda^2}{2} e^{-\lambda};$

$\sum\limits_{k=0}^{\infty} p_X(k) = e^{-\lambda} + \lambda e^{-\lambda} + \frac{\lambda^2}{2!} e^{-\lambda} + \cdots + \frac{\lambda^n}{n!} e^{-\lambda} + \cdots$

$= e^{-\lambda}\left(1 + \lambda + \frac{\lambda^2}{2!} + \cdots\right)$

$= e^{-\lambda} \cdot e^{\lambda}$

$= 1.$

4.17 $p_I(0) = .027; \quad F_I(x) = 0, \, x < 0;$
$p_I(5) = .189; \quad F_I(0) = .027;$
$p_I(10) = .441; \quad F_I(5) = .216;$
$p_I(15) = .343; \, F_I(10) = .657;$
$F_I(15) = 1.$

4.19   (a) yes; (b) no.

4.21 $K = \frac{1}{3}.$

(a) $F_{X,Y}(x, y) = 0, \qquad x < 0 \quad \text{or} \quad y < 0;$
$= \frac{1}{6}(x^2y + xy^2), \qquad 0 \le x \le 1, 0 \le y \le 2$
$= \frac{1}{6}(y + y^2), \qquad x \ge 1, 0 \le y \le 2;$
$= \frac{1}{6}(2x^2 + 4x), \qquad 0 \le x \le 1, y \ge 2;$
$= 1, \qquad x \ge 1 \quad \text{and} \quad y \ge 2.$

(b) $f_Y(y) = \frac{1}{3}(\frac{1}{2} + y), \qquad 0 \le y \le 2;$
$= 0, \qquad\qquad\qquad \text{elsewhere.}$

$f_{X|Y}(x \mid y) = \dfrac{2(x + y)}{1 + 2y}, \qquad 0 \le x \le 1;$
$= 0, \qquad\qquad x \text{ otherwise.}$

Note $f_{X|Y}(x \mid y)$ defined only for $0 \le y \le 2$.

(c) $F_{X|Y}(x \mid y) = 0, \qquad x \le 0, 0 \le y \le 2;$
$= \dfrac{2}{1 + 2y}\left(\dfrac{x^2}{2} + xy\right), \qquad 0 \le x \le 1, 0 \le y \le 2;$
$= 1, \qquad x \ge 1, 0 \le y \le 2.$

4.23   (a) .9876;

(b) $\dfrac{.383}{.9876} = .388.$

4.25   In equation 4-31, let
$B_j = \{s \mid X = x_j\}, A_i = \{s \mid Y = y_i\}.$
Then $P(B_j) = p_X(x_j) = \sum_{\text{all } i} p_{X|Y}(x_j \mid y_i)p_Y(y_i).$

4.27 $F_X(x) = P\{s \mid X \le x\} = P(X \le x) = P\{(X \le x) \cap S\};$

$F_X(x) = P\left\{(X \le x) \cap \bigcup_{i=1}^{n} B_i\right\}$

$= P\left\{\bigcup_{i=1}^{n}[B_i \cap (X \le x)]\right\}$

$= \sum_{i=1}^{n} P[B_i \cap (X \le x)]$

$= \sum_{i=1}^{n} P(X \le x \mid B_i)P(B_i)$

$= \sum_{i=1}^{n} F_{X|B_i}(x)P(B_i).$

5.1 $\frac{4}{3}$.

5.3 $4 = np$.

5.5   1.

5.7 λ.

5.8   100.

5.11   (a) $P\{|X - \mu| > 3\} \leq \frac{4}{9}$;

   (b) .0027.

5.13 $E[X] = np$;

   $E[X^2] = n^2p^2 + np(1 - p)$;

   $\sigma^2 = npq$.

5.15   5.

5.17 $\rho_{XY} = -\frac{3}{4}$.

5.20   4.

5.21 $F(s) = \int_0^\infty e^{-st}f_X(t)\,dt \quad$ if $f_X(t) = 0 \quad$ for $\quad t < 0$;

   let $s = -j\omega$;

   $F(-j\omega) = \varphi_X(\omega)$.

5.23   350,000.

6.1 $f_Y(y) = \dfrac{1}{\sqrt{\pi}} e^{-(y+1/2)^2}$.

6.3 $f_Y(y) = 1, \qquad 0 \leq y \leq 1$,

   $= 0, \qquad$ elsewhere.

6.5   (a) $\dfrac{1}{\sqrt{2\pi 5}} e^{-(x+1)^2/10}$

   (b) $f_Y(y) = 0, \qquad y < 0$,

   $= y^2, \qquad 0 \leq y \leq 1$,

   $= 2y - y^2, \qquad 1 \leq y \leq 2$,

   $= 0, \qquad y > 2$.

6.7 $f_I(i) = \dfrac{20}{\sqrt{2\pi}} e^{-200i^2}$.

6.9 $\mu_I \simeq \frac{1}{20}$;

   $\sigma_I^2 \simeq (\frac{1}{200})^2 9 + 2\left[\dfrac{10}{(200)^2}\right]^2 10$.

343

6.11 (a) $E[Y] = 10^2 + \sigma^2$;

(b) $E[Y] \simeq 10^2$;

(c) $E[Y] = 10^2 + \sigma^2$.

7.1 (a) $\dfrac{\partial Y}{\partial X_1} = 2;\ \dfrac{\partial Y}{\partial X_2} = 10;\ \dfrac{\partial Y}{\partial X_3} = -X_4;\ \dfrac{\partial Y}{\partial X_4} = -X_3;$

(b) $E[Y] \simeq 21.5$;

(c) no;

(d) $\sigma_Y{}^2 \simeq 102.8;\ X_4$;

(e) $\sigma_Y{}^2 \simeq 101.8$.

7.2 (a) $E[R_1 + R_2 \mid T = t] = 2000 + 2t$;

var $[R_1 + R_2 \mid T = t] = 200$;

(b) $E[R] = 2200$;

$\sigma_R{}^2 = 3533$;

(c) $\rho_{R_1 R_2} \simeq .898$.

7.3 (a) $\Delta Y = 7.1$;

(b) Assuming Y normal, independence, and $\sigma = \dfrac{U - L}{3.5}$; $\sigma_Y \simeq .731$;

$P(|Y - \mu| > 1) \simeq .17$.

7.4 (a) $\pm 1\ db$; (b) assuming normal $P(|Y - \mu| > 7) \simeq .00012$.

7.6 $\displaystyle\int_0^\infty \frac{1}{\sqrt{2\pi700}}\, e^{-(x-1000)^2/1400}\, dx \simeq .99286$.

7.7 $F_Y(y) = 3F(y) - 3F^2(y) + F^3(y)$;

$F_Z(z) = F^3(z)$;

7.9 $E[W] = \frac{8}{15}$;

$\sigma_W{}^2 = \frac{11}{225}$.

7.11 No.

7.13 $f_{T_p}(t) = \lambda_1 e^{-\lambda_1 t} + \lambda_2 e^{-\lambda_2 t} - (\lambda_1 + \lambda_2)e^{-(\lambda_1+\lambda_2)t}$.

8.1 (a) .87;

(b) .551.

8.3 (a) .785;

(b) .01;

(c) $\dfrac{(.05)^2(.01)}{(.05)^2(.01) + (.95)^2(.99)}$.

8.5 $P(\text{good}) = \int_{10}^{15} \frac{1}{20}[.18\lambda - 1.8]\,d\lambda + (.9)(\frac{1}{2}) + \int_{25}^{30} \frac{1}{20}[5.4 - .18\lambda]\,d\lambda$

$\simeq .675.$

8.7 $Pr(P = p_i \mid k \text{ in } n) = \dfrac{p_i^{\,k}(1 - p_i)^{n-k} a_i}{\displaystyle\sum_{j=1}^{3} a_j p_j^{\,k}(1 - p_j)^{n-k}}.$

8.9 $f_{P\mid 90;100}^{(l)} = \dfrac{(101)!}{10!\,90!}\, l^{90}(1 - l)^{10}; \quad 0 \le l \le 1.$

8.11 $f_{R\mid 9;10}(r) = (21)(20)r^{19}(1 - r); \quad 0 \le r \le 1.$

8.17 $P(\sigma^2 = 1 \mid X = \tfrac{1}{4}) = \dfrac{e^{-(\frac{1}{4})^2/2}}{e^{-(\frac{1}{4})^2/2} + \dfrac{1}{\sqrt{\frac{1}{2}}} e^{-(\frac{1}{4})^2} + \dfrac{1}{\sqrt{2}} e^{-(\frac{1}{4})^2/4}}.$

8.21 $f_{b\mid X}(\lambda \mid 2.5) = \dfrac{1/\lambda}{\ln 4 - \ln 2.5}; \quad 2.5 \le \lambda \le 4.$

9.5 $\bar{X} = \dfrac{\sum X_i}{n};$

$\hat{\mu}_2 = \dfrac{1}{n} \sum_i (X_i - \bar{X})^2 \quad \text{or} \quad s^2 = \dfrac{1}{n-1} \sum_i (X_i - \bar{X})^2.$

9.6 $\hat{\mu}_3 = \sum_{i=1}^{n} (X_i - \bar{X})^3 \dfrac{1}{n} = \dfrac{1}{n} \sum_{i=1}^{n} (X_i - \bar{X})^3.$

9.9 $f_{X_n \mid X_1, \ldots, X_{n-1}} = f_{X_n}.$

9.11 $\hat{A} = \dfrac{n + k + 1}{\sum x_i + b};$

$\hat{\hat{A}} = \dfrac{n + k}{\sum x_i + b}.$

9.13 $f_{X\mid S}(x \mid s) = \dfrac{1}{\sqrt{2\pi}} e^{-(x - s)^2/2}$

$f_{S\mid X}(s \mid x) = \dfrac{1}{\sqrt{\pi}} e^{-(s - (x + 1)/2)^2}.$

The mean and estimate is $\dfrac{x + 1}{2}$.

9.15 $\hat{s}_2 = \dfrac{\sigma_{s_1 s_2}}{\sigma_{s_1}^{2}} s_1.$

9.17 $a = \mu_x - \dfrac{m\sigma_x^2}{m^2\sigma_x^2 + \sigma_N^2}\,(m\mu_x + \mu_N);$

$b = \dfrac{m\sigma_x^2}{m^2\sigma_x^2 + \sigma_N^2}\,.$

9.19 $\epsilon_i^2 = [a + bx_i - y_i]^2;$

$\dfrac{\partial \Sigma \epsilon_i^2}{\partial a} = 0 = \dfrac{\partial \Sigma \epsilon_i^2}{\partial b}$ (produces the same a and b).

10.1 $E[L]_A = 1;$
$E[L]_B = \frac{5}{4};$
A is better.

10.3 $y_1 = \dfrac{1}{2} + \sigma^2 \ln \left[\dfrac{(L_{UA} - L_{AA})a}{(L_{AB} - L_{UB})(1-a)} \right];$

$y_2 = \dfrac{1}{2} + \sigma^2 \ln \left[\dfrac{(L_{BA} - L_{UA})a}{(L_{UB} - L_{BB})(1-a)} \right].$

10.7 Not a single stage problem.

10.11 (a) No;

(b) $\displaystyle \int_{.6}^{\infty} [6 - 100s]\,\dfrac{1}{\sqrt{2\pi}\cdot 02}\,e^{-(s-.05)^2/2(.02)^2}\,ds.$

10.13 If $f_\theta = 1$, start 0 thru 5;
if $f_\theta = 2\theta$, start 5.

10.15 (b) For all three tests, if positive drill; if not, don't drill;
(c) C is best test.

10.17 (a) either a_1 or a_4 is best; expected cost of uncertainty is 1.8;
(b) a_2 is best; expected cost of uncertainty is 1.4.

11.1 A number.

11.3 (a) Some will be $180°$ out of phase;
(b) $X(2)$ is normal with mean; cos 2, and variance: 2 cos² (2).

11.5 (a) $1 - e^{-12}\left[1 + 12 + \dfrac{(12)^2}{2} \right];$

(b) $1 - e^{-6}\left[1 + 6 + \dfrac{6^2}{2} \right];$

(c) same as (a).

11.7 $\mu_X(t) = 1 + 2t;$

$\sigma_X^2(t) = 9 + 4t^2.$

11.11 $E[X(t)] = \mu t \neq k \Rightarrow$ not stationary.

11.15 $\mu_Y = 4;$

$\sigma_Y{}^2 = \frac{19}{3}.$

11.17 (a) $p = .36;$

(b) .994.

11.19 (a) Yes; (b) no; (c) no; (d) yes.

11.21 $\Phi_{XX}(\omega) = 2\pi \, \delta(\omega) + \dfrac{3}{\omega^2 + 4}.$

11.24 $\mu_X(t) \, \mu_Y(t).$

11.25 (a) 0;

(b) $\frac{1}{2} \cos (t_2 - t_1).$

12.1 $Y(t) = \dfrac{A(1/\omega C)}{\sqrt{(1/\omega C)^2 + R^2}} \cos (\omega t - \theta)$

where $\theta = \tan^{-1} \dfrac{1}{\omega RC}.$

12.5 (b) $R_{XX}(\tau) = 1 - |\tau|, \qquad |\tau| \leq 1;$

$= 0, \qquad |\tau| > 1.$

(c) $11 - |\tau|, \qquad |\tau| \leq 1;$

$10, \qquad |\tau| > 1.$

12.9 $E[Y(t)] = f'(t);$

$\sigma_Y{}^2(t) = 200.$

12.11 (a) $y(t) = 0, \qquad t < 0;$

$= 2t, \qquad 0 \leq t \leq 1;$

$= 2, \qquad 1 \leq t \leq 3;$

$= 2(4 - t), \qquad 3 \leq t \leq 4;$

$= 0, \qquad t \geq 4.$

(b) $y(t) = 0, \qquad t < 0;$

$= 2(1 - e^{-t}), \qquad 0 \leq t \leq 3;$

$= 2(e^3 - e)e^{-t} \qquad t \geq 3.$

12.13 (a) $\Phi_{XX}(\omega) = \sqrt{\pi}\, e^{-\omega^2/4}$;

(b) $\mu_Y = 0$,

$$\Phi_{YY}(\omega) = \frac{\sqrt{\pi}(a^2 + \omega^2)e^{-\omega^2/4}}{b^2 + (2b + c^2)\omega^2 + \omega^4}.$$

12.15 Let $Y(t)$ be the output

$$\Phi_{YY}(\omega) = 1, \qquad \omega_1 \leq |\omega| \leq \omega_2;$$
$$\qquad\quad = 0, \qquad\qquad \text{elsewhere.}$$

$$R_{YY}(\tau) = \frac{2}{\pi\tau}\, [\sin \omega_2\tau - \sin \omega_1\tau].$$

12.17 1.64×10^{-12} volts^2;

the first stage.

INDEX